Analog Circuits and Signal Processing

The *Analog Circuits and Signal Processing* book series, formerly known as the *Kluwer International Series in Engineering and Computer Science*, is a high level academic and professional series publishing research on the design and applications of analog integrated circuits and signal processing circuits and systems. Typically per year we publish between 5-15 research monographs, professional books, handbooks, and edited volumes with worldwide distribution to engineers, researchers, educators, and libraries.The book series promotes and expedites the dissemination of new research results and tutorial views in the analog field. There is an exciting and large volume of research activity in the field worldwide. Researchers are striving to bridge the gap between classical analog work and recent advances in very large scale integration (VLSI) technologies with improved analog capabilities. Analog VLSI has been recognized as a major technology for future information processing. Analog work is showing signs of dramatic changes with emphasis on interdisciplinary research efforts combining device/circuit/technology issues. Consequently, new design concepts, strategies and design tools are being unveiled.Topics of interest include: Analog Interface Circuits and Systems; Data converters; Active-RC, switched-capacitor and continuous-time integrated filters; Mixed analog/digital VLSI; Simulation and modeling, mixed-mode simulation; Analog nonlinear and computational circuits and signal processing; Analog Artificial Neural Networks/Artificial Intelligence; Current-mode Signal Processing; Computer-Aided Design (CAD) tools; Analog Design in emerging technologies (Scalable CMOS, BiCMOS, GaAs, heterojunction and floating gate technologies, etc.); Analog Design for Test; Integrated sensors and actuators; Analog Design Automation/Knowledge-based Systems; Analog VLSI cell libraries; Analog product development; RFFrontends, Wirelesscommunications and Microwave Circuits; Analog behavioral modeling, Analog HDL.

Xiangyu Mao • Yan Lu • Rui P. Martins

Fully-Integrated Low-Dropout Regulators

Xiangyu Mao
University of Electronic Science and
Technology of China
Chengdu, China

Yan Lu
Tsinghua University
Beijing, China

Rui P. Martins
University of Macau
Macao, China

ISSN 1872-082X ISSN 2197-1854 (electronic)
Analog Circuits and Signal Processing
ISBN 978-3-031-84915-2 ISBN 978-3-031-84916-9 (eBook)
https://doi.org/10.1007/978-3-031-84916-9

This Springer imprint is published by the registered company Springer Nature Switzerland AG
The registered company address is: Gewerbestrasse 11, 6330 Cham, Switzerland

Preface

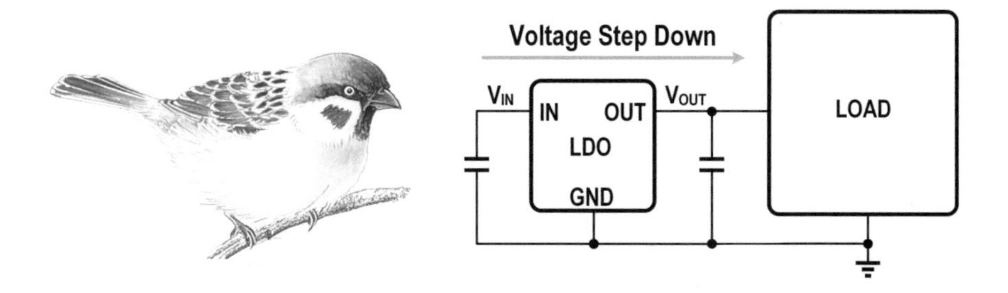

The wide utilization of low-dropout regulators (LDOs) in both high-end and low-end electronic systems, as well as in microprocessors and systems-on-a-chip (SoCs), results from their high power and current densities, fast transient response, good power supply rejection, flexibility, and cost-effectiveness. In comparison with switching-mode power converters, either inductor-based or switched-capacitor-based DC-DC converters, LDOs appear extremely compact as they have no energy storage component. Therefore, as the Chinese saying goes, "A sparrow may be small, but it has all its vital organs," an LDO is compact yet comprehensive.

In particular, fully integrated LDOs appear widely in SoCs almost always custom designed for specific applications. However, the diverse range of applications, architectures, operating conditions, and specification requirements can be perplexing for beginners and circuit engineers. This book focuses on fully integrated LDO circuit design, systematically categorizing various classic LDO architectures and key technologies while incorporating the latest research findings from top and popular publications. The aim is to help readers gain a systematic and in-depth understanding of fully integrated LDOs, enabling more accurate architecture selection and more efficient circuit implementation during design.

The book begins with a detailed introduction to the design specifications of fully integrated LDOs. Then, based on the control methods of the power transistor, we classify these LDOs into three categories: analog, digital, and switching. In the analog LDO part, we discuss key technologies and circuits, including buffers, compensation, adaptive biasing, power supply rejection ratio (PSRR) improvement, flipped-voltage follower (FVF), etc. The digital LDO part subdivides various structures, including shift register based, coarse and fine adjustment, analog-to-digital converter (ADC) based, event driven, and analog-digital hybrid, and introduces their working principles and characteristics. The switching LDO part covers hysteresis control, pulse-width modulation (PWM) control, and circuit implementation. Finally, the book explores distributed LDO designs for large-area high-current applications.

There is no single "best" architecture, only the most suitable choice for a given application. We hope that our work helps electronic engineers and students to design fully integrated voltage regulators efficiently and with high quality.

Greater Bay Area, China
October 18, 2024

Xiangyu Mao
Yan Lu
Rui P. Martins

Contents

Chapter 1
Introduction to the Low-Dropout Regulator

1.1 Fully Integrated LDO

A voltage regulator is normally a buffered reference [1]. The low-dropout voltage regulator (LDO) is a linear voltage regulator. Its widespread application in various electronic devices is due to its low power consumption, low output noise, high reliability, compact size, low cost, and ease of CMOS integration. Figure 1.1a shows a typical LDO structure. The error amplifier (EA), power transistor M_P, and feedback resistors R_{F1} and R_{F2} form a feedback loop. The power transistor M_P can be modeled as a voltage-controlled resistor or voltage-controlled current source connected in series between V_{IN} and V_{OUT}. When variations (e.g., input voltage and load current) cause V_{OUT} to change, the amplifier compares V_{OUT} against the feedback voltage V_{FB} to generate a control signal V_G that regulates the output impedance/output current, thereby maintaining a stable V_{OUT}, as illustrated in Fig. 1.1b.

The output capacitor C_L is particularly critical for an LDO as it influences its noise, stability, and transient performance. Conventional LDOs rely on off-chip capacitors in μF level, typically using ceramic capacitors. Due to the resonant frequency limitations of ceramic capacitors, the bandwidth of traditional LDOs generally remains within MHz range. In contrast, fully integrated LDOs can function without external capacitors, employing integrated capacitors on-chip or within the package. There are various ways to implement integrated capacitors, including metal-insulator-metal (MIM) capacitors, metal-oxide-metal (MOM) capacitors, metal-oxide-semiconductor (MOS) capacitors, deep trench capacitors, and in-package capacitors. The equivalent parasitic capacitor of the load can also be utilized and considered as the output capacitor. Due to the smaller output capacitor, a fully integrated LDO must achieve a sufficiently fast response to meet transient requirements. For some specific applications, the bandwidth of fully integrated LDOs can reach several hundreds of MHz.

X. Mao et al., *Fully-Integrated Low-Dropout Regulators*, Analog Circuits and Signal Processing, https://doi.org/10.1007/978-3-031-84916-9_1

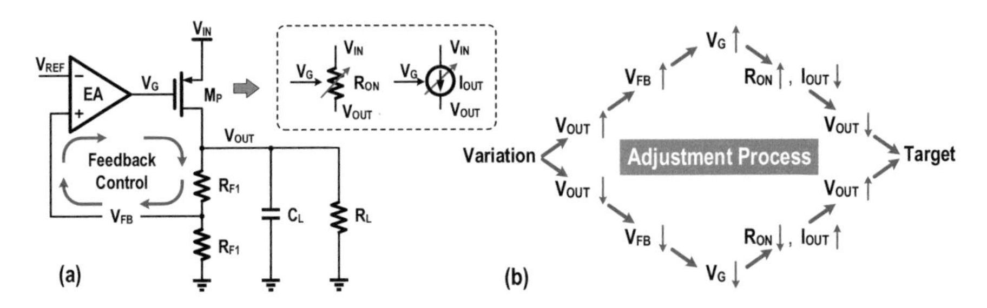

Fig. 1.1 (**a**) Typical LDO structure and (**b**) feedback adjustment process

Meanwhile, fully integrated LDO is known for various names, such as on-chip LDO, capacitor-less LDO, and capless LDO. Another variation, a cap-free LDO, is purported to operate without necessitating an output capacitor. In practical applications, since the load always has equivalent parasitic capacitance, the scenario with zero output capacitance does not exist. Therefore, accurately speaking, the cap-free LDO should be defined as capable of meeting load transient demands without an additional output capacitor while upholding loop stability and performance within a specified output capacitor range.

1.2 LDO Applications

In system-on-chip (SoC) designs, the demand for independent power domains is increasing due to increased functional modules and power efficiency requirements. However, the large number of power domains will increase the number of off-chip components and package bumps. Notably, as package bump pitches scale much slower than device scaling, the implantation overhead associated with the bumps continues to increase. Consequently, the total number of off-chip voltage domains tends to be constrained.

Figure 1.2 shows the conventional power supply and the solution with fully integrated LDOs. Compared to conventional LDOs with off-chip capacitors, fully integrated LDOs offer distinct advantages. Utilizing fully integrated LDOs can significantly reduce the number of board components and external pins [2]. Furthermore, the local fully integrated LDO being closer to the load block can reduce the IR drop and eliminate load transient voltage spikes from the bond-wire inductances.

Figure 1.3 depicts an application case of the fully integrated LDOs in an SoC design. The application purposes can be broadly categorized into three distinct groups. First, it serves as a typical step-down function to meet the operational voltage requirement of the post-stage circuit. For instance, in the 28 nm CMOS process, certain module's digital control logic circuits rely on core devices that cannot tolerate 1.8 V and require a reduction to 0.9 V. Given that the LDO's efficiency correlates with the input-output voltage, this application is relatively more suitable for

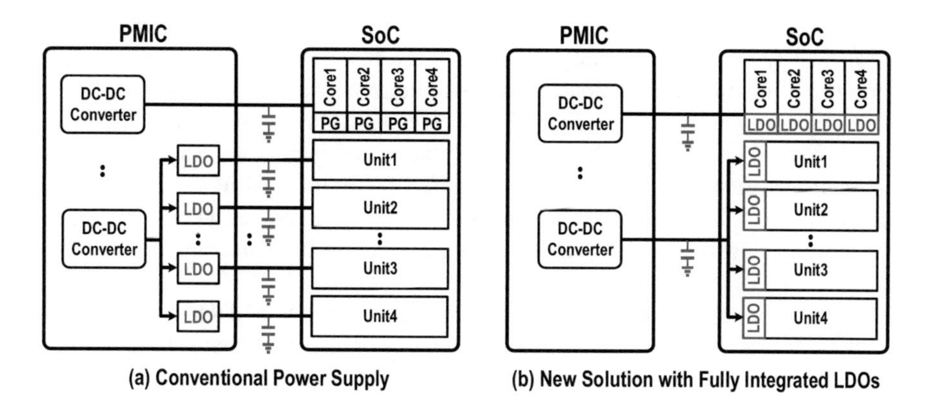

Fig. 1.2 Conventional power supply and new solution with fully integrated LDOs

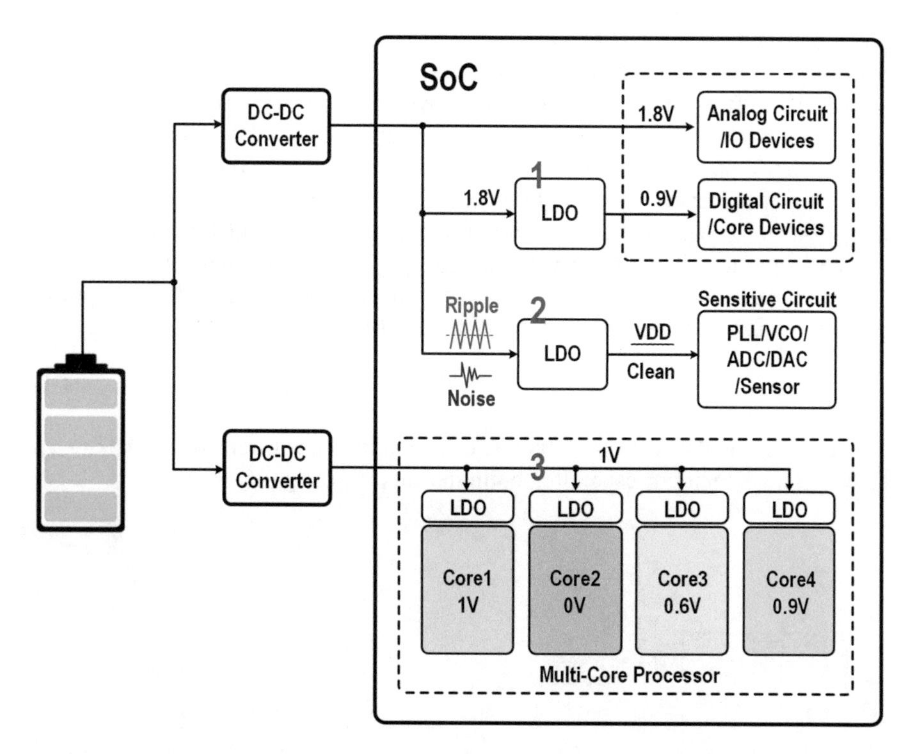

Fig. 1.3 Typical applications of fully integrated LDOs in SoC design

the scenarios with a low load current. But even for high current digital applications, using an LDO to step down the digital supply while maintaining the digital load functions will still have certain energy gain, as the digital load current reduces quite linearly with V_{DD} and then the digital load power consumption is approximately proportional to V_{DD}^2 [3], which will be discussed next.

Second, most of the analog signal processing functional modules, such as radio frequency (RF) circuits, analog-to-digital converters (ADCs), and phase-locked loop (PLL) circuits, are particularly susceptible to power supply noise. Typically, the external power supply is delivered through a DC-DC switching power converter, and its output has unavoidable ripples. Notably, since multiple modules share a common power supply, the transient response of an individual module can propagate dynamic noise across the shared common power rail, subsequently impacting the other modules. Fully integrated LDOs can offer clean power supplies to these noise-sensitive circuits, helping them perform better.

Besides the aforementioned applications, fully integrated LDOs with ampere-level load capability for large-scale digital circuits or microprocessors have become one of the research hotspots in recent years [4]. Multicore processors are widely used across many application domains, including central processing units (CPUs), graphics processing units (GPUs), and artificial intelligence (AI) processors. Core counts go up to dozens and even hundreds. Multicore processors usually share a common supply voltage. Each core has its own set of power gates (PGs), allowing it to form an independent power domain. The power supply of each core can be turned on or off to minimize the leakage power. In addition, each core can operate at a different frequency to realize dynamic frequency scaling (DFS). However, since the highest-frequency core dictates the minimum V_{IN} level, other low-frequency cores will waste extra power. A fully integrated voltage regulator can supply the local voltage domain to realize per-core dynamic voltage and frequency scaling (DVFS), significantly improving system energy efficiency, as shown in Fig. 1.4.

To analyze how much power the DVFS can save, we can calculate the power consumption P_A of the digital system with a fixed V_{IN} as its supply.

$$P_A = C_{dynA} \times V^2_{IN} \times F + I_{LEAK\_VIN} \times V_{IN} \tag{1.1}$$

Then, we calculate the system power consumption P_B with LDOs that make each core operate at their corresponding optimum supply voltage V_{OUT}:

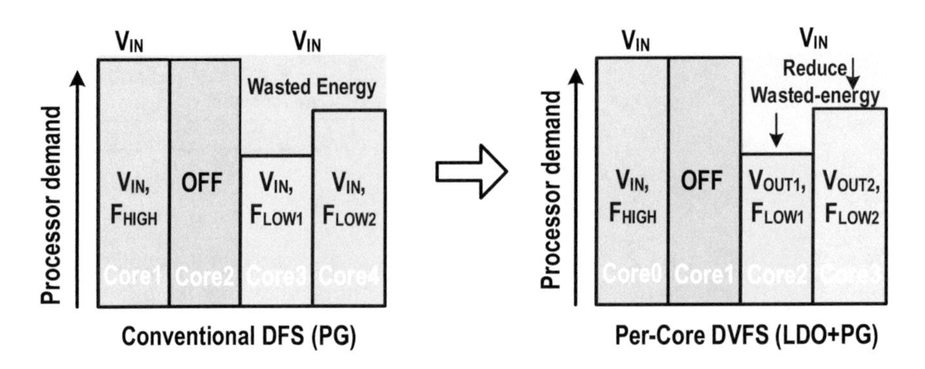

Fig. 1.4 Per-core DVFS with a share power supply V_{IN}

$$P_B = \frac{C_{dynA} \times V^2_{OUT} \times F + I_{LEAK\_VOUT} \times V_{OUT}}{h_{LDO}} \quad (1.2)$$

where $\eta_{LDO} \approx V_{OUT}/V_{IN}$. The saved power consumption is

$$P_{SAVE} = P_A - P_B = C_{dynA} \times V_{IN} \times (V_{IN} - V_{OUT}) \times F + V_{IN} \times (I_{LEAK\_VIN} - I_{LEAK\_VOUT}) \quad (1.3)$$

According to Eq. (1.3), we can deduct that although the LDO power efficiency can be very low in a large dropout voltage condition (for example, V_{IN} = 1 V, V_{OUT} = 0.5 V), it can still save a lot of power from a system perspective.

1.3 LDO Classification

LDOs maintain stable output voltage by adjusting the impedance of the power transistor or the output current. Based on different control methods of the power transistor, fully integrated LDOs can be categorized into three types: analog LDO, digital LDO, and switching LDO [5], illustrated in Fig. 1.5.

Analog LDO is the most classic, widely used, and common form of LDOs. They regulate the gate voltage V_G of the power transistor. By controlling the gate-source voltage (V_{GS}), the power transistor's impedance/output current is adjusted to stabilize the output voltage V_{OUT}. The output stage transfer function of an analog LDO is:

$$A_{VA} = \frac{\partial V_{OUT}}{\partial V_{GS}} = \frac{g_m R_o}{1 +_s R_o C_L} \quad (1.4)$$

where g_m is the transconductance of the analog power transistor related to its load current.

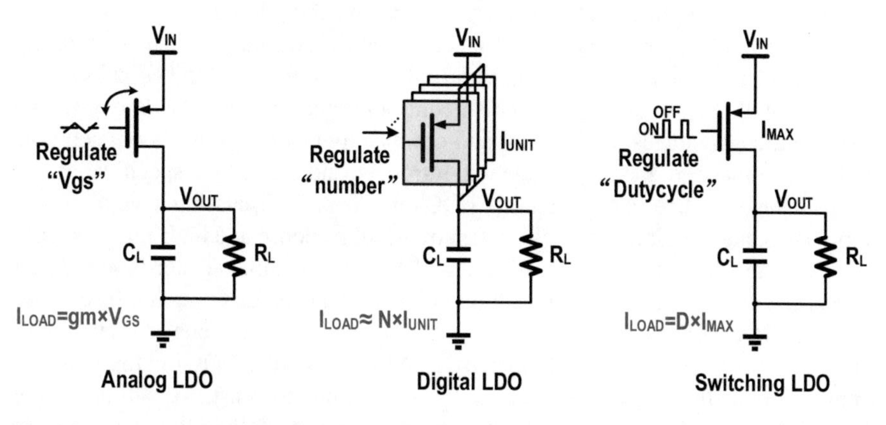

Fig. 1.5 Analog LDO, digital LDO, and switching LDO

Digital LDO, as the name suggests, controls the power transistor through digital signals. The power transistors are split into multiple small power units. For a fixed input and output voltage, the small, on-state power units can be equated to a fixed impedance resistor or a current source. By adjusting the number of parallel on-state power units, we can control the overall output impedance or output current to stabilize the output voltage V_{OUT}. The output stage transfer function is:

$$A_{VD} = \frac{\partial V_{OUT}}{\partial V} = \frac{I_{UNIT} R_o}{1 +_s R_o C_L} \tag{1.5}$$

where I_{UNIT} is the unit current conducted through a single power unit. Note that I_{UNIT} is strongly correlated with the DC values of the input voltage V_{IN} and the output voltage V_{OUT}. The digital control is discrete. To balance the output current with the load current and stabilize the output voltage with high accuracy, the control codes of the digital power stage would oscillate between several adjacent bits in steady state, resulting in the limit-cycle oscillation (LCO) phenomenon. Consequently, the output of a digital LDO usually has certain ripples.

The switching LDO stabilizes its output voltage VOUT by modulating the duty cycle D of the power transistor(s). The operation of a switching LDO can be modeled as a capacitor charging and discharging process. The load current represents the discharge current, while the charge current is determined by the strength of the power transistor. We can get the output stage transfer function of the switching LDO:

$$A_{VS} = \frac{\partial V_{OUT}}{\partial D} = \frac{I_{SW} R_o}{1 +_s R_o C_L} \tag{1.6}$$

where I_{SW} is the current conducted through the whole switching power transistor, which is also strongly related to the size of the power switch and the input-output voltage. Obviously, since the power transistors operate in a switching mode, the output of the switching LDO exhibits considerable ripples. To reduce the output ripple, larger on-chip capacitance and higher operating frequency are typically necessary, which complicates the integration of the switching LDO.

Diverse application scenarios and specific electrical characteristic index requirements lead to varied structures and control methods for LDOs. The subsequent chapters will discuss the representative control circuits for these LDO types. And the specific advantages, limitations, and optimal application domains of each type will be summarized. In Chap. 2, we will introduce and discuss the specifications of LDO. Then, we dive into the analog LDOs in Chap. 3, discussing several classic topologies and circuit techniques for fast transient response and high-power supply rejection. In Chap. 4, we start with the widely adopted shift register-based digital LDOs and discuss other formats of digital LDOs, advanced LDOs including analog-assisted digital LDOs, digital-assisted analog LDOs, computational digital LDOs, etc. In Chaps. 5 and 6, switching LDOs and distributed LDOs for high-current applications will be introduced and analyzed. Last but not least, we will add some final remarks and discussions on future directions of fully integrated voltage regulators.

References

1. G.A. Rincon-Mora, *Analog IC design with low-dropout regulators* (McGraw-Hill, New York, NY, 2009)
2. Y.-J. Lee, W. Qu, S. Singh, D.Y. Kim, K.H. Kim, S.H. Kim, J.J. Park, G.H. Cho, A 200-mA digital low drop-out regulator with coarse-fine dual loop in mobile application processor. IEEE J. Solid-State Circuits **52**(1), 64–76 (2017)
3. J.M. Rabaey, A. Chandrakasan, B. Nikolic, *Digital integrated circuits: A design perspective* (Pearson Education, 2003)
4. S.T. Kim, Y.-C. Shih, K. Mazumdar, R. Jain, J.F. Ryan, C. Tokunaga, Enabling wide autonomous DVFS in a 22 nm graphics execution core using a digitally controlled fully integrated voltage regulator. IEEE J. Solid-State Circuits **51**(1), 18–30 (2016)
5. X. Mao, Y. Lu, R.P. Martins, A scalable high-current high-accuracy dual-loop four-phase switching LDO for microprocessors. IEEE J. Solid-State Circuits **57**(6), 1841–1853 (2022)

Chapter 2
LDO Specifications

Similar to any other analog circuits, an LDO design involves multiple design parameters, considerations, and trade-offs. Table 2.1 lists the LDO design specifications, covering DC, AC, transient time, and all other domains.

2.1 Dropout Voltage

Dropout voltage ($V_{DROPOUT}$) is the input-to-output voltage difference at which the LDO is no longer capable of maintaining the desired output voltage (V_{OUT}) against further reductions in input voltage (V_{IN}) [1]. Figure 2.1 shows a simplified schematic of an LDO and illustrates its operation regions. In the normal regulation region, when input voltage V_{IN} decreases, the resistance of the power transistors decreases, while the output voltage V_{OUT} remains unchanged. As V_{IN} decreases further, the power transistors reach the minimum value, and the LDO enters the dropout region. In this region, the output voltage decreases as the input voltage reduces. The dropout voltage can be expressed as

$$V_{DROPOUT} = R_{ON} \times I_{LOAD} \tag{2.1}$$

where I_{LOAD} is the load current, and R_{ON} represents the total resistance across the entire power path, including the power transistors, on-chip power wire, and bond wires.

Figure 2.2 illustrates the operation region of an LDO under different load currents. At smaller load currents, the dropout voltage is proportionally lower. A low-dropout voltage is necessary to achieve high power efficiency, but it often requires a sizeable power transistor, especially when the load current I_{LOAD} is large. In the LDO design, it is crucial to fully consider the process-voltage-temperature (PVT) variations, parasitic resistance along the power path, and driver swing range to ensure an adequate design margin.

X. Mao et al., *Fully-Integrated Low-Dropout Regulators*, Analog Circuits and Signal Processing, https://doi.org/10.1007/978-3-031-84916-9_2

Table 2.1 Various specifications of LDO design

DC specifications			
Input voltage range	Headroom voltage	Line regulation	Shutdown current
Output voltage range	Output accuracy	Load current range	Ground current
Dropout voltage	Load regulation	Quiescent current	Current efficiency
AC specifications			
Gain	Phase margin	Noise	PSRR
Loop bandwidth	Gain margin		
Transient specifications			
Load transient response: edge time, load step, undershoot/overshoot, recovery time			
Line transient response		Dynamic voltage scaling	
Other specifications			
Area	Current density	Layout shape	Reliability consideration
Application conditions: output capacitor, operation temperature range, package parasitics			
Special functions: bypass mode, overcurrent protection			

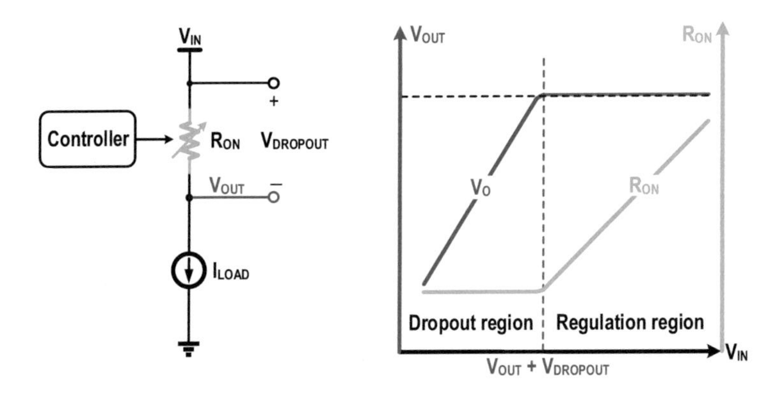

Fig. 2.1 Simplified schematic of an LDO and its operation regions

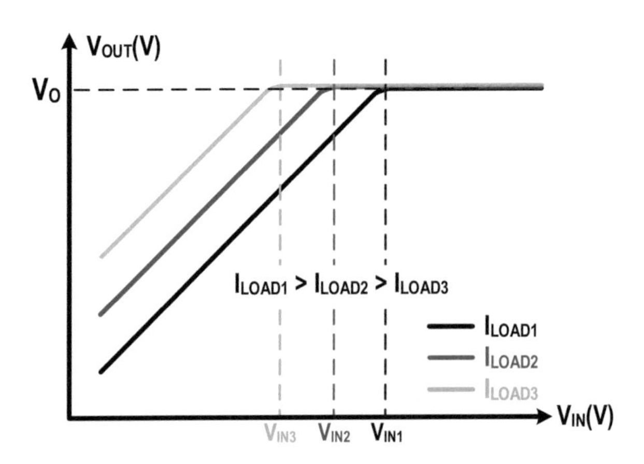

Fig. 2.2 The operation region of an LDO under different load currents

2.2 Headroom Voltage

The dropout voltage represents the minimum input-to-output voltage difference that an LDO must sustain to meet its load capability and output accuracy requirements. Apart from the dropout voltage, additional LDO specifications, such as power supply rejection (PSR) and load transient performance, may demand a higher input-to-output voltage difference for compliance. This voltage difference is defined as the headroom voltage, often employed as a condition to satisfy certain specifications.

2.3 Load Regulation

LDO can maintain a stable output voltage across varying load conditions. In practical scenarios, the output voltage V_{OUT} exhibits variations with the changes in load current. Typically, as the load current increases, the output voltage tends to decrease, as shown in Fig. 2.3. Load regulation refers to the LDO's ability to uphold a consistent voltage level despite varying load conditions [2]. It can be defined as follows:

$$\text{Load Regulation} = \Delta V_O / \Delta I_O \tag{2.2}$$

where ΔV_O is the output voltage variations and ΔI_O is the change in load current.

On the one hand, this parameter is highly dependent on the DC gain of the error amplifier (EA) in the feedback loop. A high-gain EA would strongly suppress the V_{OUT} variation versus load current. On the other hand, when employing an LDO, it is crucial to consider the IR drop along the power path. The IR drop will increase

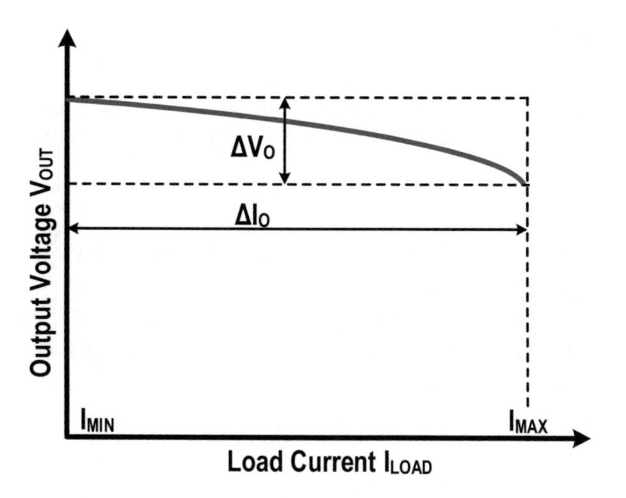

Fig. 2.3 Output voltage vs. load current

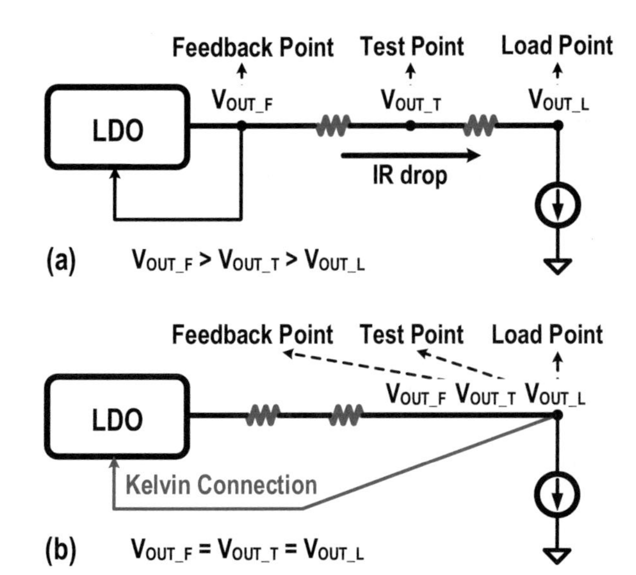

Fig. 2.4 Three point locations: (**a**) different positions and (**b**) the same position

voltage variations when the feedback point, test point, and load point are situated at different locations. An optimal approach is to utilize a Kelvin connection, which physically isolates the feedback line from the power path, ensuring that the feedback point, test point, and load point share the same position, as depicted in Fig. 2.4. For some large-scale loads, it is necessary to account for the IR drop across the load itself and select an appropriate feedback point accordingly.

2.4 Line Regulation

Line regulation is the LDO's ability to uphold a consistent voltage level despite varying input voltages. It can be defined as

$$\text{Line Regulation} = \Delta V_O / \Delta V_{IN} \tag{2.3}$$

where ΔV_O is the output voltage variations and ΔV_{IN} is the change in the input voltage, as shown in Fig. 2.5.

Numerous factors contribute to variations in the output voltage resulting from fluctuations in the input voltage, such as changes in the loop gain or offset voltage variations.

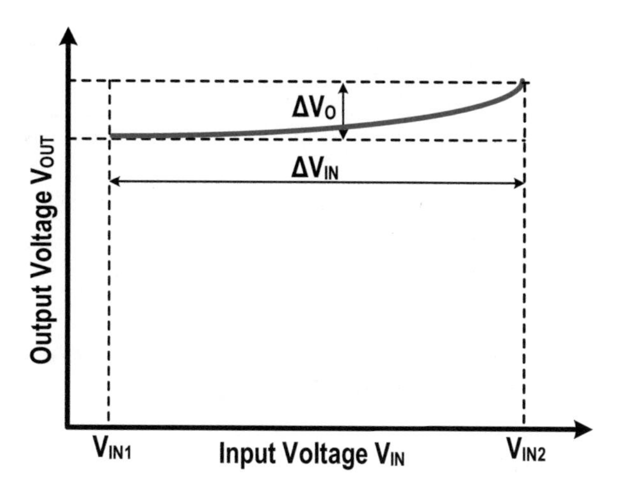

Fig. 2.5 Output voltage vs. input voltage

2.5 Accuracy

The overall accuracy needs to consider the effects of reference voltage drift, error amplifier voltage drift, external sampling resistor tolerance, temperature coefficient, and line-and-load regulation [1].

For example, let us calculate the overall accuracy of a 1-V LDO with the following operational characteristics: ±1% reference accuracy, ±60 ppm/°C temperature coefficient of the reference voltage over −40 °C to 125 °C, ±0.7% EA offset voltage drift, ±0.25% sampling resistor tolerance, 10 mV load regulation, and 6 mV line regulation. Then, the errors are listed as below:

1. The error stemming from the reference voltage is ±1%.
2. The error due to temperature can be calculated as (125 °C + 40 °C) × ±60 ppm/°C = ±0.99%.
3. The error associated with the EA offset voltage drift is ±0.7%.
4. The error arising from the sampling resistor is ±0.25%.

When measuring the line-load regulation, it is assumed that the intermediate values represent the target voltages. Then, the corresponding errors are given below:

1. The error due to the load regulation can be calculated as 100% × (0.01 V/2)/1 V = ±0.5%.
2. The error due to the line regulation can be calculated as 100% × (0.006 V/2)/1 V = ±0.3%.

Assuming that all errors occur in the same direction, the worst-case error can be calculated as

$$\text{error\_worst} = \pm(+1 + 0.99 + 0.7 + 0.25 + 0.5 + 0.3)\% = \pm 3.74\% \tag{2.4}$$

The typical error is a result of random variations, so it is reasonable to estimate the variation using a root square sum (RSS) of the errors:

$$\text{error\_typical} = \pm\sqrt{\left(1^2 + 0.99^2 + 0.7^2 + 0.25^2 + 0.5^2 + 0.3^2\right)}\% = \pm 1.695\% \quad (2.5)$$

The error distribution will be centered on the RSS error and will not exceed the worst-case error.

2.6 Quiescent Current, Ground Current, and Shutdown Current

The LDO's internal circuit consumes a certain amount of current. Ground current I_G is defined as the difference between input current I_{IN} and output current I_{OUT}, as shown in Fig. 2.6:

$$I_G = I_{IN} - I_{OUT} \quad (2.6)$$

Quiescent current I_Q specifically refers to the input current when the output current is zero:

$$I_Q = I_{IN}\big|_{I_{OUT}=0} \quad (2.7)$$

When the internal circuits of an LDO employ a fixed bias current, the ground current I_G remains constant regardless of the output current. In such cases, the quiescent current $I_Q = I_G$. However, some analog LDOs adopt an adaptive bias current in the driver and amplifier circuits. In this scenario, the ground current I_G exhibits a positive correlation with the output current I_{OUT}, leading to $I_Q <= I_G$. This technique is commonly employed to reduce quiescent current and enhance transient response performance.

When LDO is turned off, there may be a leakage current within the internal circuits and power transistors, or some specific components may remain in standby mode to comply with basic requirements. Therefore, the shutdown current is the input current when the output is disabled.

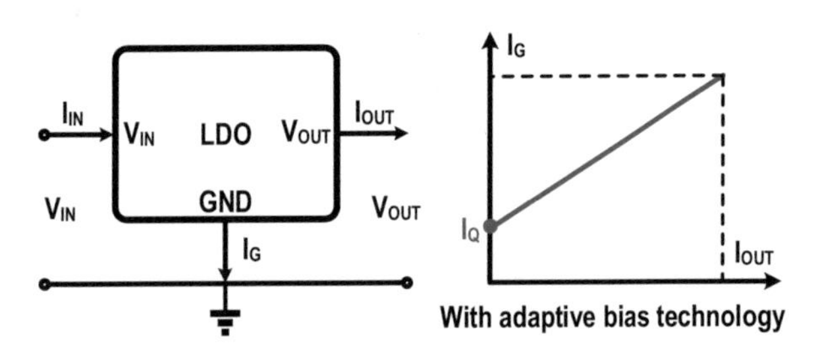

Fig. 2.6 Ground current and quiescent current of an LDO

2.7 Power Efficiency and Current Efficiency

The power efficiency of an LDO can be calculated by

$$\text{Efficiency} = \frac{V_{\text{OUT}} \times I_{\text{OUT}}}{V_{\text{IN}} \times I_{\text{IN}}} = \frac{V_{\text{OUT}}}{V_{\text{IN}}} \times \frac{I_{\text{OUT}}}{I_{\text{OUT}} + I_{\text{G}}} \approx \frac{V_{\text{OUT}}}{V_{\text{IN}}} \tag{2.8}$$

As the internal circuit consumes a tiny current and the output current I_{OUT} closely approximates the input current I_{IN}, the intrinsic factor in the power efficiency of the LDO is the difference in input-output voltage. To facilitate a more accurate comparison between different LDOs, we define the current efficiency as follows:

$$\text{Current Efficiency} = \frac{I_{\text{OUT}}}{I_{\text{IN}}} = \frac{I_{\text{OUT}}}{I_{\text{IN}} + I_{\text{G}}} \tag{2.9}$$

2.8 Load Transient Response

When the load current changes gradually, the control loop of the LDO is capable of real-time monitoring and adjusting the power transistors to maintain the output voltage. This adjustment procedure is commonly referred to as load regulation. However, when the load current undergoes rapid changes, the LDO control loop cannot swiftly track such variations due to the response time constraints, leading to temporary inconsistencies between the actual load demand and the output current. Consequently, the output voltage experiences transient deviations.

As shown in Fig. 2.7, V_{OUT} represents the output voltage, I_{LOAD} denotes the load current, and I_{OUT} is the output current. The output voltage has an undershoot when the load current steps up. Conversely, when the load current steps down, the output voltage exhibits an overshoot. This transient process is known as the load transient response.

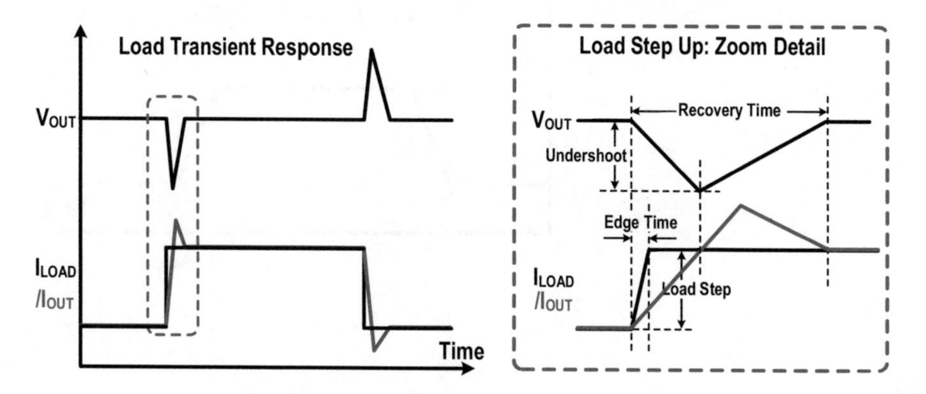

Fig. 2.7 Load transient response

The amplitude of load current variation is referred to as the load step, while the duration of load current change is termed as the edge time. The corresponding decrease in output voltage is known as undershoot or droop, whereas the excessive increase in output voltage is referred to as overshoot. The output voltage will stop falling or rising when the output current I_{OUT} is equal to the load current I_{LOAD}. The interval between the occurrence of a transient event and the moment the output voltage returns to the target voltage is denoted as the recovery time or setting time.

The load transient performance is one of the most critical indicators of an LDO. Smaller undershoot/overshoot and shorter recovery time denote better transient performance. The amplitude of overshoot/undershoot is directly related to the load step, edge time, output capacitor, and loop bandwidth. And the recovery time is usually related to the −3 dB bandwidth of the control loop. When multiple loops exist in an LDO, the slowest one will determine the recovery time.

2.9 Line Transient Response

The slow changes in the input voltage that result in a DC variation in the output voltage are referred to as line regulation. On the other hand, when the input voltage undergoes rapid changes, causing transient fluctuations in the output voltage, it is termed as line transient, as illustrated in Fig. 2.8.

Considering that the adjustment speed of the external power supply is typically in the ~μs range, LDOs are generally capable of tracking such changes. The

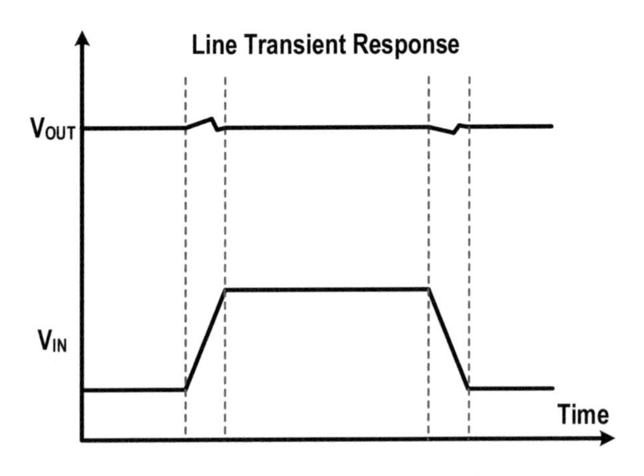

Fig. 2.8 Line transient response

magnitude of overshoot/undershoot in line transient is related to the input voltage step value, the edge time, the load current, and the output capacitor. The magnitudes are usually much smaller in line transient compared to those observed in load transient responses.

2.10 Dynamic Voltage Scaling

The output voltage V_{OUT} changes with the reference voltage V_{REF}, called dynamic voltage scaling (DVS) [3]. As discussed in Chap. 1, DVS is a critical power management technology where we adjust supply voltage in response to the load demand. In digital systems, the voltage adjustment is usually coordinated with the adjustment of the operating frequency, known as dynamic voltage frequency scaling (DVFS). The power consumption of digital circuits can be expressed as

$$P = C_{\text{dynA}} \times V_{\text{DD}}^2 \times F + I_{\text{LEAK\_VIN}} \times V_{\text{DD}} \tag{2.10}$$

While maintaining performance, reducing the operating voltage V_{DD} and operating frequency F of the digital circuit can significantly reduce its power consumption. Since the load demand of digital circuits fluctuates rapidly, the speed of V_{OUT} voltage regulation is very important, and a fast DVS can significantly reduce the system energy loss, as illustrated in Fig. 2.9. On the other hand, we also need to ensure that V_{OUT} tracks smoothly with minimal overshoot or undershoot.

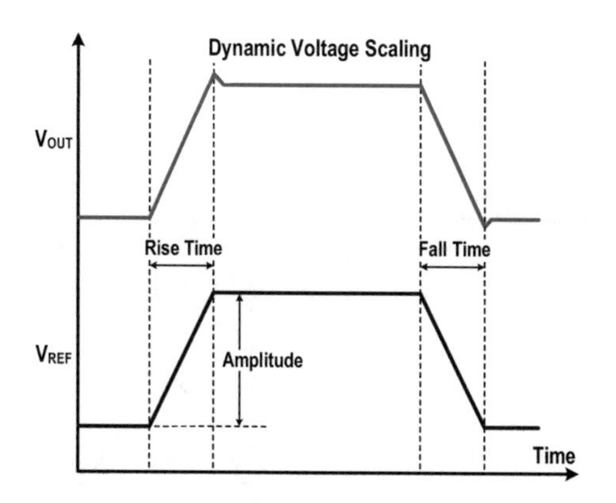

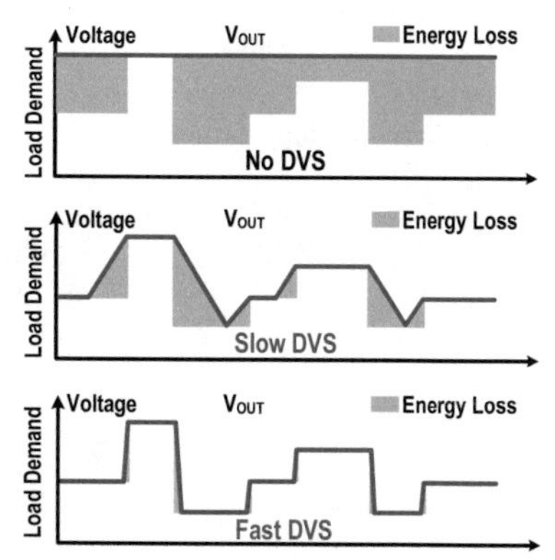

Fig. 2.9 Dynamic voltage scaling

2.11 Power Supply Rejection Ratio

The parameter power supply rejection (PSR) or power supply rejection ratio (PSRR) is widely used to quantify the capability of a circuit to suppress any power supply variations to its output signal, as shown in Fig. 2.10. PSRR can be defined as follows:

$$\mathrm{PSRR} = 20 \times \log\left(V_{\mathrm{O,RIPPLE}} / V_{\mathrm{IN,RIPPLE}}\right) \tag{2.11}$$

The PSRR value is related to the loop gain and ripple transfer gain of the LDO [4]. With fixed load current and fixed input-output voltage, the LDO exhibits different loop gains at different frequencies, and the ripple transmission gain is also different. Therefore, the LDO's ability to suppress input ripple is strongly dependent on the frequency, as illustrated in Fig. 2.11.

The performance of sensitive analog circuits, such as ADCs, DACs, and VCOs, is directly influenced by the quality of their power supplies. Many high-efficiency switch-mode power supplies have output ripple, which can be effectively

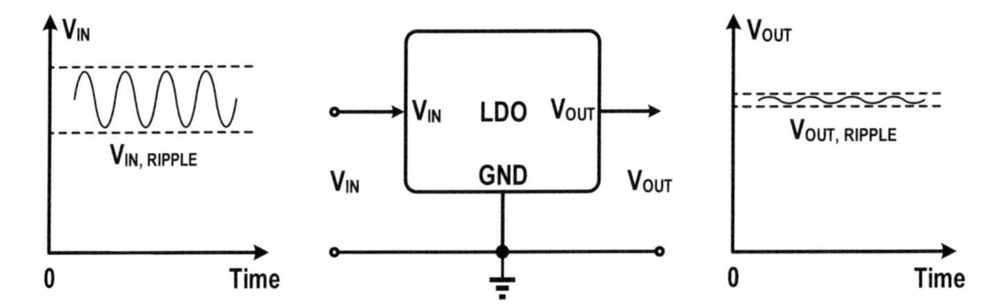

Fig. 2.10 Ripple suppression

Fig. 2.11 Power supply rejection

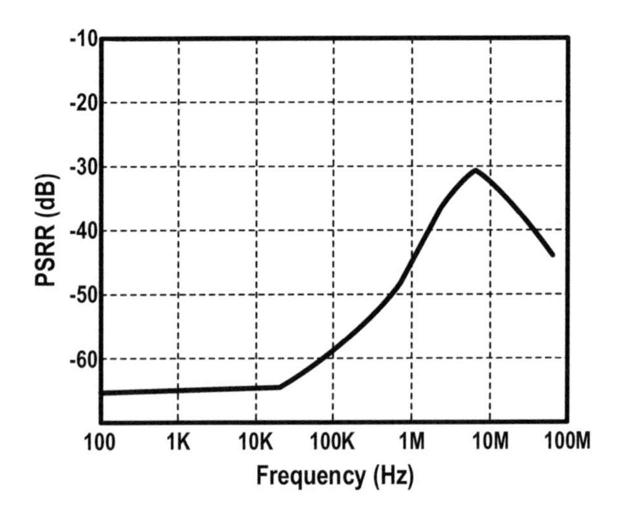

suppressed by employing LDOs in the post-stage. Considering that the operating frequency of a DC-DC converter typically lies in the MHz range, the PSR of the LDO in the 100 kHz to 10 MHz range is particularly important.

2.12 Output Noise Voltage

Noise is a physical phenomenon that occurs with resistors and transistors, encompassing thermal noise, shot noise, and flicker noise (1/f noise). Shot noise and flicker noise are generated by transistors, while resistors and the resistive component of MOSFETs produce thermal noise. Thermal noise and shot noise are intrinsically random in nature, and its power is flat over frequency. Flicker noise arises from trapped charges at the MOSFET gates. It follows Poisson's distribution with 1/f roll-off in power versus frequency. At low frequencies, the flicker noise is higher and dominates until its power becomes smaller than the thermal noise [5], as illustrated in Fig. 2.12.

The output noise of an LDO refers to the noise generated only by the LDO itself under the conditions of a constant output current and a ripple-free input voltage. Typically, the output noise in an LDO is specified in two fashions. One is "integrated output noise-in μV_{RMS}," which is root mean square (RMS) value of the spectral noise density integrated over a specific frequency range (typically 10 Hz or 100 Hz to 100 kHz). The second is a plot of noise density versus frequency, and the unit is μV/sqrt (Hz) (Fig. 2.13).

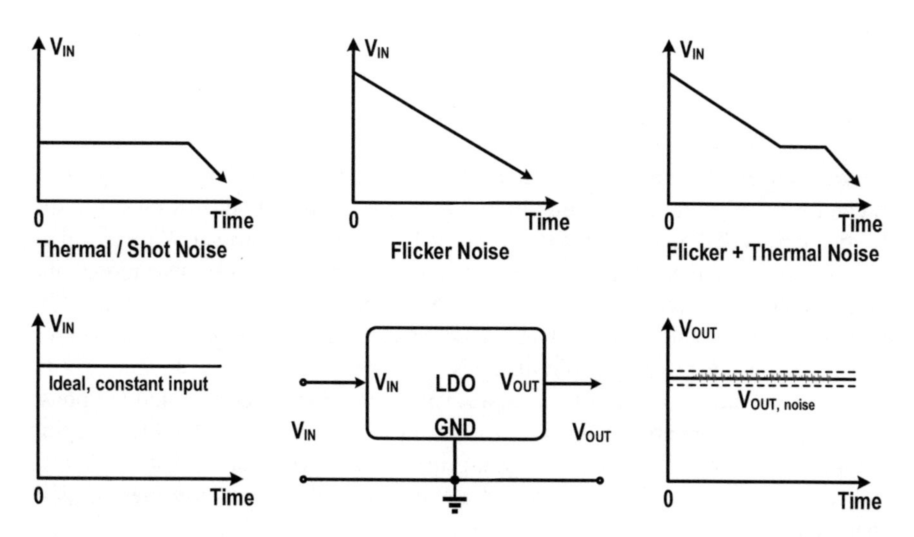

Fig. 2.12 The output noise of an LDO

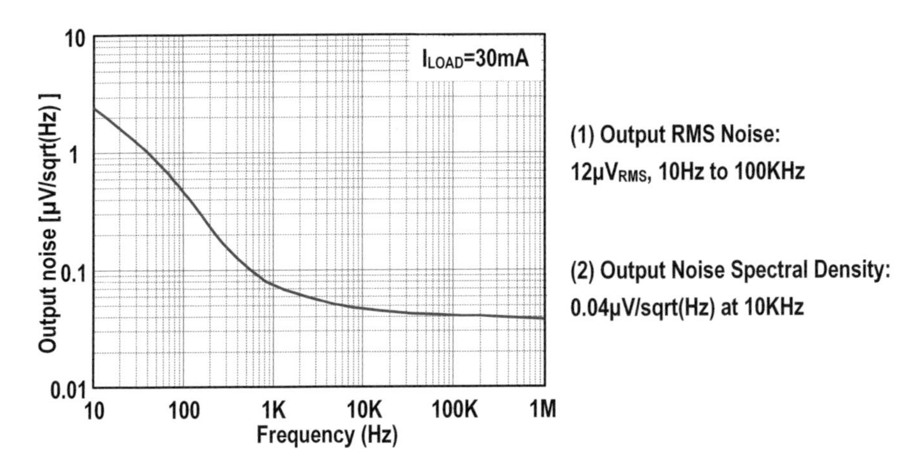

Fig. 2.13 Output noise specified in two ways: RMS noise and noise spectral density

2.13 Loop Stability

Stability is the most crucial indicator for an LDO. Although it is not explicitly reflected in the functional specifications, it serves as a prerequisite for the LDO to achieve other performance metrics. Only when the loop is stable will the other specifications become meaningful. Figure 2.14 illustrates the block diagram of a typical analog LDO. It comprises four main components: an amplifier, a diver, a power stage, and feedback resistors. The transfer functions of these four blocks are denoted as $H_A(s)$, $H_D(s)$, $H_P(s)$, and $H_F(s)$, respectively.

The overall transfer function of an LDO is

$$H(s) = H_A(s) \times H_D(s) \times H_P(s) \times H_F(s) \tag{2.12}$$

We often employ the Bode plot to analyze the loop stability. For an analog LDO, it can be directly simulated using the stability (STB) simulation. However, the stability analysis of digital LDOs and switching LDOs is more intricate. We can utilize periodic stability (PSTB) simulations or construct an equivalent linear model and perform STB simulations or utilize MATLAB to obtain the Bode plot based on the transfer function. The high-resistance nodes (nodes a and b in Fig. 2.15) in the loop are selected as the nodes for STB analysis.

The Bode plot consists of two separate plots: the magnitude plot and the phase plot. It provides valuable insights into a system's frequency characteristics, helping to understand its stability and analyzing its overall performance. There are four indicators that we need to pay special attention to: DC gain, phase margin, gain margin, and bandwidth, as shown in Fig. 2.16.

The phase margin in a Bode plot represents the amount of additional phase shift that can be tolerated before the system becomes unstable. It is measured at the frequency where the magnitude plot crosses the 0 dB (unity gain) line. A positive phase

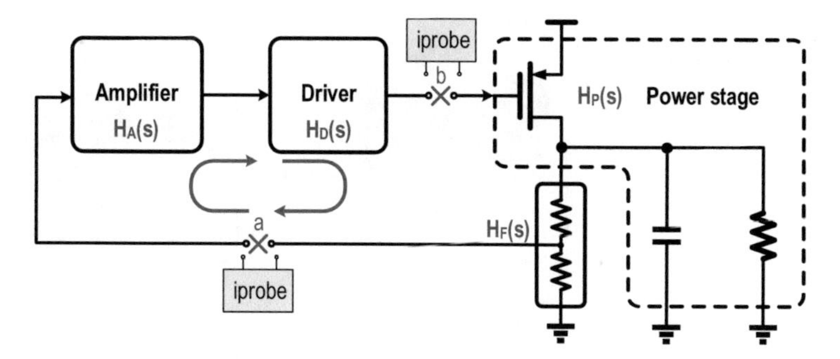

Fig. 2.14 Block diagram of a typical analog LDO

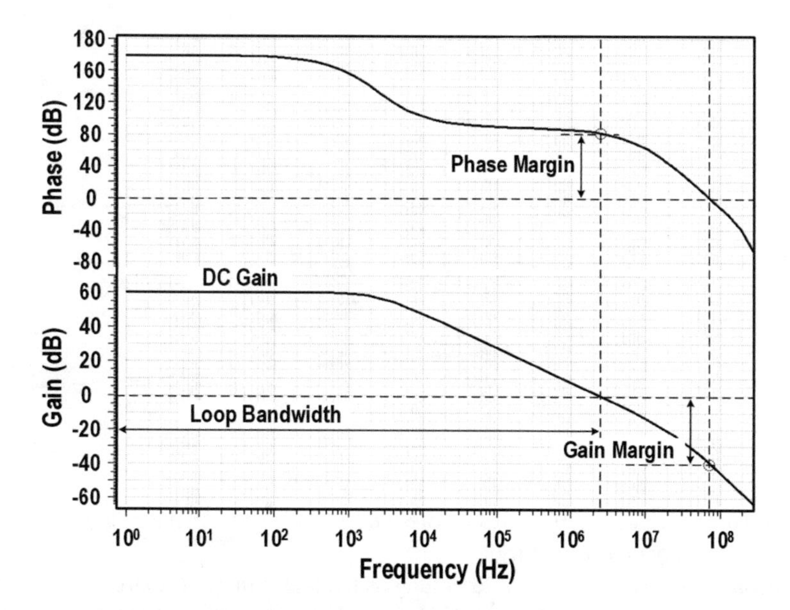

Fig. 2.15 The simulated Bode plot

margin indicates a stable system, while a negative phase margin suggests an unstable system. Specifically, the phase margin quantifies the amount of additional phase shift required to reach the instability threshold.

The gain margin, on the other hand, is determined by examining the magnitude plot at the frequency where the phase shift is −180 degrees. It represents the amount of additional gain that can be introduced before the system becomes unstable. In other words, the gain margin quantifies the amount of additional gain required to reach the instability threshold.

The bandwidth in a Bode plot stands for the unity gain bandwidth (UGB), which represents the frequency at which the magnitude plot crosses the 0 dB line. The UGB is related to the loop's response speed, where a fast response usually requires a large loop bandwidth. However, it should be noted that a wide loop bandwidth

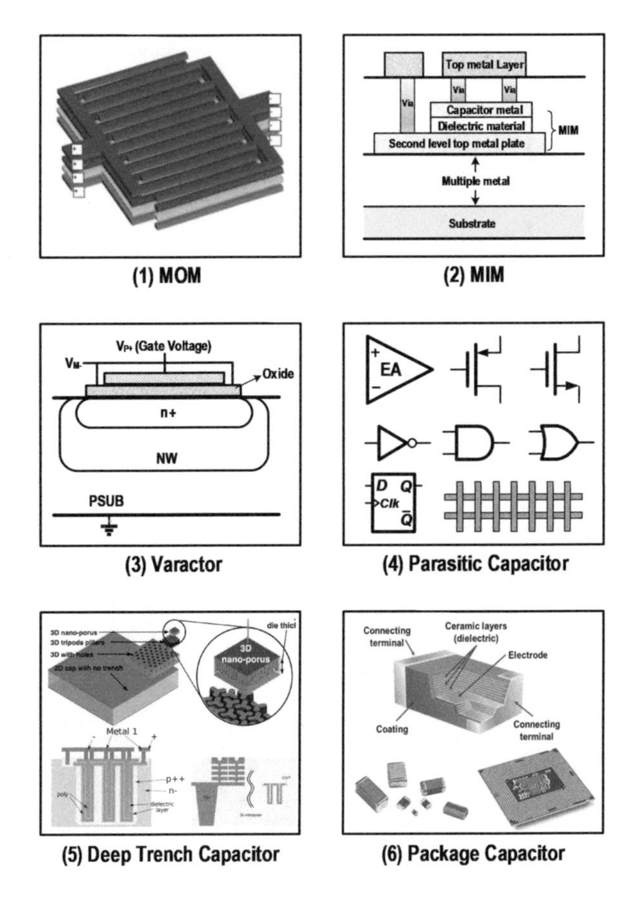

Fig. 2.16 Six types of capacitors: (1) MOM, (2) MIM, (3) varactor, (4) parasitic capacitor, (5) deep trench capacitor, and (6) package capacitor

means that the influence of parasitic poles and zeros is greater, making it more challenging to achieve loop stability.

Theoretically, the loop is stable as long as the phase margin is larger than 0° and the gain margin is greater than 0 dB. However, considering the transient ring, process-voltage-temperature (PVT) variations, manufacturing mismatch, and other factors, it is evident that this is far from enough. Then, how to design a robust and stable LDO?

A large phase margin and gain margin indicate a more stable system with better robustness against disturbances or uncertainties. Additionally, a larger phase margin results in fewer oscillations during transient responses. Based on our previous design experience, the phase margin is typically required to be greater than 45°, and the gain margin is required to be greater than 10 dB. Suppose it is challenging to meet the phase margin greater than 40° to ensure sufficient margin; then the numerical value of the sum of phase margin plus gain margin is required to exceed 40 to ensure sufficient stability margins.

An LDO is initially designed for stability in the typical-typical (TT) corner to ensure that the four-stability metrics (DC gain, phase margin, gain margin, and

Table 2.2 PVT verification case

No.	Items	Content	Amount
1	R_L	R_{L_MIN} to R_{L_MAX}	20
2	C_L	C_{L_MIN}, C_{L_MAX}	2
3	MOS	SS, FF, SF, FS	4
4	Res	RES_SS, RES_FF	2
5	Cap	CAP_SS, CAP_FF	2
6	Input-Output	(V_{IN_MIN}, V_{OUT_MIN}) (V_{IN_DROP1}, V_{OUT_MIN}) (V_{IN_DROP2}, V_{OUT_MAX}) *V_{IN_DROP}=V_{OUT} + $V_{DROPOUT}$	3
7	Temperature	−40 °C, 125 °C	2
8	I_{BIAS}	I_{BIAS_MIN}, I_{BIAS_MAX}	2
Total: **7680**			

bandwidth) meet the requirements across the entire load range, and then the analog design environment (ADE) XL is employed for PVT verification. Table 2.2 lists the PVT cases, which can be adjusted according to the actual conditions.

2.14 Output and Compensation Capacitor

The output and compensation capacitors of fully integrated LDOs can be divided into six types:

(1) Metal-oxide-metal capacitors

Metal-oxide-metal (MOM) capacitor consists of interdigitated metal fingers that form a multi-finger capacitor structure, also known as the "finger cap," as shown in Fig. 2.16(1). The standard metal wiring lines and vias serve as the capacitor plates, and the lateral capacitive coupling effect between plates generates the desired capacitance. For a higher capacitance density, multiple metal layers can be connected in parallel using vias, creating a vertical metal wall or mesh structure. Typically, the lowest metal layers (such as M1–M5) with the smallest line width and spacing are used in MOM capacitors to maximize capacitance density.

(2) Metal-insulator-metal capacitors

Metal-insulator-metal (MIM) capacitor is another class of compact capacitors. They are typically constructed using three plates, like a sandwich, and thus are also known as the "sandwich cap": the top metal layer, the second top metal layer, and a special dielectric material in the middle, as shown in Fig. 2.16(2). The dielectric material affects the capacitance density, leakage current, and breakdown voltage characteristics of the MIM capacitor. The square capacitance of MIM capacitors is small, generally ~10 nF/mm^2. However, some MIM capacitors using special processes, such as Intel's SuperMIM capacitor, can reach 376 nF/mm^2.

(3) Metal-oxide-semiconductor (MOS) capacitors and varactor capacitors

Metal-oxide-semiconductor (MOS) transistors can be used as MOS capacitors. The gate acts as the top electrode, the drain and source connection make up the bottom plate, and the gate oxide serves as the dielectric layer. It is important to note that the bias voltage applied at the gate needs to be higher than the threshold voltage of the transistors; otherwise, there is no capacitive characteristic.

Varactor capacitors, depicted in Fig. 2.16(3), resemble diodes. The capacitance value of varactor depends on the bias voltage on the gate. Unlike MOS capacitors, the varactor capacitors have no significant threshold characteristic. The capacitance value first increases linearly with the bias voltage, subsequently stabilizing.

(4) Parasitic capacitors

In addition to the MOM, MIM, MOS CAP, and varactor cap, the parasitic capacitance of the load itself should not be ignored. These parasitic components include the equivalent parasitic capacitance of the transistors, the routing metal parasitic capacitance, and the inter-decoupling capacitors, as shown in Fig. 2.16(4). Ideally, for digital loads, only the equivalent parasitic capacitance is used as the output capacitance. Any additional new capacitance will increase the area cost of the LDO.

(5) Deep trench capacitors

A deep-trench capacitor (DTC) is a three-dimensional vertical capacitor formed by etching a deep trench (DT) into a silicon substrate. Compared to other on-die capacitor solutions, such as MIM or MOS capacitors, DTCs provide higher capacitance per unit area. The capacitance density of DTC capacitors can reach approximately 1200 fF/μm^2 [6]. In a 2.5 D/3D packaging, DTCs are typically integrated into the silicon interposer.

(6) In-package capacitors

Another type of capacitor that can be employed is the package capacitor. Package capacitors are typically multilayer ceramic capacitors (MLCCs), and their capacitance can reach the μF level. When on-chip capacitance is insufficient and no dedicated decoupling capacitor capacitance is available, incorporating in-package capacitance should be considered. However, it is noteworthy that compared to DTCs and on-die capacitors, MLCCs exhibit a lower resonant frequency, typically ranging from a few MHz to tens of MHz. Figure 2.17 shows the location of these six types of capacitors under advanced packaging technology.

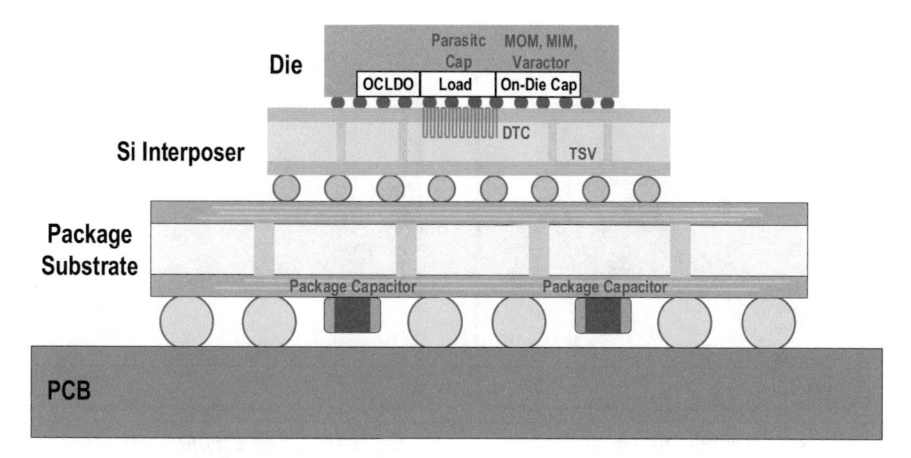

Fig. 2.17 The location of these six types of capacitors

2.15 Integration Method

During integration, to minimize the interference with the LDO, the ground of the LDO should not be directly connected to the load. Instead, it is recommended to connect it directly to the package bump, or if silicon area and I/O resource are sufficient, provide a dedicated ground path for the LDO.

Typically, LDOs need to be placed close to the load and connected through the top metal layer to mitigate the IR drop on the power delivery path. The area occupied by the LDO is usually much smaller than the load area. For low-current LDOs (say <150 mA), particularly for analog LDOs, centralized integration is more suitable (Fig. 2.18a). However, if the load current is substantial, centralized integration may result in a significant IR drop. In such cases, a distributed power stage can be employed, where the power transistors are arranged in long strips and distributed on one side of the load, as illustrated in Fig. 2.18b. This distributed power stage architecture can leverage more top metal resources and effectively reduce the IR drop.

Nevertheless, the integrated approach illustrated in Fig. 2.18a, b only has a single detection point. When the load area is further increased (>1 mm2), the load current is substantially higher (>1A); especially when there are local transients, there is still a large lateral IR drop. Moreover, the single sensing point cannot respond to the local transients in time, leading to severe local voltage drops. To address this issue,

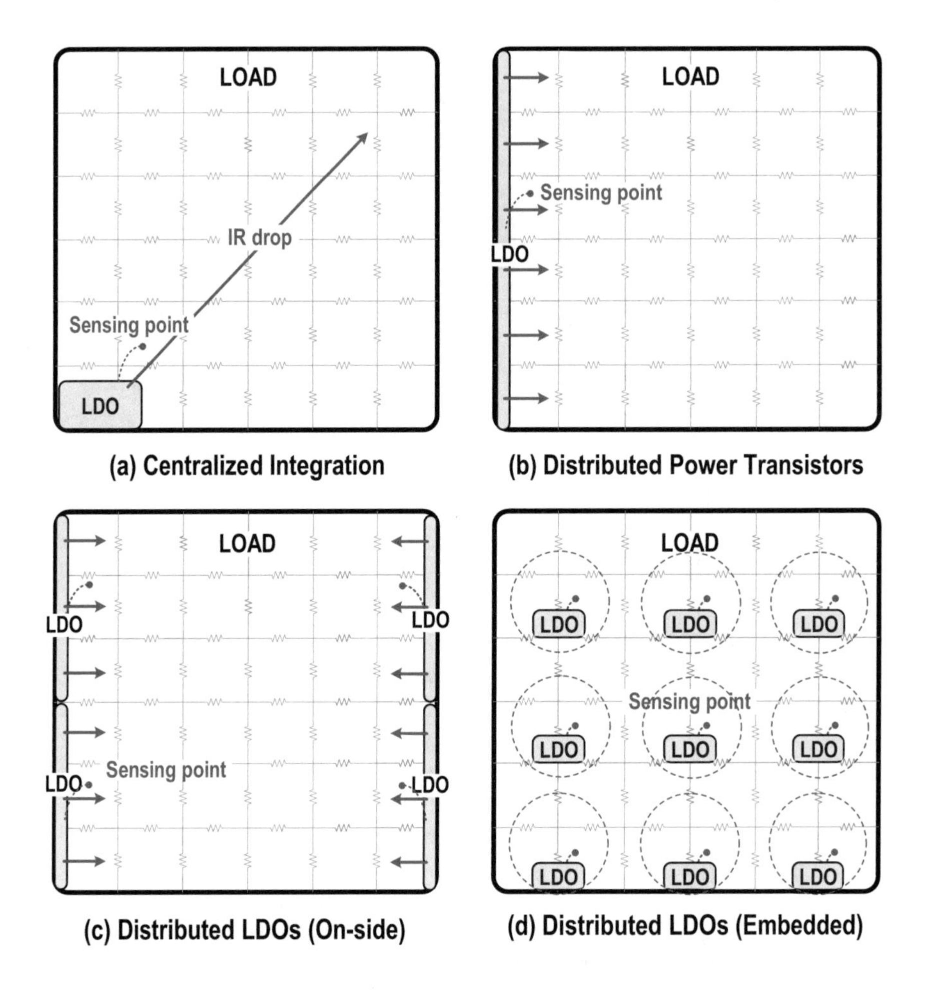

Fig. 2.18 Integration methods. (**a**) Centralized integration. (**b**) Distributed power transistors. (**c**) On-side distributed LDO. (**d**) Embedded distributed LDO

a large LDO can be partitioned into multiple smaller LDOs, each local LDO with a separate sensing point and tuning loop. The distributed power network helps to reduce the current redistribution through the grid and thus can decrease the IR drop. In addition, due to the multiple sensing points and their short distances to the load circuits, the distributed regulators can respond much faster to sudden local load transients [7].

Distributed LDOs can have two integration forms: one is that they are integrated on both sides of the load, and the other is that they are embedded inside the load. With the embedded integration, the power supply regulator is closer to the load, resulting better performance, but the integration is also the most challenging issue.

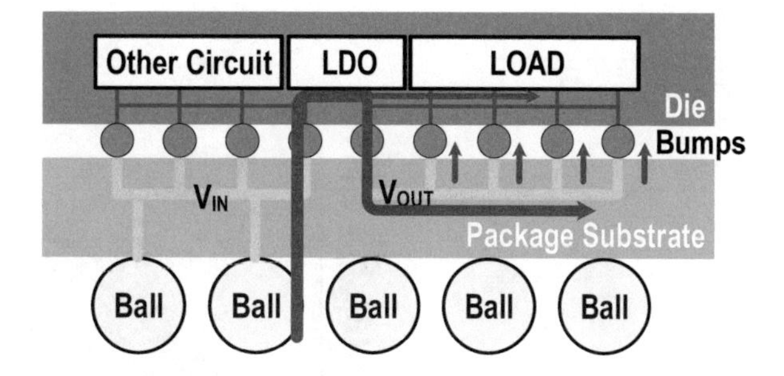

Fig. 2.19 Package connection to reduce the impedance

Furthermore, since multiple regulators with different offsets share the same input-output power grid, we should pay attention to the current-sharing/balancing problem. Severe unbalanced currents may lead to local hotspots and will also limit the local transient response.

Since the on-chip metal resources are limited, we can leverage package connections to reduce the impedance of the power delivery network (PDN), as illustrated in Fig. 2.19. By utilizing the package connections effectively, the overall impedance profile of the PDN can be improved, mitigating potential voltage droops and ensuring a stable supply voltage across the chip.

2.16 Reliability Consideration

When designing and integrating LDO circuits, especially in advanced processes, we may encounter two reliability problems: electromigration (EM) and self-heating effect (SHE).

EM is the movement of atoms due to the current flow through a material. If the current density is high enough, the heat dissipated within the material will repeatedly detach the atoms from the structure and move them [8]. This can lead to changes in metal resistance and eventually cause open-circuit failures. To mitigate the EM issues, it is essential to ensure that the wires carrying potentially large current density have proper widths to hold them. We need to strictly abide by the constraints of design rules. It should be noted that the EM capability of metal wires is strongly temperature dependent. As the temperature increases from 85 °C to 125 °C, the maximum DC current that can pass through a metal wire may decrease by up to nine times. Therefore, it is crucial to thoroughly consider the operating conditions of the LDO to ensure sufficient EM margin.

Another reliability issue is the self-heating effect (SHE). The heat generated by carrier collisions is accumulated in the well, causing the local temperature to rise [9]. With the continuous shrinkage of technology, the fin field effect transistor (FinFET) structure makes it harder for heat to dissipate. The smaller the process, the more significant the temperature rise will be, as shown in Fig. 2.20. The effects of SHE exacerbate signal and power metal EM reliability failures. The LDO control

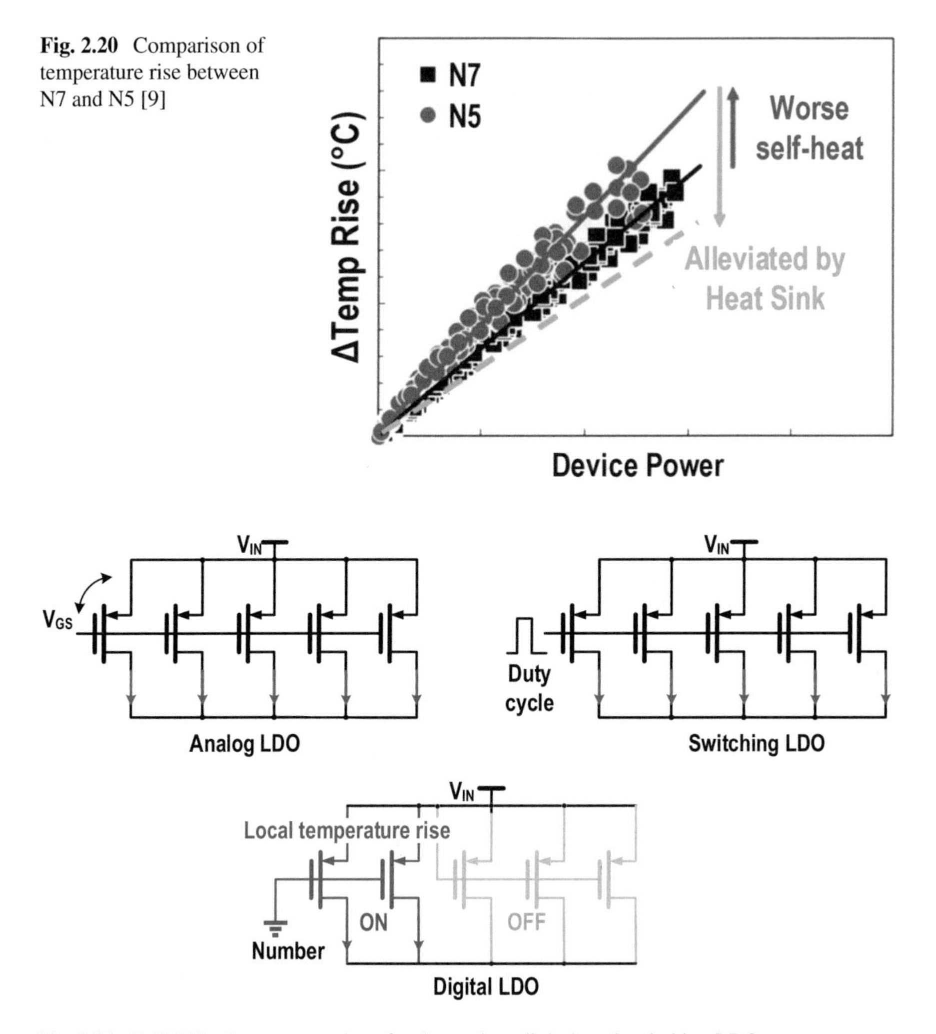

Fig. 2.20 Comparison of temperature rise between N7 and N5 [9]

Fig. 2.21 Reliability issue comparison for the analog, digital, and switching LDOs

methods also have an impact on reliability. Figure 2.21 illustrates the reliability comparison between three LDO types: analog, switching, and digital.

The analog LDO regulates the gate voltage of the power transistors, and the transistors do not fully conduct. The load current is distributed across all power transistors, facilitating the dissipation of heat among all power units. Switching LDOs engage all power switches simultaneously, resulting in a dispersion of both load current and heat across all power switches. However, digital LDOs operate differently from these two control methods. With a gate voltage of 0 V, the p-type power transistor operates at maximum conduction capacity. Consequently, the load current and heat tend to concentrate on part of power transistors.

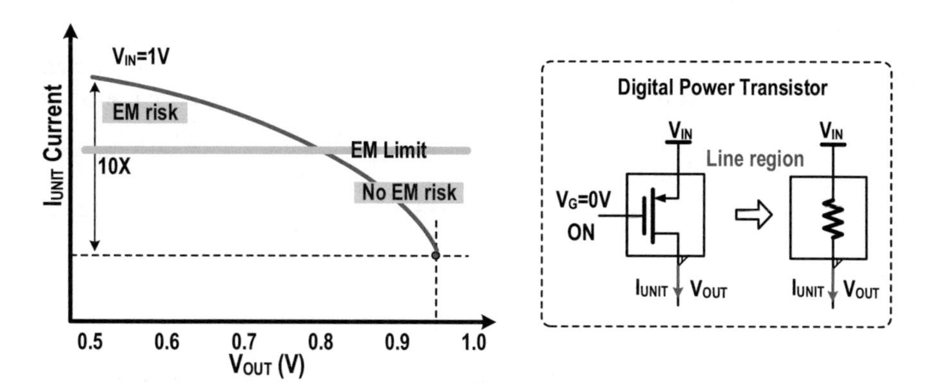

Fig. 2.22 As the V_{DS} voltage increases, I_{UNIT} increases accordingly

In digital LDO, the power transistor operates in the linear region. As the input-output voltage difference increases, the current flowing through the transistor increases rapidly with the rising V_{DS} voltage, leading to a high risk of EM [10], as illustrated in Fig. 2.22.

Furthermore, when V_{DS} and I_{UNIT} increase simultaneously, the power dissipation on the conducting power transistors increases quadratically, potentially forming a localized hotspot and exacerbating the risk of EM. Therefore, these reliability concerns need to be fully considered during the design of digital LDOs.

2.17 Figure of Merit

Various LDO architectures exist, and how do we compare their performance? Many research papers utilize figure of merit (FoM) values in comparison tables [11]. The most commonly used FoM is defined as follows:

$$FoM = \frac{C_{\mathrm{L}} \times V_{\mathrm{DROOP}}}{\Delta I_{\mathrm{L}}} \times \frac{I_{\mathrm{Q}}}{\Delta I_{\mathrm{L}}} = T_{\mathrm{R}} \times \frac{I_{\mathrm{Q}}}{\Delta I_{\mathrm{L}}} \tag{2.13}$$

$$T_{\mathrm{R}} = \frac{C_{\mathrm{L}} \times V_{\mathrm{DROOP}}}{\Delta I_{\mathrm{L}}} \tag{2.14}$$

where C_L denotes the output capacitance, I_{L} represents the load transient step amplitude, V_{DROOP} is the output voltage drop, I_{Q} refers to the quiescent current, and TR represents the response time.

In the same load transient scenario, the smaller the FoM value, the better the LDO performance, which means the faster the response speed, the smaller the output capacitance requirement, the less the output voltage drop, or the lower the quiescent current.

According to the description in Sect. 2.8, edge time is an important factor affecting transient performance. The prerequisite for the formula to hold is that the edge time is much smaller than TR. Taking this factor into account, there are also some improved versions of the FoM value definition, such as FoM2 in [12]:

$$\mathrm{FoM}_2 = I_\mathrm{Q} \times \left(\frac{C_\mathrm{L} \times V_\mathrm{DROOP}}{\Delta I_\mathrm{L}^2} + \frac{0.5}{\mathrm{SR}} \right) \tag{2.15}$$

$$\mathrm{SR} = \frac{\Delta I_\mathrm{L}}{\Delta t} \tag{2.16}$$

The impact of load current slew rate (SR) on droop magnitude is taken into consideration. There is another FoM in [13], mainly for the case of extremely small output capacitance. It is given by

$$\mathrm{FoM}_3 = K \times \left(\frac{I_\mathrm{Q} \times V_\mathrm{DROOP}}{\Delta I_\mathrm{L}} \right) \tag{2.17}$$

References

1. G. Morita, Understand low-dropout regulator (LDO) concepts to achieve optimal designs, ADI, Wilmington. [Online]. Available: https://www.analog.com/en/resources/analog-dialogue/articles/understand-ldo-concepts.html
2. B.S. Lee, Understanding the terms and definitions of LDO voltage regulators. TI Designs (1999). [Online]. Available: https://www.ti.com/lit/an/slva079/slva079.pdf?ts=1738985776133&ref_url=https%253A%252F%252Fwww.google.com%252F
3. F. Li, Q. Fang, J. Wu, Y. Jiang, R.I. Mak, R.P. Martins, M.K. Law, A93.4% peak efficiency CLOAD-free multi-phase switched-capacitor DC-DC converter achieving a fast DVS up to 222.5 mV/ns. IEEE J. Solid-State Circuits, Early Access (2023)
4. A. Paxton, LDO basics: power supply rejection ratio, TI Designs. [Online]. Available: https://www.ti.com/lit/ta/ssztad7/ssztad7.pdf?ts=1715834248-673
5. S. Pithadia, A. Verma, LDO Noise Demystified, TI Designs. [Online]. Available: https://www.ti.com/lit/an/slaa412b/slaa412b.pdf?ts=1715857664887
6. S. Felix, S. Morton, S. Stacey, J. Walsh, Wafer-level stacking of high-density capacitors to enhance the performance of a large multicore processor for machine learning applications, in *IEEE International Solid-State Circuits Conference (ISSCC)*, (Digest of Technical Papers, 2023), pp. 424–425
7. J.F. Bulzacchelli et al., Dual-loop system of distributed microregulators with high DC accuracy, load response time below 500 ps, and 85-mV dropout voltage. IEEE J. Solid-State Circuits **47**(4), 863–874 (2012)
8. What is electromigration? Synopsys. [Online]. Available: https://www.synopsys.com/-glossary/what-is-electromigration.html
9. J.C. Liu, S. Mukhopadhyay, A. Kundu, et al., A reliability enhanced 5nm CMOS technology featuring 5th generation FinFET with fully-developed EUV and high mobility channel for mobile SoC and high performance computing application, in *2020 IEEE International Electron Devices Meeting (IDEM)*, (2020), pp. 179–182

10. R. Muthukaruppan et al., A digitally controlled linear regulator for per-core wide-range dvfs of atom cores in 14 nm Tri-gate CMOS featuring non-linear control, adaptive gain and code roaming, in *Proceeding 43rd IEEE European Solid State Circuits Conference*, (2017), pp. 275–278
11. P. Hazucha, T. Karnik, B.A. Bloechel, et al., Area-efficient linear regulator with ultra-fast load regulation. IEEE J. Solid-State Circuits **40**(4), 933–940 (2005)
12. X. Liu, H.K. Krishnamurthy, T. Na, S. Weng, K.Z. Ahmed, K. Ravichandran, J. Tschanz, V. De, A modular hybrid LDO with fast load-transient response and programmable PSRR in 14nm CMOS featuring dynamic clamp tuning and time-constant compensation, in *IEEE International Solid-State Circuits Conference (ISSCC)*, (Digest of Technical Papers, 2019), pp. 234–235
13. J. Guo, K.N. Leung, A 6-μW chip-area-efficient output-capacitorless LDO in 90-nm CMOS technology. IEEE J. Solid-State Circuits **45**(9), 1896–1905 (2010)

Chapter 3
Analog LDO

3.1 Introduction

As its name implies, an analog LDO stabilizes the output voltage by modulating the gate voltage of the power transistor through analog control mechanisms. A typical analog LDO architecture comprises five key components: an error amplifier, a compensation network, a drive circuit, power transistors, and a feedback network, as shown in Fig. 3.1. The error amplifier is used to amplify the discrepancy between the feedback voltage V_{FB} and the reference voltage V_{REF}. The output signal V_{EA} is then processed by the drive circuit, which controls the power transistor's gate voltage, adjusting its on-resistance. The error amplifier, drive circuit, power transistor, and feedback network form a negative feedback loop. The compensation network is incorporated to ensure the stability of this loop under various operating conditions.

It is worth noting that the complexity of the LDO's control circuit can be tailored based on the specific requirements. In its simplest form, an NMOS source follower can be considered as an LDO. The control circuit of an analog LDO can be highly compact. Fully integrated analog LDOs are widely used for application requirements ranging from a few mA to hundreds of mA. Specifically, analog control offers continuous regulation with no output ripple and can suppress noise from the input power supply, making it suitable for loads with high-power quality demands such as analog-to-digital converters (ADCs), radio frequency (RF) circuits, and sensors [1].

The following sections provide a detailed overview of the core components and key technologies of LDOs, including error amplifiers, compensation, buffers, adaptive biasing, and power supply ripple rejection (PSRR) analysis. Throughout these sections, we will also highlight representative papers that utilize these classic technologies and structures. Finally, the last two sections will discuss flipped-voltage follower (FVF)-based LDOs and NMOS LDOs in detail.

X. Mao et al., *Fully-Integrated Low-Dropout Regulators*, Analog Circuits and Signal Processing, https://doi.org/10.1007/978-3-031-84916-9_3

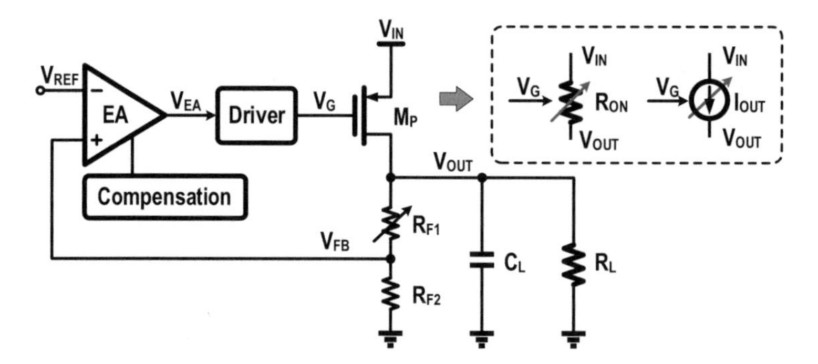

Fig. 3.1 The block diagram of a typical analog LDO

3.2 Single-Stage Amplifiers

The analysis of analog LDO necessitates extensive small-signal analysis. Figure 3.2 illustrates the small-signal model of a PMOS transistor [2]. In this model, S, D, G, and B nodes represent the source, drain, gate, and body terminals, respectively. The parasitic capacitances between these terminals are denoted as C_{GS}, C_{GD}, C_{SB}, C_{DB}, and C_{GB}.

Table 3.1 presents the parasitic capacitance values of a PMOS transistor in a 65 nm CMOS process. The PMOS transistor operates in saturation region. Notably, C_{GS} constitutes the primary component of the gate parasitic capacitance.

The transconductance g_m is defined as the ratio of the change in drain current to the change in gate-to-source voltage V_{GS}. The output resistance r_o accounts for the channel length modulation effect and represents the output impedance of the transistor. In addition, g_{mb} is defined as the change in drain current divided by the change in body-to-source voltage (V_{BS}), which arises due to the body effect.

Single-stage amplifiers form the foundational basis for the study of LDOs. Figure 3.3 presents three fundamental single-stage amplifier structures along with their simplified small-signal models [2]. In Fig. 3.3a, the input signal is applied to the gate of transistor M_1, and the output is taken from the drain, defining it as a common-source (CS) amplifier. The gain of this amplifier is expressed as

$$A_V = \frac{\partial V_{OUT}}{\partial V_{IN}} = -g_m \left(R_D // r_o \right) = -g_m \frac{R_D r_o}{R_D + r_o} \tag{3.1}$$

The common-source stage is an inverting amplifier; thus, for every change ΔV_{IN} at the input, the output changes by $-g_m(R_D//r_o)\Delta V_{IN}$.

In Fig. 3.3b, the input signal is applied to the gate of the transistor, while the output is taken from the source, forming a common-drain (CD) amplifier. Its gain is given by

$$A_V = \frac{\partial V_{OUT}}{\partial V_{IN}} = \frac{g_m \left(R_S // r_o \right)}{1 + \left(g_m + g_{mb} \right) \left(R_S // r_o \right)} \tag{3.2}$$

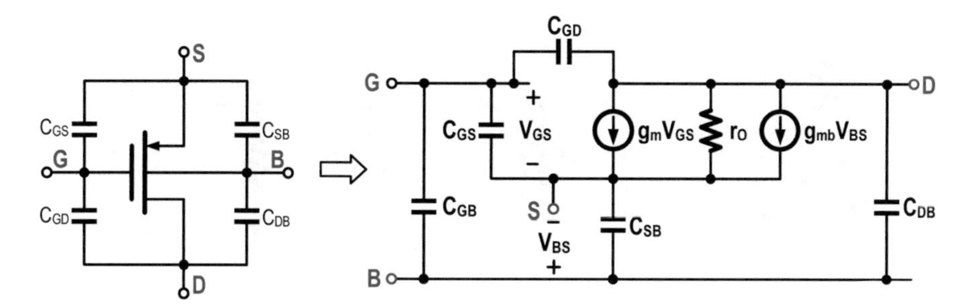

Fig. 3.2 The small-signal model of a PMOS transistor

Table 3.1 The parasitic capacitance values of a PMOS transistor

PMOS	C_{GS}	C_{GD}	C_{GB}	C_{SB}	C_{DB}
$W = 5$ μm	5.1 fF	0.98 fF	0.25 fF	0.39 fF	0.013 fF
$L = 100$ nm	Total: $C_{GG} = 6.33$ fF				

Usually, $g_m \gg g_{mb}$, $g_m(R_S//r_o) \gg 1$, and $A_V \approx 1$. Therefore, the common-drain amplifier is also known as a source follower, as the output signal closely follows the input signal with approximately the same amplitude.

In Fig. 3.3c, the input signal is applied to the source of the transistor, the gate is tied to a fixed bias voltage V_B, and the output is taken from the drain. This configuration is defined as a common-gate (CG) amplifier, and its gain is given by

$$A_V = \frac{\partial V_{OUT}}{\partial V_{IN}} = (g_m + g_{mb})\left[1 + \frac{1}{(g_m + g_{mb})r_o}\right](R_D // r_o) \approx (g_m + g_{mb})(R_D // r_o) \quad (3.3)$$

The common-gate stage functions as a non-inverting amplifier.

In Fig. 3.3, R_D (R_S) represents the load resistor in the basic single-stage amplifiers. In practical circuits, MOSFETs are commonly used as load resistors in addition to actual resistors. Figure 3.4 presents six typical configurations of MOSFETs utilized as load resistors and the equivalent resistance R_{EQ} observed from the node [3].

Figure 3.4a represents the impedance seen from the gate of the MOSFET, where the equivalent resistance is an infinite (an open circuit). In Fig. 3.4b, the current mirror is used as a load, with the gate voltage of the NMOS set to V_B and the source grounded. The R_{EQ} seen from the drain is r_o.

Figure 3.4c describes a source follower configuration, where the drain is connected to the power supply, the gate is connected to V_B, and the impedance seen from the source is $r_o \,||\, \frac{1}{g_m} \,||\, \frac{1}{g_{mb}} \,||$. Figure 3.4d illustrates a diode-connected method. Since the source is grounded, there is no body effect, and the impedance seen from the drain is $r_o \,||\, \frac{1}{g_m}$.

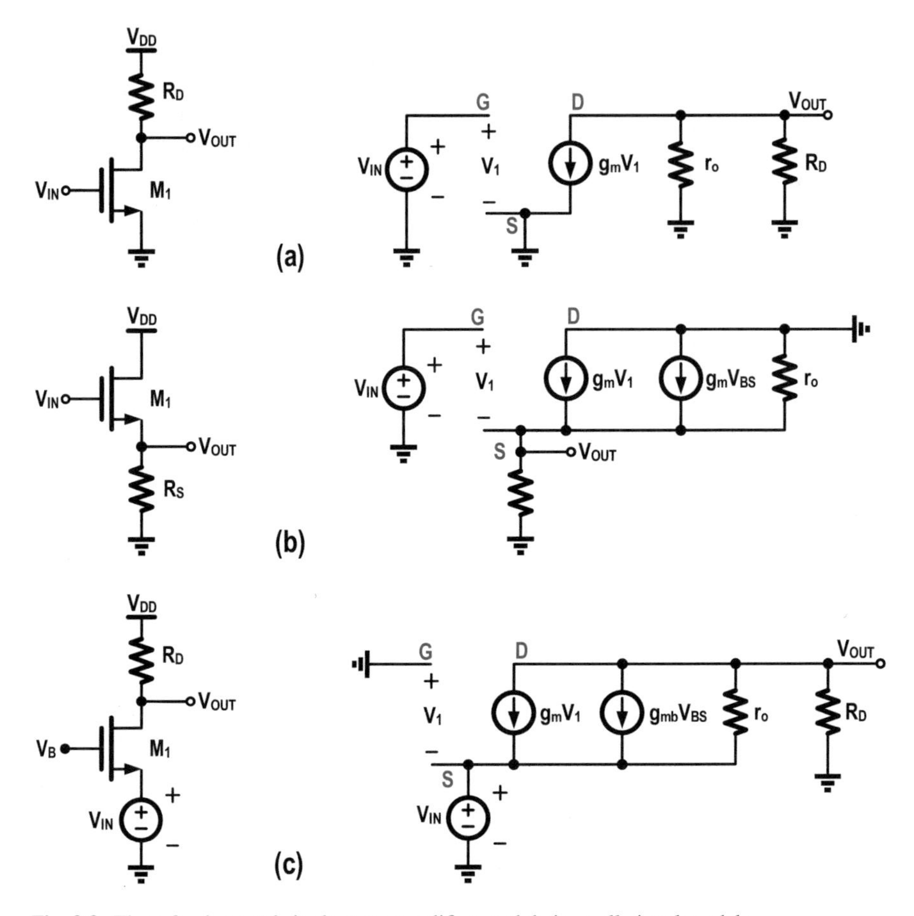

Fig. 3.3 Three fundamental single-stage amplifiers and their small-signal models

Next, in Fig. 3.4e, a resistor R_S is connected in series at the source of the NMOS. The impedance seen from the drain will be significantly greater than R_S, with an equivalent resistance of $r_o + R_S + (g_m + g_{mb})r_oR_S$, amplified by a factor of at least $(g_m + g_{mb})r_o$ times. For Fig. 3.4f, a resistor R_D is added in series with the drain of the NMOS, and the impedance seen from the source will be significantly less than R_D, with an equivalent impedance of $\frac{R_S + r_o}{1+\left(g_m+g_{mb}\right)r_o}$, effectively reducing the impedance by a factor of $(g_m + g_{mb})r_o$.

Figure 3.4 uses NMOS as an example, and similar impedance formulas apply for PMOS. By combining the five resistor configurations (b–f) with the three basic single-stage operational amplifiers (shown in Fig. 3.3), various single-stage amplifier topologies can be obtained.

Table 3.2 lists some typical single-stage amplifier circuits along with their gains and equivalent output impedances.

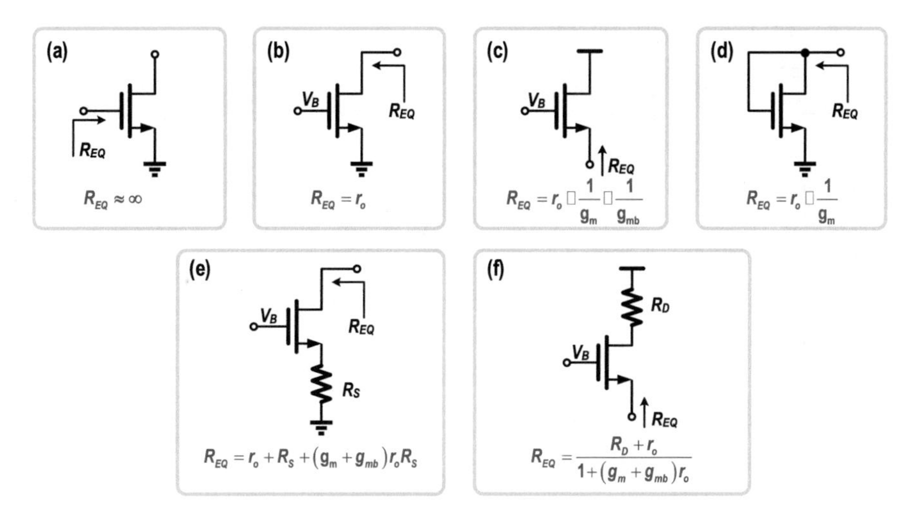

Fig. 3.4 Six configurations of MOSFETs and the equivalent resistance [3]

3.3 Error Amplifiers

The error amplifier is a crucial component in the design of analog LDO. It is mainly used to amplify the differential signal between the feedback signal of the output voltage and the reference voltage. The performance of the error amplifier has a significant impact on nearly all characteristics of LDO. In this chapter, we present several commonly used error amplifiers.

Figure 3.5 illustrates two basic single-stage, single-ended output operational transconductance amplifiers (OTAs). The core part consists of only five MOSFETs. M_3 and M_4 form the differential input pair, M_2 supplies the bias current for this differential pair, and M_5 and M_6 serve as the current mirror loads. For optimal operation, M_2–M_5 should function in the saturation region. The gain of the operational amplifier is

$$A_V \approx g_{m4} \times \left(r_{o6} // r_{o4}\right) \tag{3.4}$$

where g_{m4} is the transconductance of M_4, and r_{o4} and r_{o6} are the output impedances of M_4 and M_6, respectively. The output impedance of the operational amplifier is

$$R_O \approx r_{o6} // r_{o4} \tag{3.5}$$

The output swing of the operational amplifier in Fig. 3.5 is constrained by the input common-mode voltage. Figure 3.6 illustrates an alternative operational transconductance amplifier. In this design, M_3 and M_4 serve as the input differential pair, converting the differential voltages into differential currents. M_5, M_7, and M_1 are diode connected, which relay the differential current information pass to the output stage. The gain of the OTA is

$$A_V \approx g_{m4} \times \left(r_{o8} // r_{o2}\right) \tag{3.6}$$

Table 3.2 Typical single-stage amplifier circuits

No.	Structure	Gain	Ro
1	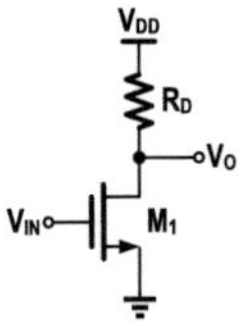 VDD, RD, VO, VIN, M1	$A_V = -g_{m1}\dfrac{R_D r_{o1}}{R_D + r_{o1}}$	$\dfrac{R_D r_{o1}}{R_D + r_{o1}}$
2	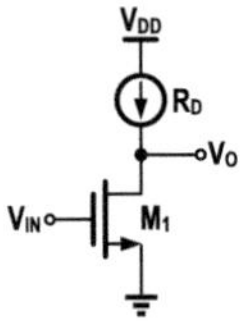 VDD, RD, VO, VIN, M1	$A_V = -g_{m1} r_{o1}$	r_{o1}
3	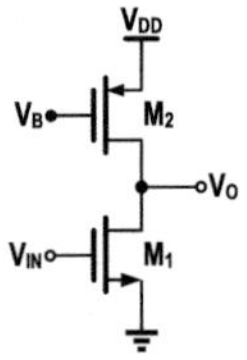 VDD, VB, M2, VO, VIN, M1	$A_V = -g_m \dfrac{r_{o1} r_{o2}}{r_{o1} + r_{o2}}$	$\dfrac{r_{o1} r_{o2}}{r_{o1} + r_{o2}}$
4	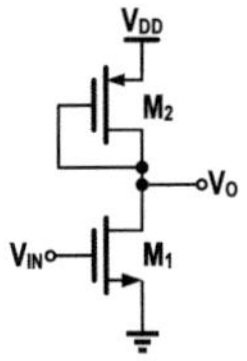 VDD, M2, VO, VIN, M1	$A_V = -g_{m1}\dfrac{1}{g_{m2} + \dfrac{1}{r_{o1}} + \dfrac{1}{r_{o2}}}$	$\dfrac{1}{g_{m2} + \dfrac{1}{r_{o1}} + \dfrac{1}{r_{o2}}}$
5	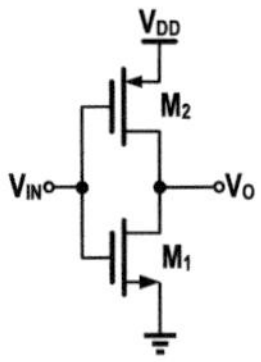 VDD, M2, VIN, VO, M1	$A_V = -(g_{m1} + g_{m2})\dfrac{r_{o1} r_{o2}}{r_{o1} + r_{o2}}$	$\dfrac{r_{o1} r_{o2}}{r_{o1} + r_{o2}}$
6	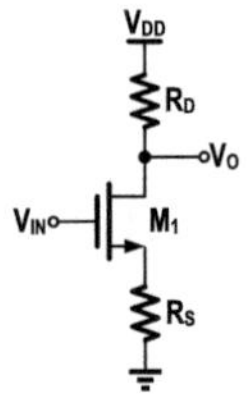 VDD, RD, VO, VIN, M1, RS	$A_V = \dfrac{-g_{m1} r_o R_D}{R_D + R_S + r_o + (g_{m1} + g_{mb1}) R_S r_o}$	$\left[R_S + r_o + (g_m + g_{mb}) R_S r_o\right] \Box R_D$

(continued)

Table. 3.2 continued

7	V_{DD}, V_{IN}, M_1, V_O, R_S	$A_V = \dfrac{g_{m1}}{\dfrac{1}{R_S} + \dfrac{1}{r_{o1}} + (g_{m1} + g_{mb1})}$	$\dfrac{1}{\dfrac{1}{R_S} + \dfrac{1}{r_{o1}} + (g_{m1} + g_{mb1})}$
8	V_{DD}, V_{IN}, M_1, V_O, I_1	$A_V = \dfrac{g_{m1}}{\dfrac{1}{r_{o1}} + (g_{m1} + g_{mb1})}$	$\dfrac{1}{\dfrac{1}{r_{o1}} + (g_{m1} + g_{mb1})}$
9	V_{DD}, V_{IN}, M_1, V_O, V_B, M_2	$A_V = \dfrac{g_{m1}}{\dfrac{1}{r_{o2}} + \dfrac{1}{r_{o1}} + (g_{m1} + g_{mb1})}$	$\dfrac{1}{\dfrac{1}{r_{o2}} + \dfrac{1}{r_{o1}} + (g_{m1} + g_{mb1})}$
10	V_{DD}, R_D, V_O, V_B, M_1, V_{IN}, +, −	$A_V =$ $(g_{m1} + g_{mb1})\dfrac{R_D r_{o1}}{R_D + r_{o1}} + \dfrac{R_D}{R_D + r_{o1}}$	$\dfrac{R_D r_{o1}}{R_D + r_{o1}}$
11	V_{DD}, R_D, V_O, V_B, M_1, R_S, V_{IN}, +, −	$A_V =$ $\dfrac{[(g_{m1} + g_{mb1}) r_{o1} + 1] R_D}{r_{o1} + (g_{m1} + g_{mb1}) r_{o1} R_S + R_S + R_D}$	$[R_S + r_o + (g_m + g_{mb}) R_S r_o] \square R_D$
12	V_{DD}, V_{B3}, M_4, V_{B2}, M_3, V_O, V_{B1}, M_2, V_{IN}, M_1	$A_V \approx$ $g_{m1}[(g_{m2} r_{o2} r_{o1}) \square (g_{m3} r_{o3} r_{o4})]$	$\approx (g_{m2} r_{o2} r_{o1}) \square (g_{m3} r_{o3} r_{o4})$

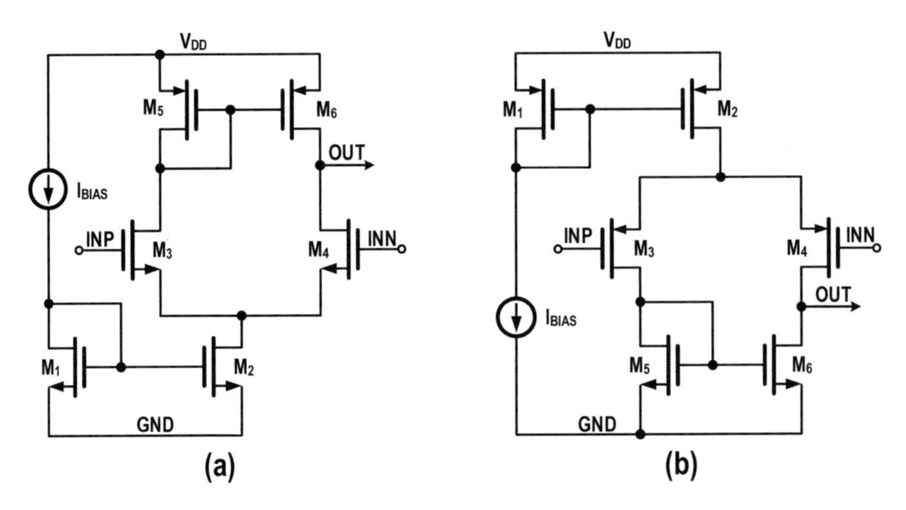

Fig. 3.5 Basic differential amplifier with (**a**) NMOS differential input pair and (**b**) PMOS differential input pair

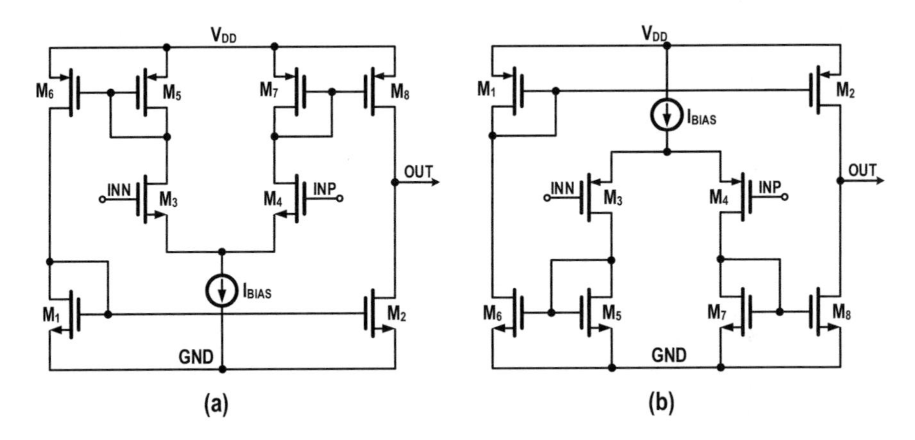

Fig. 3.6 Operational transconductance amplifier with (**a**) NMOS differential input pair and (**b**) PMOS differential input pair

The output impedance of the OTA is

$$R_O \approx r_{o8} // r_{o2} \tag{3.7}$$

Taking Fig. 3.6a as an example, the maximum swing of the output signal can be

$$V_{DSAT2} < V_R < V_{DD} - V_{DSAT8} \tag{3.8}$$

It is significantly larger than the swing range in Fig. 3.5a, but this improvement comes at the cost of adding two additional current paths, thereby increasing the total current consumption.

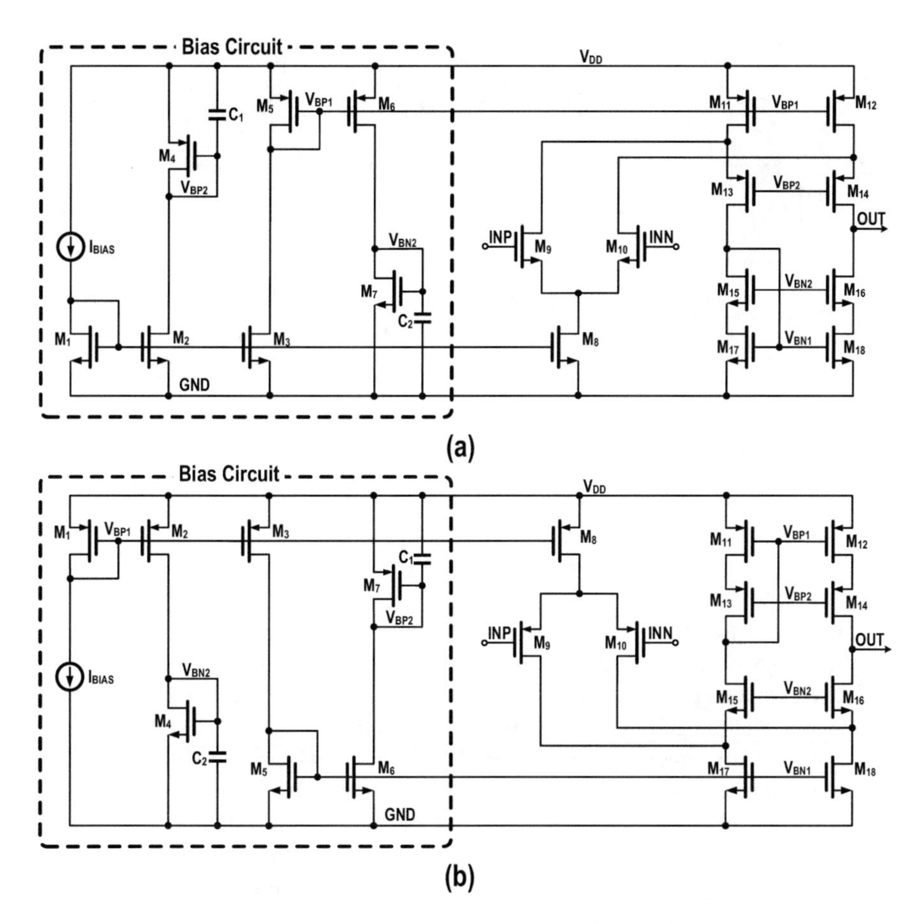

Fig. 3.7 Folded cascode operational amplifier with (**a**) NMOS differential input pair and (**b**) PMOS differential input pair

Figure 3.7 illustrates the folded cascode amplifier. The bias circuit on the left generates bias voltages V_{BN2} and V_{BP2} for the cascode transistors. Transistors M_4, M_{15}, and M_{16} are NMOS transistors, while M_7, M_{13}, and M_{14} are PMOS transistors, which helps mitigate the effects of process, voltage, and temperature (PVT) variations. Using Fig. 3.7a as an example, the minimum supply voltage V_{DDMIN} is

$$V_{DDMIN} \geq V_{DSAT11} + V_{CM} - V_{GS9} + V_{DSAT9} \tag{3.9}$$

while in Figs. 3.6 and 3.7, the minimum power supply voltage is

$$V_{DDMIN} \geq V_{GS7} + V_{CM} - V_{GS4} + V_{DSAT4} \tag{3.10}$$

The folded cascode amplifier reduces the minimum power supply voltage requirement, making it more suitable for low-voltage applications and providing a higher voltage design margin. The folded cascode amplifier has a higher DC gain:

$$A_V \approx g_{m9} \times \left(g_{m14} r_{o14} r_{o12} \, / / \, g_{m16} r_{o16} r_{o18} \right) \tag{3.11}$$

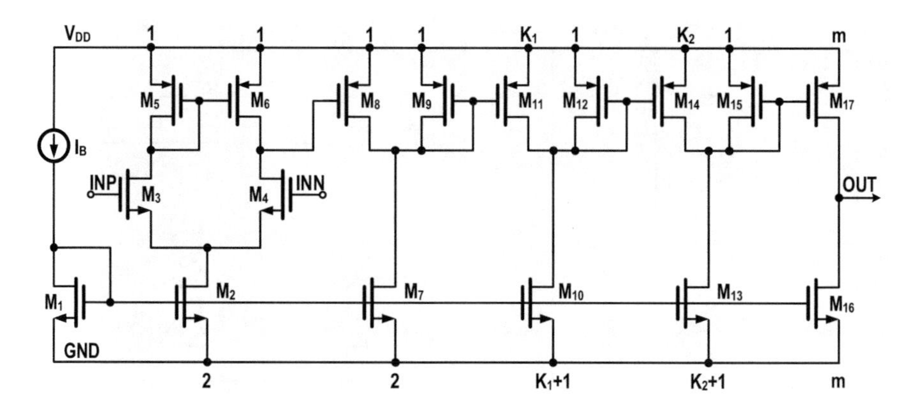

Fig. 3.8 The amplifier with multiple small-gain stage in [4]

The output impedance is

$$R_{\mathrm{O}} \approx g_{\mathrm{m}14} r_{\mathrm{o}14} r_{o12} \,//\, g_{\mathrm{m}16} r_{\mathrm{o}16} r_{\mathrm{o}18} \tag{3.12}$$

To simplify the compensation of LDO, the operational amplifier in LDOs typically employs a single-stage design, as shown in Figs. 3.5, 3.6, and 3.7. However, with the advanced process, the shrinkage in device feature size reduces the channel resistances, and the gain of the amplifier has diminished.

Figure 3.8 illustrates an amplifier with multiple small-gain stages in [4]. This structure improves the loop gain and bandwidth, enabling a fast transient response LDO design. Since the output impedance of each stage is low and the poles are at high frequency (much higher than the loop bandwidth), no additional on-chip compensation circuit is required. On the other hand, the multistage structure also increases static current consumption.

3.4 Buffer Circuits

Due to the requirements for low-dropout voltage and high current output, the device size of power transistors is typically very large. The gate parasitic capacitance of these power transistors often reaches tens to hundreds of *p*F.

In the two-stage LDO structure depicted in Fig. 3.9a, the error amplifier (EA) output directly drives the power transistor. According to the single-stage amplifier designs shown in Figs. 3.5, 3.6, and 3.7, ensuring sufficient gain necessitates a high output impedance. This high impedance makes it challenging to effectively drive large power transistors, resulting in a low gate voltage slew rate, thus severely compromising the transient response performance of the LDO. Moreover, in nanometer processes (<65 nm and below), the gate of low-voltage devices typically exhibits considerable leakage current, which affects the DC operating point of the op amp and induces an offset voltage, consequently impacting the accuracy of the output voltage. Since leakage current varies significantly with temperature, calibrating the offset voltage through a fixed trimming becomes impossible.

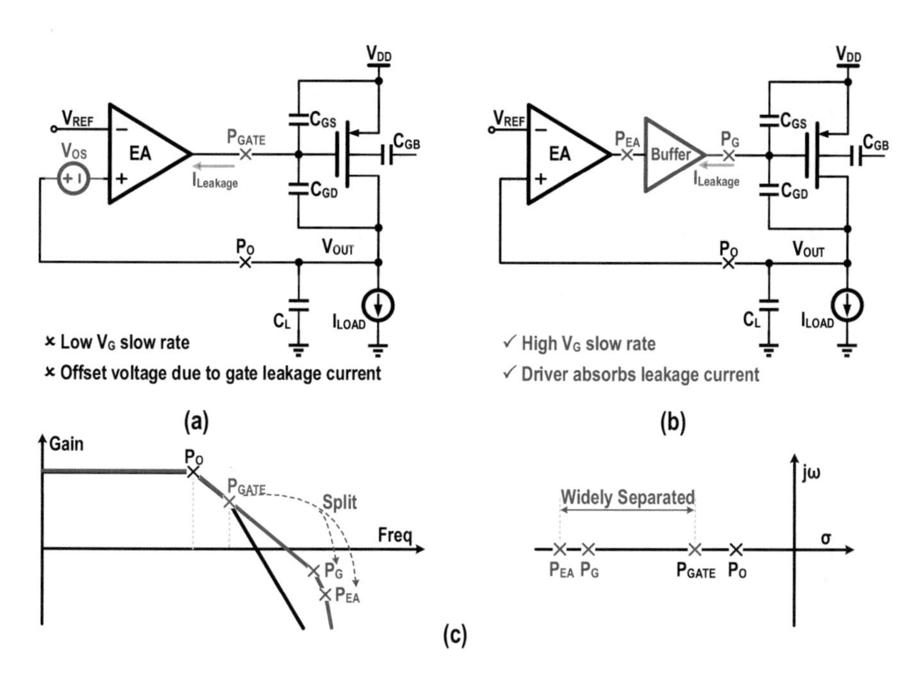

Fig. 3.9 The role of the buffer (**a**) LDO without buffer, (**b**) LDO with buffer, (**c**) pole separation by inserting the buffer

A buffer circuit is indispensable for fast transients or gate leakage current applications. As illustrated in Fig. 3.9b, the buffer receives the output signal V_{EA} from the EA and enhances the driving capability before driving the power transistor. The buffer output exhibits lower impedance and higher output current, facilitating a higher slew rate, which can rapidly adjust the gate voltage V_G during fast transients. In addition, the gate leakage current of the power transistor is absorbed by the buffer circuit, preventing it from impacting the operational state of the error amplifier. This effectively eliminates the offset voltage induced by leakage current.

In the frequency domain, the insertion of the buffer causes the original low-frequency gate pole P_{GATE} to split into two high-frequency poles P_{EA} and P_G, as depicted in Fig. 3.9c. If the output pole is the dominant pole, the separation assists in improving the loop stability. If the output pole is a nondominant pole, P_{EA} can still be repositioned appropriately through compensation, increasing the flexibility of the compensation design.

3.4.1 Common-Source Buffer

Figure 3.10 illustrates the common-source buffer configurations. In Fig. 3.10a, the gate of M_1 is connected to the output signal V_{EA} from the error amplifier, while the drain is connected to V_G and the current source I_1. When V_{EA} increases, V_G decreases, making it an inverting buffer.

In Fig. 3.10b, the addition of current mirrors M_2 and M_3 transforms it into a non-inverting buffer. When V_{EA} increases, V_G also increases. For Fig. 3.10a, the output impedance is

$$R_O = r_{o1} \tag{3.13}$$

Increasing the current I_1 can reduce the output impedance R_O.

Figure 3.11 illustrates a diode-connected common-source buffer. In Fig. 3.11a, the current source I_1 is replaced by a diode-connected MOSFET M_2. The output impedance is

$$R_O = r_{o1} // r_{o2} // \frac{1}{g_{m2}} \approx \frac{1}{g_{m2}} \tag{3.14}$$

Compared to Fig. 3.10a, the output impedance R_O is significantly reduced. The diode connection of M_2 and the power transistor M_P form a current mirror, making the buffer current proportional to the output current. Under heavy load conditions, the output pole shifts to a higher frequency. Consequently, the buffer current increases, R_O decreases, and the parasitic pole of the gate also shifts to a higher frequency, aiding in loop stabilization. Under light-load conditions, the buffer current decreases, which helps reduce the quiescent current.

In Fig. 3.11b, current mirrors M_2 and M_3 are incorporated to form a non-inverting buffer. In the current mirror shown in Fig. 3.11a, b, the slew rate of V_G is limited by $C_G \times 1/g_{m2}$. To further enhance the slew rate, [5] proposed a supercurrent mirror structure. Under steady-state conditions, $V_G = V_B$, and V_{BOOST} regulates M_7 such that $I_{BOOST} = I_{IN}$. When V_{EA} increases rapidly, the resulting increase in I_{IN} quickly reduces the currents in M_6 and M_5, causing V_{BOOST} to rise. This leads to an increase in I_{BOOST} and a corresponding decrease in V_B and V_G.

The lower right corner of Fig. 3.11c presents an equivalent model for calculating the output impedance. M_2–M_7 can be regarded as a two-stage operational amplifier (op amp), with M_3–M_6 comprising the first stage, and M_2–M_7 forming the second stage (output stage). V_B serves as both the input and output of the op amp. The op

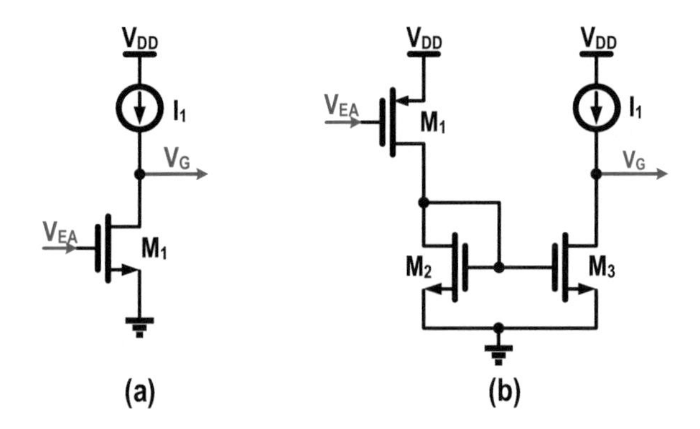

Fig. 3.10 Common-source buffer with a fixed biasing current: (**a**) inverting buffer and (**b**) non-inverting buffer

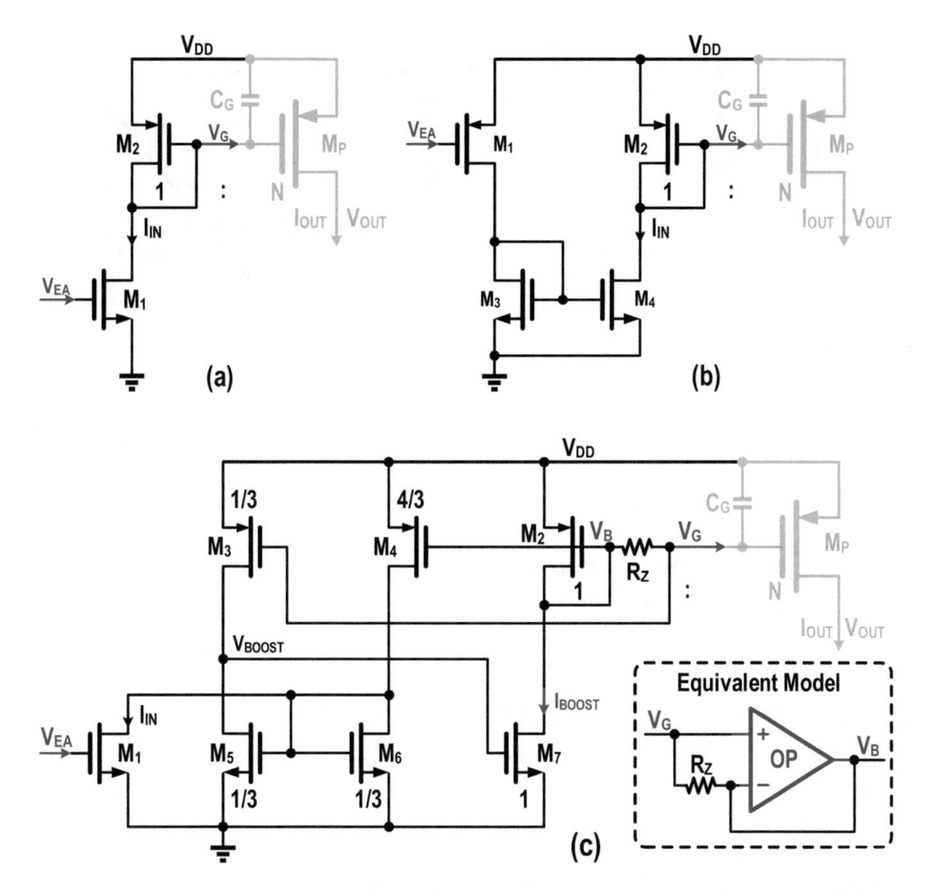

Fig. 3.11 Diode-connected common-source buffer: (**a**) inverting buffer, (**b**) non-inverting buffer, (**c**) supercurrent mirror

amp (composed of M_2–M_7) forms a unity-gain negative feedback loop. Negative feedback can significantly reduce the output impedance. The output impedance of the supercurrent mirror is

$$R_O \approx R_Z + \frac{1}{g_{m3}g_{m7}\left(r_{o3}//r_{o5}\right)} \tag{3.15}$$

For applications with wide-range load currents, the buffer current will also fluctuate significantly with the load current. Under heavy-load conditions, if the gate pole is already far beyond the bandwidth, further increase in buffer current will not enhance loop stability or transient performance. Instead, it may result in higher current consumption and adversely affect the operating point of the EA and the drive circuit. In such scenarios, adding a resistor R_1 can limit the maximum value of the drive current, as illustrated in Fig. 3.12. The bias current I_1 is used to establish the

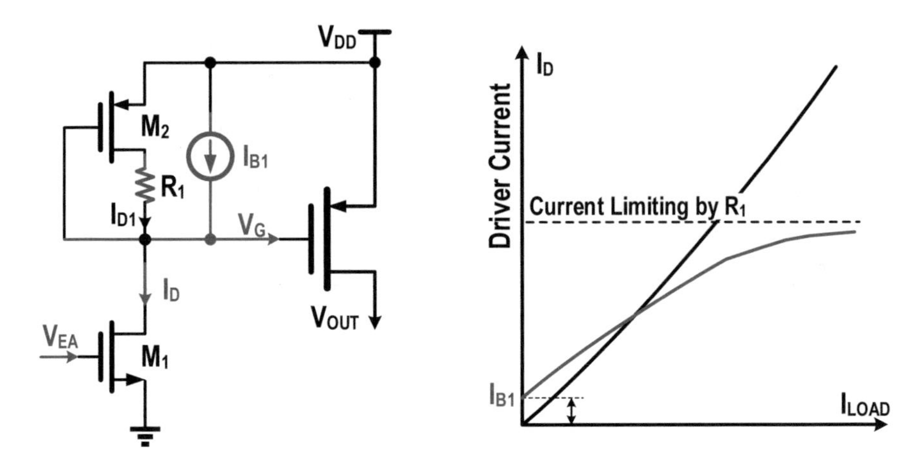

Fig. 3.12 A current mirror buffer with maximum and minimum current limits

minimum buffer current, ensuring that the buffer circuit functions correctly under extremely light-load condition.

3.4.2 Source Follower Buffer

The source follower is also a commonly used buffer. Figure 3.13 illustrates five types of source follower buffers. Figure 3.13a shows the basic structure [6], where the gate of M_1 is connected to V_{EA}, and the source is the buffer's output, V_G. I_1 provides the bias current. The output impedance of this structure is

$$R_O = r_{o1} // \frac{1}{g_{m1}} \approx \frac{1}{g_{m1}} \tag{3.16}$$

Increasing the W/L ratio of transistor M_1 or enhancing the DC bias current can increase the transconductance of M_1, thereby reducing the output impedance. However, increasing the bias current will elevate the total quiescent current of the LDO, consequently diminishing its current efficiency. In addition, if a larger M_1 is employed, the gate parasitic capacitance of M_1 will increase.

To further reduce the output impedance, Fig. 3.13b presents a super source follower (SSF) structure [7]. In this configuration, an M_2 path is connected in parallel with the output, V_G. The components M_1, I_2, and M_2 form a negative feedback loop, effectively reducing the output impedance. The output impedance of the SSF is

$$R_O \approx \frac{r_{o2}}{g_{m1}r_{o1} \times g_{m2}r_{o2}} = \frac{1}{g_{m1}r_{o1} \times g_{m2}} \tag{3.17}$$

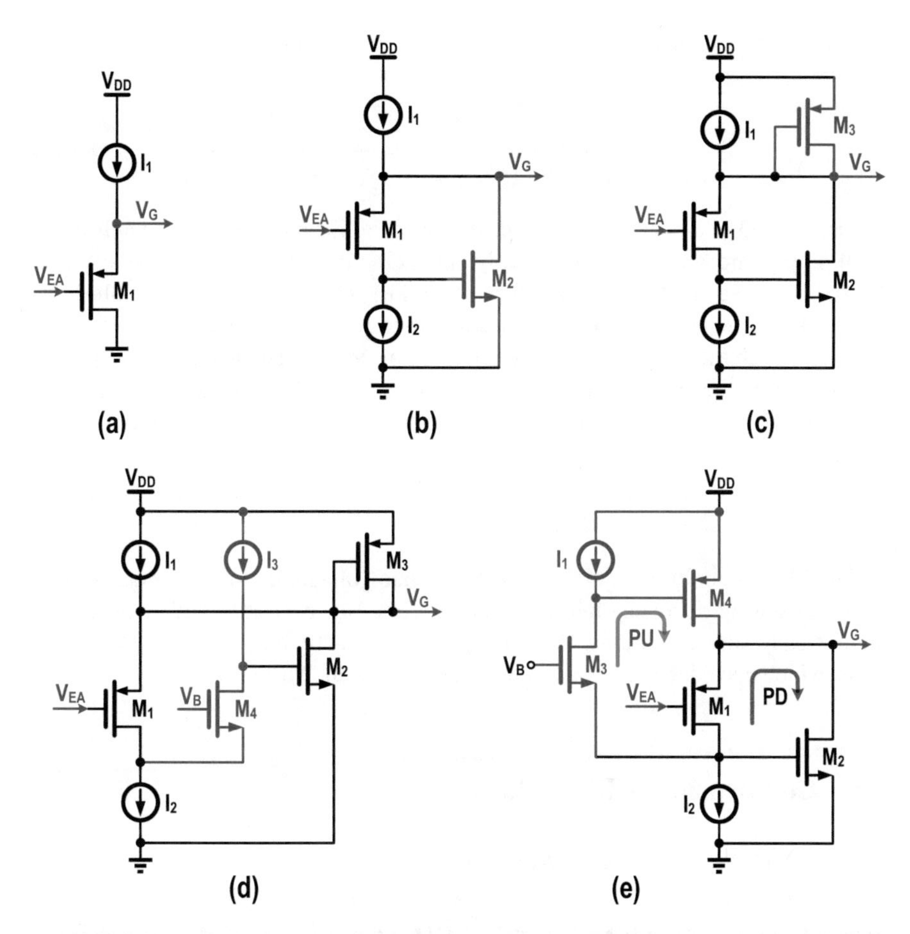

Fig. 3.13 Source follower buffers: (**a**) basic structure [6], (**b**) super source follower (SSF) [7], (**c**) SSF with adaptive biasing [8], (**d**) enhanced SSF with adaptive biasing [9], (**e**) pull-up/pull-down buffer [10]

Figure 3.13c illustrates an enhancement of the SSF structure by incorporating a diode-connected transistor M_3 in parallel with V_G [8]. M_3 and the power transistor M_P form a current mirror. Consequently, the current through M_2 increases in response to rising load current. Under heavy-load conditions, this mechanism further reduces the output impedance and shifts the parasitic gate pole of the power transistor to a higher frequency. The output impedance in Fig. 3.13c is

$$R_O \approx \frac{r_{o2}}{g_{m1}r_{o1} \times g_{m2}r_{o2}} \parallel \frac{1}{g_{m3}} = \frac{1}{g_{m1}r_{o1}g_{m2} + g_{m3}} \tag{3.18}$$

Based on the structure shown in Fig. 3.13c, [9] introduces a common-gate amplifier stage comprising M_4 and I_3. This additional amplifier stage enhances the DC gain and further reduces output impedance. The output impedance of this enhanced super source follower (E-SSF) is

$$R_{\mathrm{O}} \approx \frac{r_{\mathrm{o2}}}{g_{\mathrm{m1}} r_{\mathrm{o1}} \times g_{\mathrm{m4}} r_{\mathrm{o4}} \times g_{\mathrm{m2}} r_{\mathrm{o2}}} \| \frac{1}{g_{\mathrm{m3}}} = \frac{1}{g_{\mathrm{m1}} r_{\mathrm{o1}} g_{\mathrm{m4}} r_{\mathrm{o4}} g_{\mathrm{m2}} + g_{\mathrm{m3}}} \tag{3.19}$$

From Fig. 3.13b–d, the super source follower significantly reduces the pull-down (PD) output impedance, thereby greatly enhancing the buffer's pull-down capability. In Fig. 3.13e, the SSF composed of M_1 and M_2 similarly reduces the pull-down impedance. Concurrently, M_3 and current source I_1 form a common-gate amplifier stage. Through the negative feedback provided by M_1, M_3, and M_4, the pull-up (PU) impedance of the buffer is significantly reduced, thereby enhancing the pull-up capability [10]. The output impedance of the configuration in Fig. 3.13e is

$$R_{\mathrm{O}} \approx \frac{1}{g_{\mathrm{m1}} r_{\mathrm{o1}} g_{\mathrm{m2}}} \| \frac{1}{g_{\mathrm{m1}} r_{\mathrm{o1}} g_{\mathrm{m3}} r_{\mathrm{o3}} g_{\mathrm{m4}}} = \frac{1}{g_{\mathrm{m1}} r_{\mathrm{o1}} \left(g_{\mathrm{m2}} + g_{\mathrm{m3}} r_{\mathrm{o3}} g_{\mathrm{m4}} \right)} \tag{3.20}$$

The PU/PD buffer can quickly adjust the gate voltage, improving transient performance during rising and falling load transients. However, it should be noted that under steady-state conditions, M_2 and M_4 are significantly influenced by PVT variations, which considerably impact the buffer's quiescent current.

3.5 Compensation Technology

3.5.1 *Simple Miller Compensation*

Figure 3.14 illustrates a basic two-stage LDO and its transient response waveform. There are at least two poles:

$$p_1^{'} = \frac{-1}{R_1 C_1} \tag{3.21}$$

$$p_2^{'} = \frac{-1}{R_{\mathrm{O}} C_{\mathrm{L}}} \tag{3.22}$$

$$R_{\mathrm{O}} = R_{\mathrm{L}} // r_{\mathrm{on}} \tag{3.23}$$

where r_{on} is the output impedance of the power tube. When two poles are in proximity, the phase will rapidly drop by 180°, leading to an insufficient phase margin (<45°), which can induce ringing during load transients. If the phase margin is further reduced to <0°, the loop will become unstable. The compensation circuit's role is to separate these poles or introduce left-half-plane zeros, thereby stabilizing the loop and ensuring adequate phase margin.

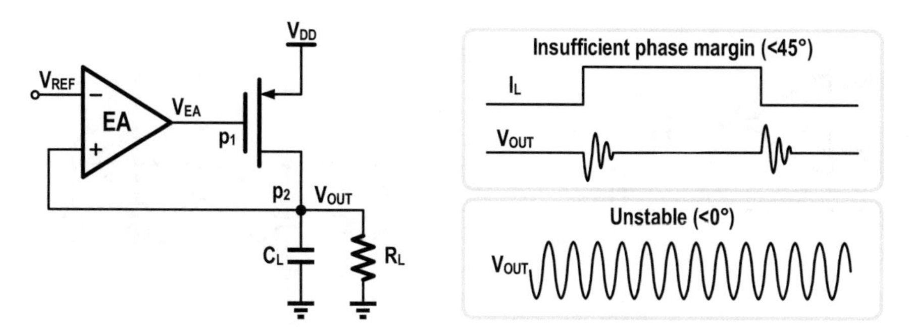

Fig. 3.14 Two-staged LDO without compensation

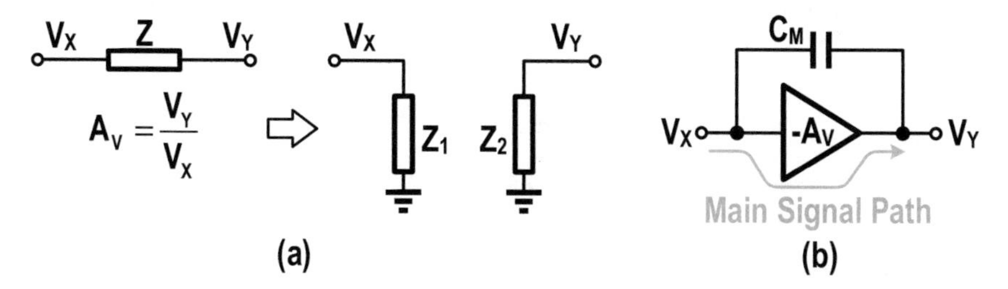

Fig. 3.15 (**a**) Miller's theorem. (**b**) Typical application of Miller's theorem

Miller compensation is a commonly used compensation method. Here, we first introduce Miller's theorem. As illustrated in Fig. 3.15a, the impedance Z is connected across the X and Y nodes. There are additional signal paths between X and Y, and the voltage gain is given by $A_V = V_Y/V_X$.

In this context, the "floating" impedance Z can be equivalently represented by two grounded impedances, Z_1 and Z_2 [2], where

$$Z_1 = \frac{Z}{1 - A_V} \tag{3.24}$$

$$Z_2 = \frac{Z}{1 - A_V^{-1}} \tag{3.25}$$

Figure 3.15b illustrates a typical application of Miller's theorem, where the capacitor C_M is used to split the poles at points X and Y.

As depicted in Fig. 3.16, the compensation capacitor C_M bridges the gate and drain of the power transistor M_P. This compensation circuit is referred to as simple Miller compensation (SMC) in some literature [11, 12]. Based on the small-signal model on the right side of Fig. 3.16, we can derive the transfer function:

$$H(s) = \frac{-g_{m1} g_{mL} R_1 R_O \left(1 - sC_M / g_{mL}\right)}{\begin{array}{l}1 + s\left[R_1\left(C_1 + C_M\right) + R_O\left(C_L + C_M\right) + g_{m1} R_1 R_O C_M\right] \\ + s^2 R_1 R_O \left[C_1 C_L + C_M C_1 + C_M C_L\right]\end{array}} \tag{3.26}$$

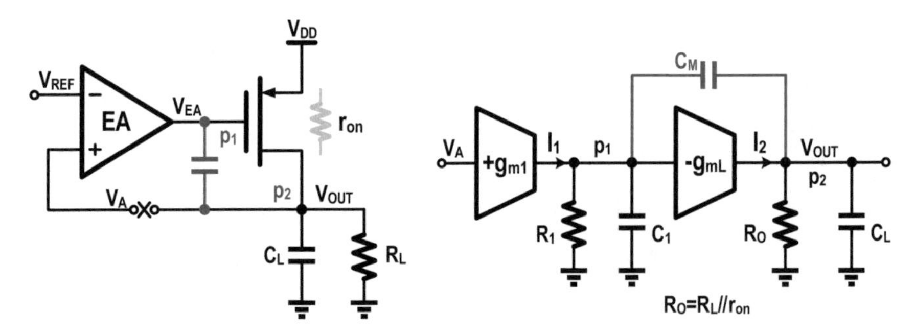

Fig. 3.16 A two-stage LDO with simple Miller compensation

Here, we assume that $g_{m1}R_1 >> 1$, $g_{mL}R_O >> 1$, $C_L >> C_1$, and $C_{M1} >> C_1$. The transfer function can be simplified to

$$H(s) \approx \frac{-g_{m1}g_{mL}R_1R_O(1 - sC_M/g_{mL})}{(1 + sC_Mg_{mL}R_1R_O)\left(1 + s\dfrac{C_L}{g_{mL}}\right)} \tag{3.27}$$

The new poles are

$$p_1 \approx \frac{-1}{g_{mL}R_1R_OC_M} \tag{3.28}$$

$$p_2 \approx \frac{-g_{mL}}{C_L} \tag{3.29}$$

Figure 3.17 presents the Bode plots for the uncompensated and Miller compensated cases. The dominant pole p_1 is significantly shifted to lower frequencies, while the nondominant pole is moved to higher frequencies. The separation between these two poles improves the phase margin.

3.5.2 *Right-Half-Plane Zero Cancellation*

It should be noted that according to (3.20), a right-half-plane (RHP) zero appears as

$$z_1 = \frac{g_{mL}}{C_M} \tag{3.30}$$

An RHP zero increases the phase shift (like a left-half-plane (LHP) pole) and increases the magnitude (identical to an LHP zero). Consequently, an RHP zero presents two worst-case scenarios with respect to stability. This RHP zero arises due to the feedforward small-signal path. Reducing the feedforward current can eliminate the RHP zero.

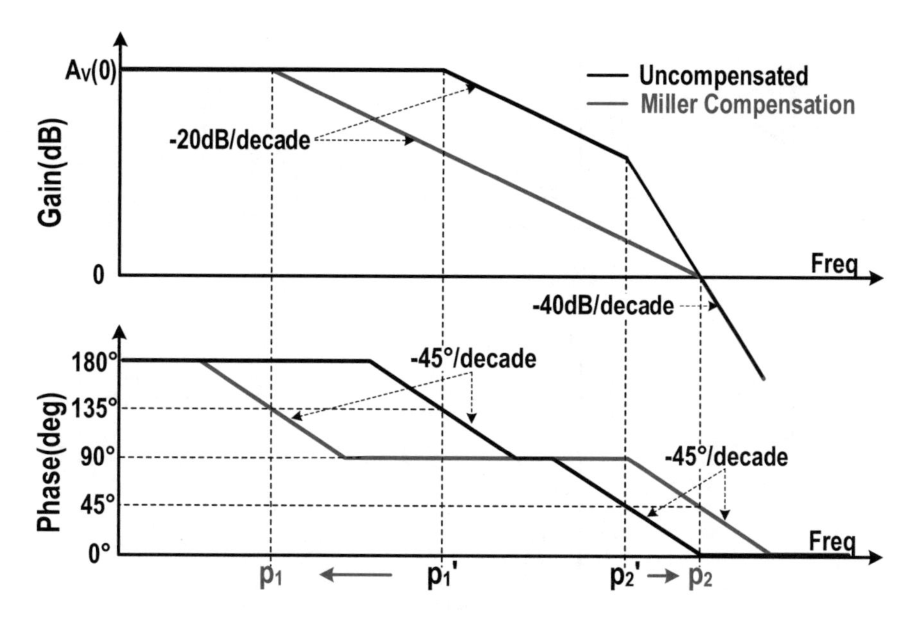

Fig. 3.17 The Bode plots for the uncompensated and Miller compensation cases

Figure 3.18 illustrates several methods for RHP zero cancellation [13–15]: (a) SMC using a nulling resistor (SMCNR); (b) SMC using a voltage buffer (SMCVB); (c) SMC using a current buffer (SMCCB); and (d) SMC using multipath zero cancellation (MZC).

As illustrated in Fig. 3.18a, a nulling resistor is connected in series with C_{M} to increase the impedance of the capacitor path. The resulting transfer function is given by

$$H(s) \approx \frac{-g_{\mathrm{m1}} g_{\mathrm{mL}} R_1 R_{\mathrm{O}} \left[1 + sC_{\mathrm{M}}\left(R_{\mathrm{M}} - \frac{1}{g_{\mathrm{mL}}}\right)\right]}{\left[1 + sC_{\mathrm{M}}\left(R_{\mathrm{M}} + g_{\mathrm{mL}} R_1 R_{\mathrm{O}}\right)\right]\left[1 + s\frac{C_L\left(R_1 + R_{\mathrm{M}}\right)R_{\mathrm{O}}}{R_{\mathrm{M}} + g_{\mathrm{mL}} R_1 R_{\mathrm{O}}}\right]} \tag{3.31}$$

According to Eq. (3.31), the poles and zero are

$$p_1 \approx \frac{-1}{\left(R_{\mathrm{M}} + g_{\mathrm{mL}} R_1 R_{\mathrm{O}}\right) C_{\mathrm{M}}} \tag{3.32}$$

$$p_2 \approx \frac{-\left(R_{\mathrm{M}} + g_{\mathrm{mL}} R_1 R_{\mathrm{O}}\right)}{R_{\mathrm{O}}\left(R_1 + R_{\mathrm{M}}\right) C_{\mathrm{M}}} \tag{3.33}$$

$$z_1 \approx \frac{-1}{\left(R_{\mathrm{M}} - \frac{1}{g_{\mathrm{mL}}}\right) C_{\mathrm{M}}} \tag{3.34}$$

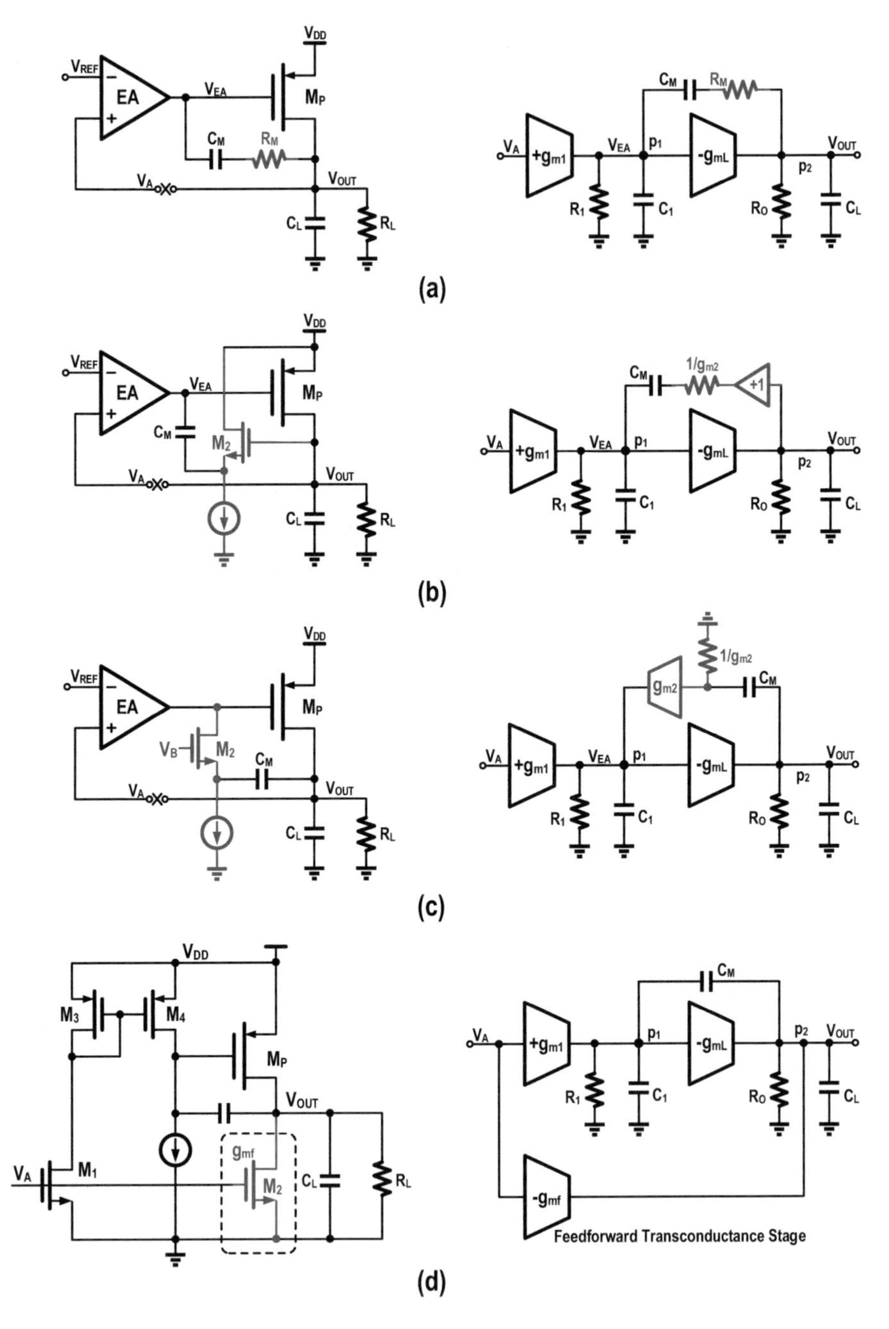

Fig. 3.18 RHP zero cancellation: (**a**) SMC using nulling resistor (SMCNR); (**b**) SMC using voltage buffer (SMCVB); (**c**) SMC using current buffer (SMCCB); (**d**) SMC using multipath zero cancellation (MZC)

When $R_M > 1/g_{mL}$, the right-half-plane zero transitions to a left-plane zero. By selecting an appropriate R_M value, it can offset the nondominant pole p_2, thereby increasing the bandwidth and improving the phase margin.

Figure 3.18b shows an SMC using a voltage buffer for RHP zero cancellation. A source follower M_2 in series with a capacitor is inserted to interrupt the feedforward current path. The resulting transfer function is

$$H(s) \approx \frac{-g_{m1}g_{mL}R_1R_O\left[1+s\frac{C_M}{g_{m2}}\right]}{R_OC_LC_M\left(\frac{1}{g_{m2}}+R_1\right)s^2+\left[\left(\frac{1}{g_{m2}}+g_{mL}R_OR_1\right)C_M+R_OC_L\right]s+1} \tag{3.35}$$

The poles and zero are

$$p_1 \approx \frac{-1}{g_{mL}R_OR_1C_M} \tag{3.36}$$

$$p_2 \approx \frac{-g_{mL}}{C_L} \tag{3.37}$$

$$z_1 \approx \frac{-g_{m2}}{C_M} \tag{3.38}$$

Figure 3.18c illustrates an SMC using a current buffer for RHP zero cancellation. The common-gate amplifier stage, composed of M_2, converts the output voltage swing into a current. The resulting transfer function is

$$H(s) \approx \frac{-g_{m1}g_{mL}R_1R_O\left[1+s\frac{C_M}{g_{m2}}\right]}{R_OC_LC_M\frac{1}{g_{m2}}s^2+\left[\left(1+g_{mL}R_1\right)R_OC_M+\frac{C_M}{g_{m2}}+R_OC_L\right]s+1} \tag{3.39}$$

The poles and zero are

$$p_1 \approx \frac{-1}{g_{mL}R_OR_1C_M} \tag{3.40}$$

$$p_2 \approx \frac{-g_{m2}R_1g_{mL}}{C_L} \tag{3.41}$$

$$z_1 \approx \frac{-g_{m2}}{C_M} \tag{3.42}$$

The nondominant pole is observed to shift to a higher frequency. Several works have adopted SMCCB technology and introduced improvements to reduce the required compensation capacitance. For instance, [16] employs capacitor multiplier technology to decrease the need for the compensation capacitor. Similarly, [17] utilizes a current amplifier to lower the value of C_M.

Figure 3.18d demonstrates the method of eliminating the RHP zero using MZC [18]. A feedforward transconductance stage (FTS) is incorporated to generate an out-of-phase small-signal current, canceling the feedforward small-signal current at high frequencies. The resulting transfer function is

$$H(s) \approx \frac{-g_{m1} g_{mL} R_1 R_O \left[1 + s \frac{C_M (g_{mf} - g_{m1})}{g_{m1} g_{mL}} \right]}{(1 + s C_M g_{mL} R_1 R_O) \left(1 + s \frac{C_L}{g_{mL}} \right)} \tag{3.43}$$

The poles and zero are

$$p_1 \approx \frac{-1}{g_{mL} R_O R_1 C_M} \tag{3.44}$$

$$p_2 \approx \frac{-g_{mL}}{C_L} \tag{3.45}$$

$$z_1 \approx \frac{-g_{m1} g_{mL}}{C_M (g_{mf} - g_{m1})} \tag{3.46}$$

MZC does not substantially change the pole splitting effect in SMC, and when $g_{mf} \geq g_{m1}$, it can cancel out the RHP zero and produce a left-half-plane zero.

When the error amplifier has a cascode structure, the cascode Miller compensation can also be employed [19, 20]. As depicted in Fig. 3.19a, one plate of the compensation capacitor is connected to the source of M_8. This configuration is equivalent to inserting a common-gate amplifier stage between V_{EA} and V_M.

In SMC, the nondominant pole shifts to a higher frequency, which can be equated to a reduction in the output impedance to $1/g_{mL}$ at high frequency, as illustrated in Fig. 3.19b. With cascode Miller compensation, the equivalent impedance at high frequency decreases further due to the gain of the common-gate stage:

$$R_O \approx \frac{1}{g_{m8} (r_{ds8} \parallel r_{ds4}) g_{mL}} \tag{3.47}$$

Therefore, compared to SMC, the cascode Miller compensation can shift the nondominant pole to a higher frequency:

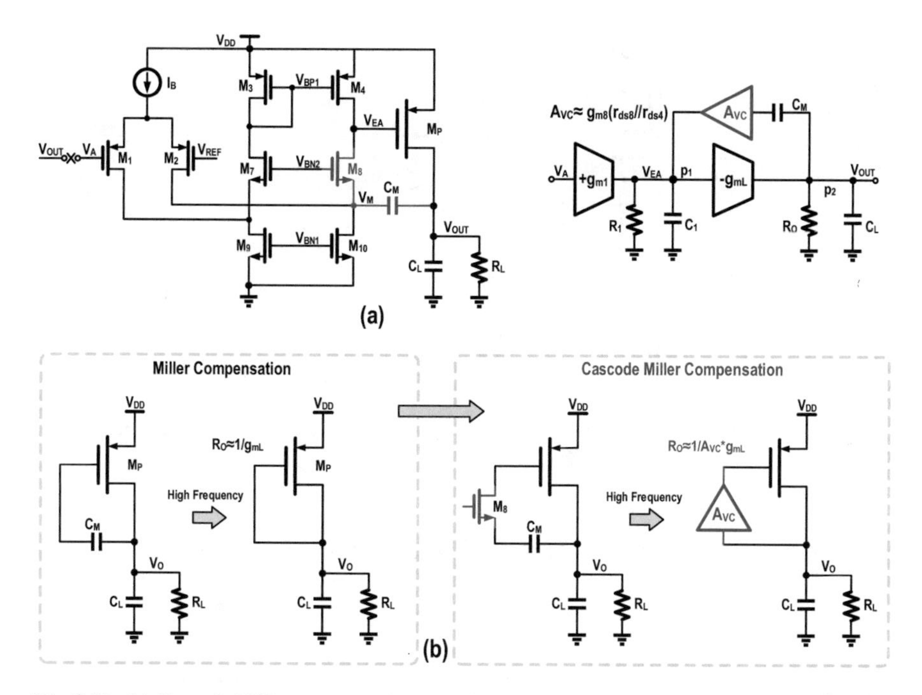

Fig. 3.19 (**a**) Cascode Miller compensation, and (**b**) illustration of how the output pole is increased

$$p_2 = \frac{-g_{\mathrm{m8}}\left(r_{\mathrm{ds8}} \parallel r_{\mathrm{ds4}}\right)g_{\mathrm{mL}}}{C_{\mathrm{L}}} \tag{3.48}$$

3.5.3 Nested Miller Compensation

In Sect. 3.4, we introduced the role of buffers in LDOs. By using a buffer with low output impedance, the parasitic pole p_{G} of the power transistor gate can be shifted to a higher frequency. When p_{G} is much higher than the unity-gain bandwidth (UGB), the parasitic pole no longer affects loop stability. In this case, p_{G} can be ignored during loop compensation design, allowing the three-stage LDO structure to be simplified to a two-stage configuration for compensation analysis. This significantly reduces the design complexity of the LDO and helps improve the slew rate of gate control signal, thus enhancing transient performance (Fig. 3.20).

If multistage amplifiers are used, or if the driver's output impedance is insufficient to shift the parasitic gate pole to a higher frequency, multistage compensation is required. Generally, a simplified LDO structure should not exceed three stages whenever possible. Two-stage amplification plus output stage is typically sufficient to meet the practical requirements of an LDO. Additional gain stages complicate the circuit structure and significantly increase the complexity of frequency

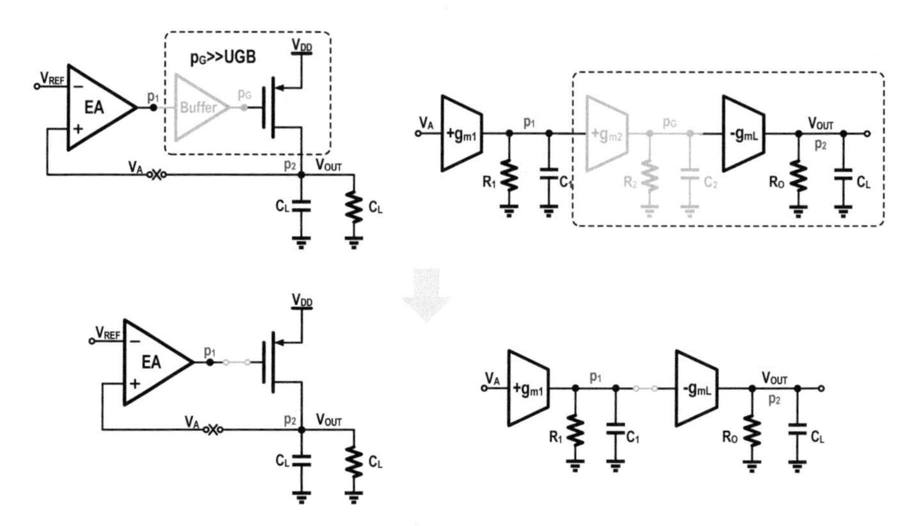

Fig. 3.20 When $p_G >> \text{UGB}$, the three-stage LDO structure can be analyzed as equivalent to a two-stage structure

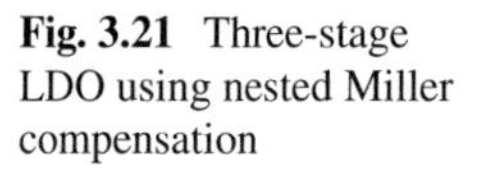
Fig. 3.21 Three-stage LDO using nested Miller compensation

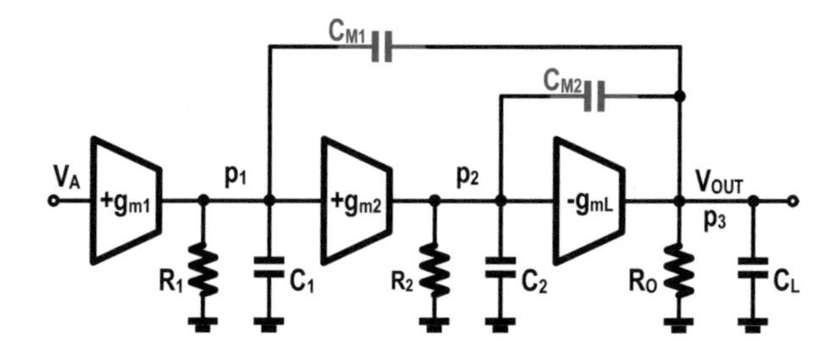

compensation. Alternatively, the structure proposed in [4] can be adopted, which utilizes multiple small-gain stages. This approach effectively reduces the output impedance at each stage and shifts the parasitic poles to higher frequencies.

Figure 3.21 depicts the nested Miller compensation scheme [21–23] applied to the three-stage LDO. The external compensation capacitor C_{M1} is connected across the second and third stages, while the internal compensation capacitor C_{M2} is connected across the third stage. The resulting transfer function is

$$H(s) \approx \frac{-g_{m1}g_{m2}g_{mL}R_1R_2R_O\left(1 - s\frac{C_{M2}}{g_{mL}} - s^2\frac{C_{M1}C_{M2}}{g_{m2}g_{mL}}\right)}{\left(1 + sC_{M1}g_{m2}g_{mL}R_1R_2R_O\right)\left[1 + s\frac{C_{M2}\left(g_{mL} - g_{m2}\right)}{g_{m2}g_{mL}} + s^2\frac{C_LC_{M2}}{g_{m2}g_{mL}}\right]} \quad (3.49)$$

The dominant pole is

$$p_1 \approx \frac{-1}{R_1C_{M1}g_{m2}g_{mL}R_2R_O} \quad (3.50)$$

The two nondominant poles are governed by the second-order term in the denominator of Eq. (3.49). The capacitor C_{M2} is employed to separate pole p_2 and the output pole p_3.

As analyzed in reference [4], for nested Miller compensation (NMC), it is crucial that $g_{mL} >> g_{m1}$ and $g_{mL} >> g_{m2}$. If $g_{mL} - g_{m2}$ is not sufficiently large, the damping factor of the complex pole will be inadequate. If $g_{mL} < g_{m2}$, a right-half-plane pole will emerge, leading to instability in the loop. Additionally, we observe the presence of right-half-plane zeros in the system. Only when $g_{mL} >> g_{m1}$ and $g_{mL} >> g_{m2}$ will the small-signal output current be significantly larger than the feedforward current. Under these conditions, the impact of the RHP zero on system stability becomes negligible.

To mitigate the influence of the RHP zero, nested Miller compensation can employ various techniques, including nulling resistor (NR), voltage buffer (VB), current buffer (CB), and multipath zero cancellation (MZC). Figure 3.22 illustrates the implementation of a nulling resistor [21].

The new transfer function is

$$H(s) \approx \frac{-g_{m1}g_{m2}g_{mL}R_1R_2R_O\left\{1+s\left[C_{M1}R_M + C_{M2}\left(R_M - \frac{1}{g_{mL}}\right) + s^2\frac{C_{M1}C_{M2}\left(g_{mL}R_M - 1\right)}{g_{m2}g_{mL}}\right]\right\}}{\left(1+sC_{M1}g_{m2}g_{mL}R_1R_2R_O\right)\left[1+s\frac{C_{M2}\left(g_{mL}-g_{m2}\right)}{g_{m2}g_{mL}} + s^2\frac{\left(1-g_{m2}R_M\right)C_LC_{M2}}{g_{m2}g_{mL}}\right]} \tag{3.51}$$

According to Eq. (3.51), when $R_M = 1/g_{mL}$, the RHP zero disappears, leaving only the LHP zero:

$$z_1 = \frac{-1}{R_M C_{M1}} \tag{3.52}$$

As R_M is further increased, the complex pole vanishes when $R_M = 1/g_{m2}$. However, it is crucial to note that R_M must not exceed $1/g_{m2}$, as this would result in the

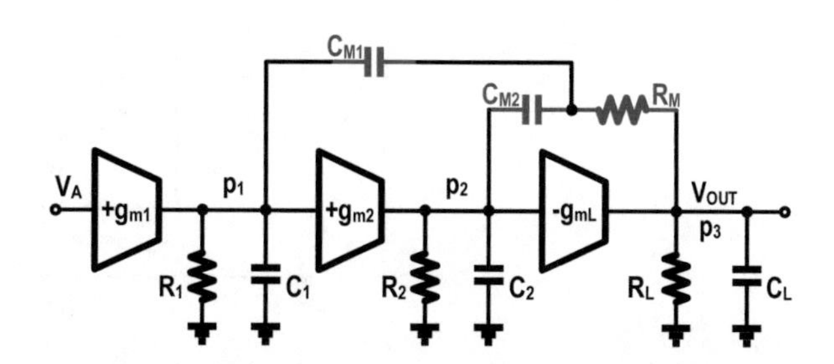

Fig. 3.22 The nested Miller compensation using a nulling resistor

emergence of right-half-plane poles. Consequently, the condition $g_{mL}> > g_{m2}$ is essential, providing a margin for R_M adjustment and tolerance for resistance variations.

Figure 3.23 illustrates three structures that employ feedforward transconductance stage (FTS) technology to eliminate the RHP zero [11]. The incorporation of an FTS generates an out-of-phase small-signal current that counteracts the high-frequency feedforward small-signal current. These techniques can effectively mitigate the detrimental effects of the RHP zero on the system's stability and frequency response.

In addition to the aforementioned nesting structures, an alternative approach known as reverse nested Miller compensation (RNMC) [24] is presented in Fig. 3.24. This technique modifies the second stage into an inverting amplifier, utilizing the diode-connected common-source structure (described in Sect. 3.4.1). Resistors R_1 and R_2 are employed to set the maximum and minimum drive current values, respectively.

In the conventional nested Miller compensation (NMC), the internal Miller capacitor is typically connected to the third stage, while RNMC connects it across

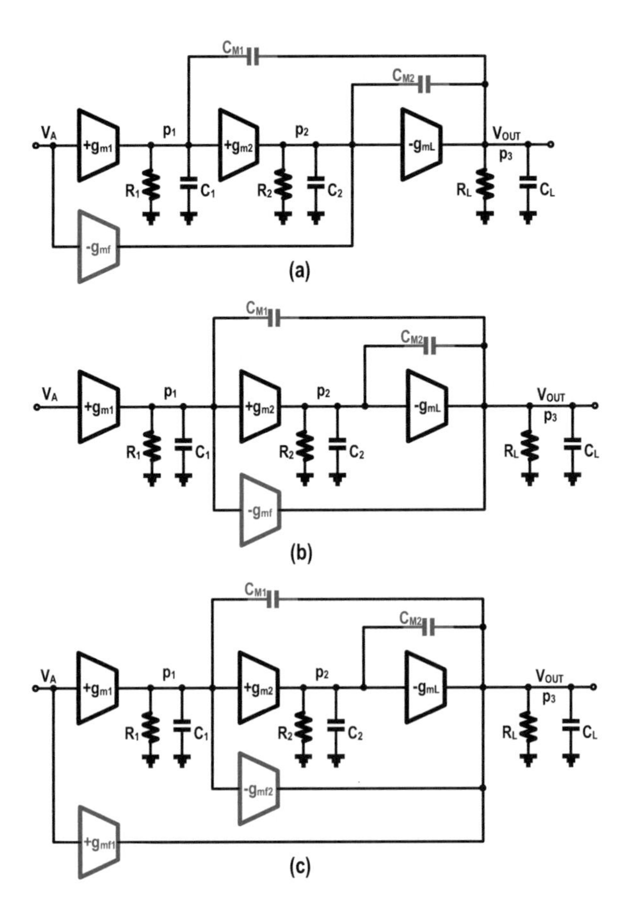

Fig. 3.23 Three feedforward transconductance-stage structures to eliminate the RHP zero

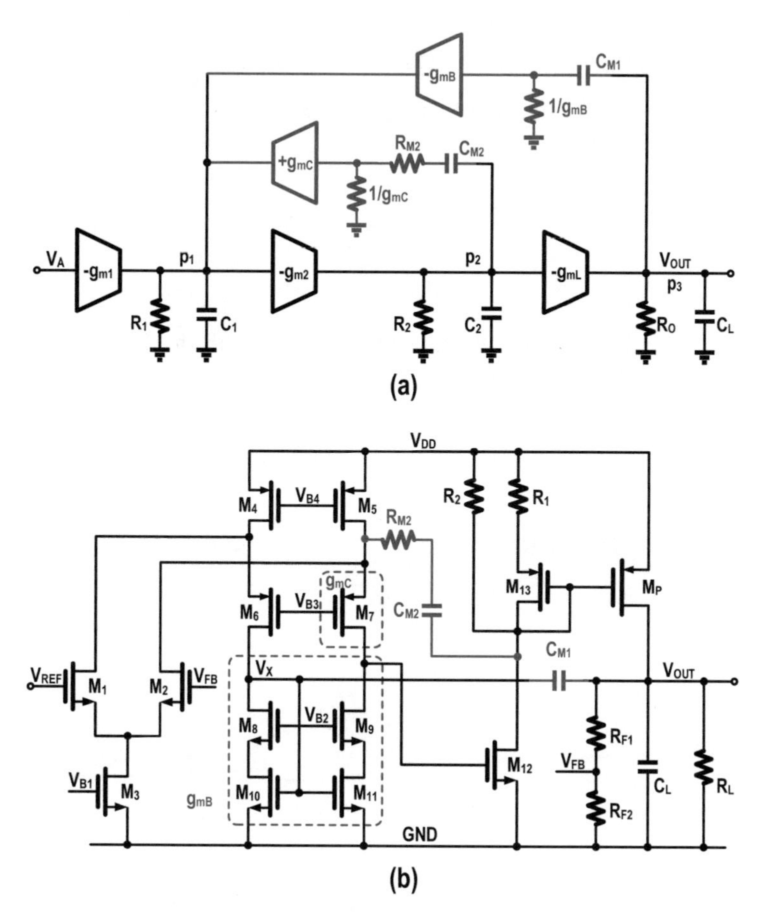

Fig. 3.24 The reverse nested Miller compensation in [24]

the second stage. The external Miller compensation uses a current buffer (M_{10}, M_{11}) to generate a left-half-plane zero. The internal Miller compensation utilizes a cascode Miller configuration with a series resistor R_{M2}, producing another left-half-plane zero.

The resulting system comprises three poles and two zeros, thereby ensuring loop stability. Notably, the RNMC technology implemented in [24] does not add any additional active components and therefore does not increase the power consumption.

3.5.4 Damping-Factor-Control Frequency Compensation

According to the previously derived NMC transfer function (3.49), assume that $g_{mL} >> g_{m1}$, g_{m2}, and $C_{M2} >> C_1$, C_2. We can obtain

$$H(s) \approx \frac{-g_{\mathrm{m1}} g_{\mathrm{m2}} g_{\mathrm{mL}} R_1 R_2 R_{\mathrm{O}}}{\left(1 + sC_{\mathrm{M1}} g_{\mathrm{m2}} g_{\mathrm{mL}} R_1 R_2 R_{\mathrm{O}}\right)\left[1 + s\frac{C_{\mathrm{M2}}}{g_{\mathrm{m2}}} + s^2 \frac{C_L C_{\mathrm{M2}}}{g_{\mathrm{m2}} g_{\mathrm{mL}}}\right]} \tag{3.53}$$

The dominant pole is

$$p_1 \approx \frac{-1}{R_1 C_{\mathrm{M1}} g_{\mathrm{m2}} g_{\mathrm{mL}} R_2 R_{\mathrm{O}}} \tag{3.54}$$

We compare the second-order function in (3.53) with the following standard second-order function:

$$F(s) = 1 + s\left(\frac{2\zeta}{p_{\mathrm{c}}}\right) + s^2\left(\frac{1}{p_{\mathrm{c}}}\right)^2 \tag{3.55}$$

where ζ is the damping factor, and p_{c} is the center frequency of nondominant poles:

$$p_{\mathrm{C}} \approx \sqrt{\frac{g_{\mathrm{m2}} g_{\mathrm{mL}}}{C_L C_{\mathrm{M2}}}} \tag{3.56}$$

$$\zeta = \frac{1}{2}\sqrt{\frac{g_{\mathrm{mL}} C_{\mathrm{M2}}}{g_{\mathrm{m2}} C_{\mathrm{L}}}} \tag{3.57}$$

$$Q = \frac{1}{2\zeta} = \sqrt{\frac{g_{\mathrm{m2}} C_{\mathrm{L}}}{g_{\mathrm{mL}} C_{\mathrm{M2}}}} \tag{3.58}$$

If the damping factor ζ is too small, a frequency "peak" will occur, and the separated zero cannot effectively cancel the pole. C_{M2}, C_{L}, and g_{mL} influence the damping factor ζ. When the load capacitor C_{L} is large, a larger C_{M2} is needed to ensure an appropriate damping factor. Larger C_{M2} and C_{L} will result in the nondominant pole frequency p_{c} appearing at a relatively low-frequency range, thereby reducing the bandwidth of the NMC [25].

Therefore, NMC is not suitable for scenarios that require low power consumption, large load capacitor, and high bandwidth. In such cases, damping-factor-control frequency compensation (DFCFC) is a more appropriate choice [26]. Figure 3.25b presents the small-signal model of the three-stage LDO and the comparison of the Bode plots for NMC and DFCFC. This figure intuitively illustrates the role of DFC.

According to the previous analysis, the poor compensation bandwidth of the NMC is mainly due to the presence of C_{M2}. If C_{M2} is removed, the nondominant pole will revert to a higher frequency, thus expanding the NMC bandwidth. However, the two nondominant poles may form a complex conjugate pole pair. Without C_{M2}, the damping factor is small (Q is high), leading to a frequency peak near the unity-gain bandwidth. If this peak exceeds 0 dB, it will affect the phase margin and gain margin, resulting in loop instability. In addition, the transconductance g_{mL} of the output

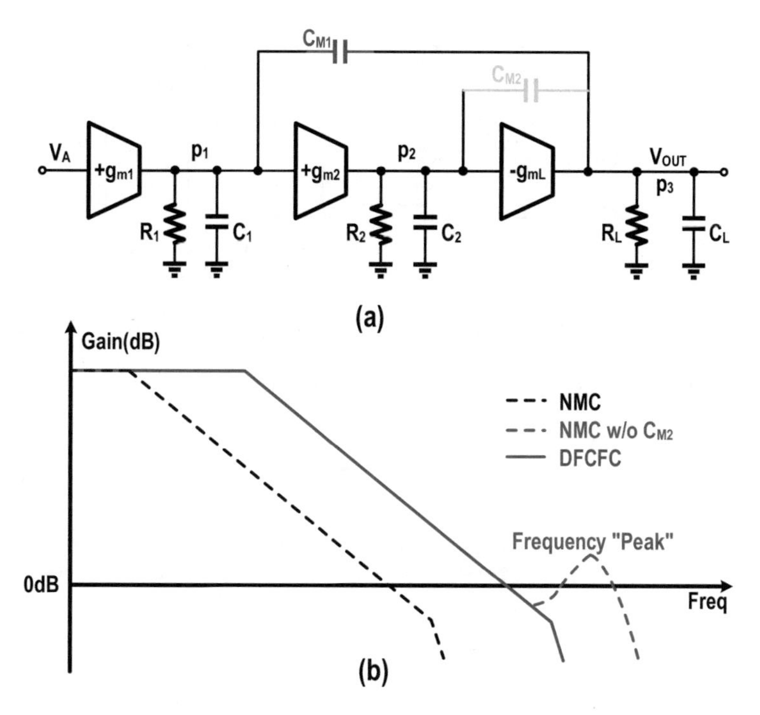

Fig. 3.25 (**a**) Small-signal model of the three-stage LDO. (**b**) Bode plot comparison of NMC and DFC

power transistor will vary with the load. At light loads, a reduced g_{mL} will further exacerbate the "frequency peak" issue.

When $\zeta \approx 0.707$, this peak will not occur [26]. In LDO design, it is not necessary to completely eliminate the peak; as long as the peak amplitude is low enough to not affect the phase margin and stability, it is acceptable. In [27], using DFCFC, when $\zeta \approx 0.5$, a sufficiently high phase margin has been achieved. On the other hand, the damping factor ζ should not be too large, since an overly high damping factor would cause the complex pole to split into two separate real poles, which would negatively affect the loop bandwidth and phase margin.

Figure 3.26 illustrates an LDO regulator design using DFCFC [25]. The main circuit comprises three parts: the first-stage error amplifier, the second-stage Gm-boosting circuit, and the third-stage power stage. The compensation circuit includes a Miller capacitor C_{M1} across the second and third stages and the DFCFC circuit in the middle. The key component of the DFC is transistor M_9, which receives input from EA' output, V_1. However, considering the process variation of M_9 and that the DC voltage of input V_1 may change with the load, the drain voltage of M_9 may rise to V_{DD} or drop to ground, thus affecting the DFCFC effect. Thus, a local feedback circuit (M_{10}–M_{15}) is used to set M_9's drain DC operating point.

In addition to DFCFC and Miller compensation, the feedback network components R_1, R_2, and C_F create a pole-zero pair (Fig. 3.27) that can enhance the phase margin. The zero-pole pairs in the feedback network are

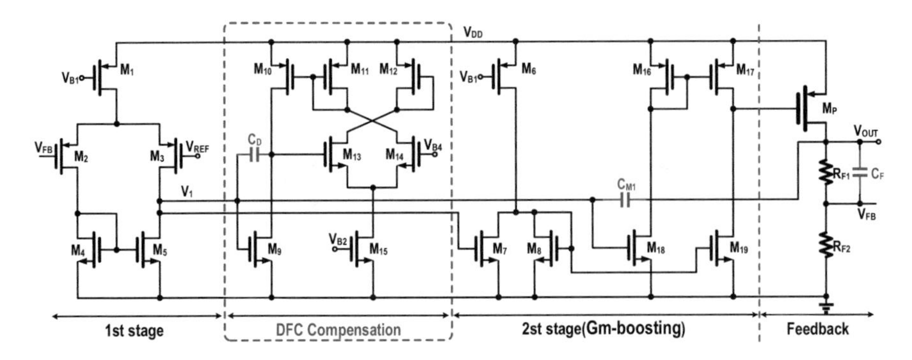

Fig. 3.26 Schematic of the three-stage LDO with DFCFC in [25]

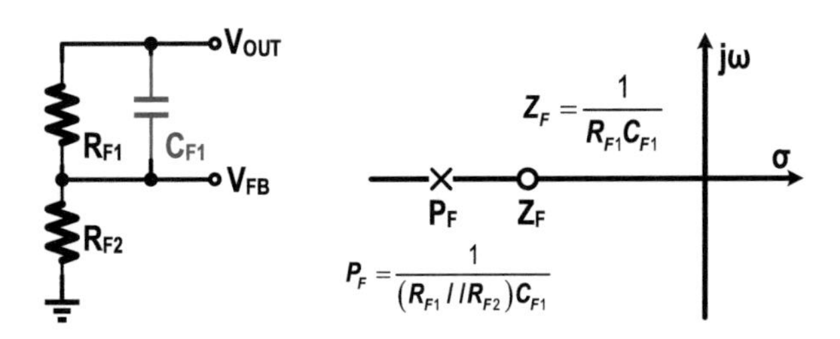

Fig. 3.27 The pole-zero pairs of feedback network

$$p_{\mathrm{F}} = \frac{1}{C_{\mathrm{F1}}\left(R_{\mathrm{F1}} \parallel R_{\mathrm{F2}}\right)} \tag{3.59}$$

$$z_{\mathrm{F}} = \frac{1}{C_{\mathrm{F1}}R_{\mathrm{F1}}} \tag{3.60}$$

The frequency of the zero is lower than that of the pole, which can be leveraged to increase bandwidth or improve phase margin.

Figure 3.28 illustrates the small-signal model of the LDO. By ignoring the effects of some high-frequency zeros, the transfer function can be expressed as

$$H(s) \approx \frac{-g_{\mathrm{m1}}g_{\mathrm{m2}}g_{\mathrm{mL}}R_1R_2R_{\mathrm{O}}}{\left(1+sC_{\mathrm{M1}}g_{\mathrm{m2}}g_{\mathrm{mL}}R_1R_2R_{\mathrm{O}}\right)\left[1+s\dfrac{C_2\left(C_{\mathrm{M1}}g_{\mathrm{mL}}+C_{\mathrm{L}}g_{\mathrm{mD}}\right)}{g_{\mathrm{m2}}g_{\mathrm{mL}}C_{\mathrm{M1}}}+s^2\dfrac{C_2C_{\mathrm{L}}}{g_{\mathrm{m2}}g_{\mathrm{mL}}}\right]} \tag{3.61}$$

When the load is light, g_{ML} decreases. Based on the previous analysis, the frequency peak is exacerbated under light-load conditions. Assuming that $C_{\mathrm{M1}}g_{\mathrm{mL}} << C_{\mathrm{L}}g_{\mathrm{m3}}$, we obtain

$$p_{\mathrm{C}} \approx \sqrt{\frac{g_{\mathrm{m2}}g_{\mathrm{mL}}}{C_{\mathrm{L}}C_2}} \tag{3.62}$$

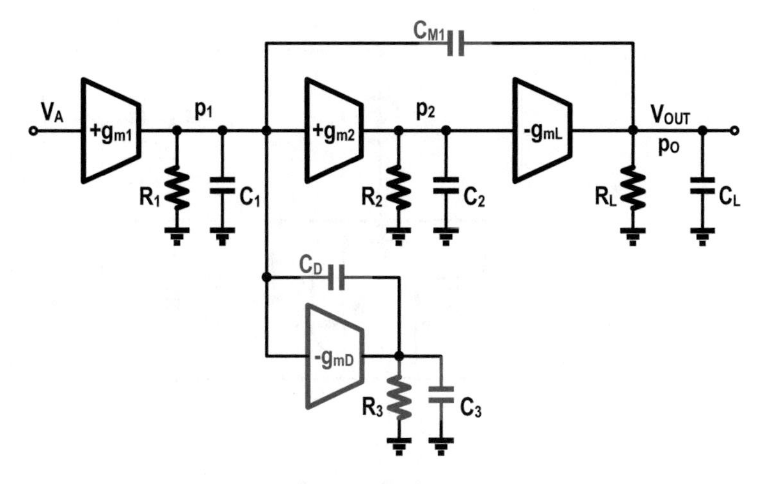

Fig. 3.28 Small-signal model of the LDO with DFCFC in [25]

$$\zeta = \frac{1}{2}\sqrt{\frac{C_2 C_L}{g_{m2} g_{mL}}} \times \left(\frac{g_{mD}}{C_{M1}}\right) \quad (3.63)$$

It can be observed that the elimination of C_{M2} (since $C_{M2} >> C_2$) shifts the non-dominant pole to a higher frequency. The damping factor can be adjusted by g_{mD}, and g_{mD} does not affect the loop DC gain.

In addition, according to Eq. (3.62), the secondary pole can be further shifted to a higher frequency by increasing g_{m2}. Although g_{m2} also affects the damping factor, DFCFC can mitigate the negative impact by increasing g_{mD}.

Figure 3.29 illustrates the g_m-boosting circuit. M_7 and M_{19}, and M_8 and M_{18}, form current mirrors with a ratio of K. The output equivalent transconductance can be calculated as follows:

$$G_M = \frac{I_2}{V_1} = \frac{I_A - I_B}{V_1} = \frac{-g_{m7} \times V_1 \times \frac{1}{g_{m8}} \times K \times g_{m8} - g_{m7} \times V_1 \times K}{V_1} = -2K \times g_{m7} \quad (3.64)$$

As Eq. (3.61) shows, the transconductance is amplified by a factor of 2 K. The push-pull output stage formed by M_{17} and M_{19} can charge and discharge the gate capacitance more efficiently during transient response.

The work in [28] proposed a Q-reduction compensation scheme to address the frequency peaking issue. Figure 3.30 illustrates the schematic of the LDO with Q-reduction. M_1–M_5 form the first-stage error amplifier, while M_6–M_9 make up the second non-inverting gain stage. M_P is the power transistor. The feedforward path is constituted by M_2, M_4, and M_9. The Miller capacitor C_M is connected across the first-stage output V_1 and V_{OUT}. The capacitor C_D is used for Q-reduction compensation, aiming to increase the damping factor. In contrast to [25], [28] does not include an additional damping factor gain stage; instead, C_D is connected between V_D and V_2.

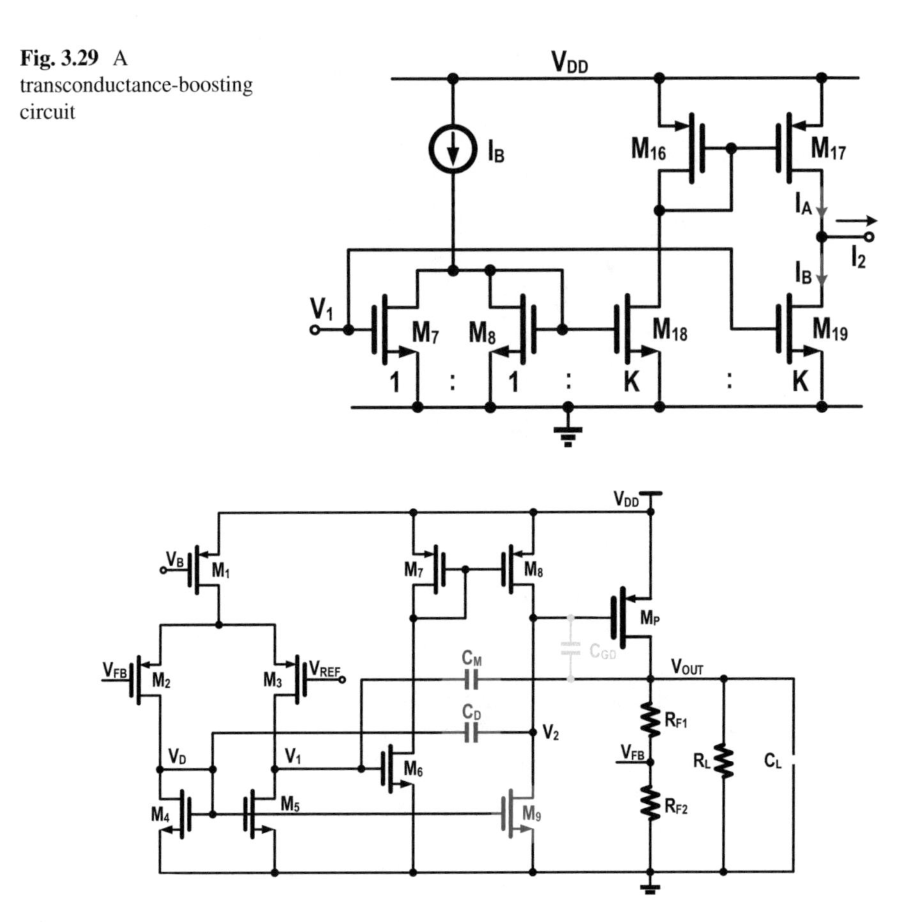

Fig. 3.29 A transconductance-boosting circuit

Fig. 3.30 Schematic of the three-stage LDO with Q-reduction in [28]

Figure 3.31 depicts the small-signal model of the LDO. The transfer function is given below:

$$H(s) \approx \frac{-A_{\mathrm{DC}}\left\{\begin{array}{l} 1+s\left(C_{\mathrm{D}}R_{\mathrm{D}}+\dfrac{C_{\mathrm{M}}g_{\mathrm{mf}}}{g_{\mathrm{m1}}g_{\mathrm{m2}}}-\dfrac{C_{\mathrm{GD}}}{g_{\mathrm{mL}}}\right)- \\ s^{2}\left[\dfrac{C_{\mathrm{M1}}\left(C_{\mathrm{GD}}+C_{2}\right)}{g_{\mathrm{m2}}g_{\mathrm{mL}}}+\dfrac{C_{\mathrm{GD}}C_{\mathrm{D}}R_{\mathrm{D}}}{g_{\mathrm{mL}}}\right] \end{array}\right\}}{\left(1+\dfrac{s}{p_{1}}\right)\left[\begin{array}{l} 1+s\dfrac{C_{\mathrm{M}}C_{\mathrm{GD}}\left(g_{\mathrm{mL}}-g_{\mathrm{m2}}\right)+C_{\mathrm{D}}C_{L}g_{\mathrm{m2}}+C_{\mathrm{M}}C_{\mathrm{D}}g_{\mathrm{m2}}g_{\mathrm{mL}}R_{\mathrm{D}}}{g_{\mathrm{m2}}g_{\mathrm{mL}}C_{\mathrm{M1}}}+ \\ s^{2}\dfrac{\left(C_{\mathrm{GD}}+C_{2}+C_{\mathrm{D}}\right)C_{\mathrm{L}}}{g_{\mathrm{m2}}g_{\mathrm{mL}}} \end{array}\right]} \tag{3.65}$$

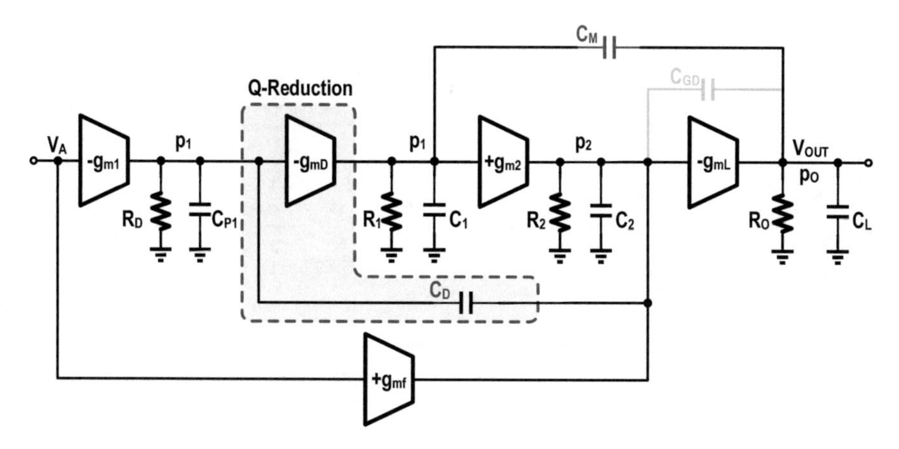

Fig. 3.31 Small-signal model of the LDO with Q-reduction in [29]

The dominant pole is

$$p_1 \approx \frac{-1}{R_1 C_{M1} g_{m2} g_{mL} R_2 R_O} \tag{3.66}$$

For the moderate to maximum output current, $g_{mL} >> g_{m1}$ and g_{m2}. The transfer function can be simplified to

$$H(s) \approx \frac{-A_{DC}\left(1+s\frac{C_M g_{mf}}{g_{m2} g_{mL}}\right)}{\left(1+\frac{s}{p_1}\right)\left[1+s\frac{(C_{GD}+C_D)}{gm2}+s^2\frac{(C_{GD}+C_2+C_D)C_L}{g_{m2} g_{mL}}\right]} \tag{3.67}$$

$$p_2 \approx \frac{-g_{m2}}{C_{GD}+C_D} \tag{3.68}$$

$$p_3 \approx \frac{-(C_{GD}+C_D)g_{mL}}{(C_{GD}+C_D+C_2)\times C_L} \approx \frac{-g_{mL}}{C_L} \tag{3.69}$$

$$z_1 \approx \frac{-g_{m1} g_{m2}}{C_M g_{mf}} \tag{3.70}$$

In this condition, Q-reduction compensation can be considered as reverse nested Miller compensation.

For the low to moderate output current, g_{mL} is not too much larger than g_{m1} and g_{m2}. The frequency of the complex pole in [28] is

$$p_{C} \approx \sqrt{\frac{g_{m2}g_{mL}}{C_{L}\left(C_{2}+C_{GD}\right)}} \tag{3.71}$$

$$Q=\sqrt{\frac{\left(C_{2}+C_{GD}\right)C_{L}}{g_{m2}g_{mL}}} \times \left[\frac{C_{M}g_{m2}g_{mL}}{C_{M}C_{GD}\left(g_{mL}-g_{m2}\right)+C_{D}C_{L}g_{m2}+C_{M}C_{D}g_{m2}g_{mL} \times \frac{1}{g_{mD}}}\right] \tag{3.72}$$

According to Eq. (3.56), a smaller Q value (indicating a higher damping factor) can be achieved by increasing C_D or reducing g_{mD}. By decreasing the aspect ratio of M_4 and M_5, the Q value can be reduced. This also helps in minimizing the random offset voltage of the input stage.

3.5.5 Dynamic Pole-Zero Compensation

In the previous analysis, LDOs are typically modeled as multistage amplifiers. However, in practice, the compensation design of LDOs is often much more complex than that of multistage amplifiers. LDOs often need to supply a wide range (>1000×) of output currents, which significantly impacts the output impedance, causing the output pole to shift across a broad frequency band.

The output pole acts as a nondominant pole in the Miller compensation. To maintain stability, the loop bandwidth will be limited by this nondominant pole. For a wider bandwidth, pole-zero cancellation technology is necessary, which involves introducing a zero in the left half-plane to offset the influence of the nondominant pole.

For the nondominant pole with a wide range of variation, a fixed zero point can only optimize bandwidth and phase margin within a specific frequency range. For instance, when the zero is positioned at a relatively low frequency, it benefits stability and loop bandwidth expansion under light-load conditions. However, under heavy-load conditions, this might cause the loop bandwidth to become excessively wide, which could lead to higher-frequency poles deteriorating the phase margin. In such cases, the zero needs to be moved to a higher frequency region. Ideally, the introduced zero can be adjusted according to load changes, thereby maintaining good bandwidth and phase margin across a wide load range.

The technology that adjusts the zero dynamically with the load is known as dynamic compensation [31]. It is also referred to as active zero compensation [32] or pole-zero tracking technique [33]. Figure 3.32 illustrates an implementation of active zero in Miller compensation [30].

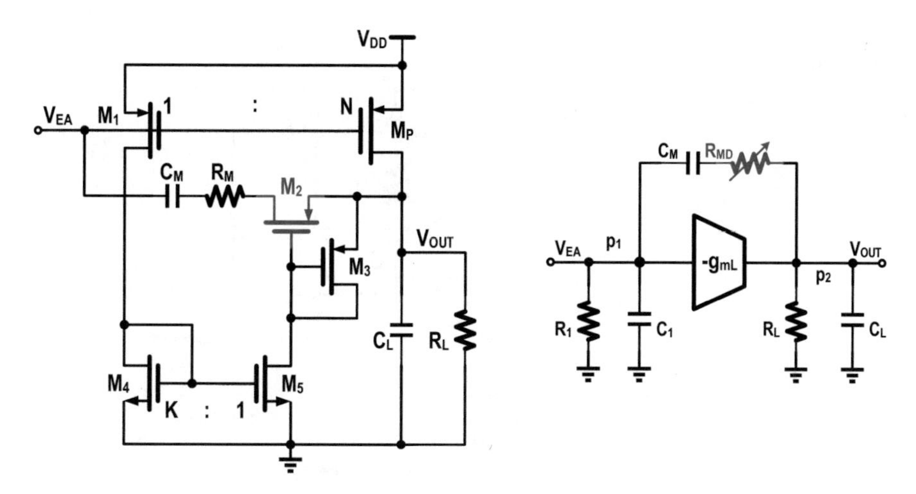

Fig. 3.32 The active zero in the Miller compensation [30]

Unlike traditional Miller compensation, in [30], transistor M_2 is connected in series with the nulling resistor. Since the V_{GS} of M_2 matches that of M_3, M_2's impedance can be adjusted by the current through M_3:

$$r_{o2} = \frac{1}{\mu \text{Cox}(W/L)_2 (V_{GS2} - V_{TH})} = \frac{(W/L)_3}{(W/L)_2 \times g_{m3}} \tag{3.73}$$

Here, M_1 and M_P, and M_4 and M_5, are current mirrors. Therefore, the current flowing through M_3 is proportional to the load current I_L. Using Miller compensation with a nulling resistor, we have

$$z_1 = \frac{-1}{(R_{MD} - g_{mL})C_M} \tag{3.74}$$

$$R_{MD} = R_M + r_{o2} = R_M + \frac{(W/L)_3}{(W/L)_2 \times g_{m3}} \tag{3.75}$$

R_{MD} is inversely proportional to the load current. With a light load, the zero point moves to a lower frequency following the output pole, while with a heavy load, it shifts to a higher frequency.

In a broader sense, dynamic compensation not only involves adjusting zeros dynamically but can also modulate the poles in the loop. As illustrated in Fig. 3.33, the work in [34] segments the load range and adjusts the compensation capacitor based on the load current, thus achieving a variable pole-zero pair and improving the phase margin.

In addition, Fig. 3.34 shows a wide-range adaptive nested Miller compensation (WRA-NMC) in [35]. R_M is a nulling resistor used to introduce a zero in the left half-plane, improving the phase margin. The gain of amplifier A_{V3} varies with the

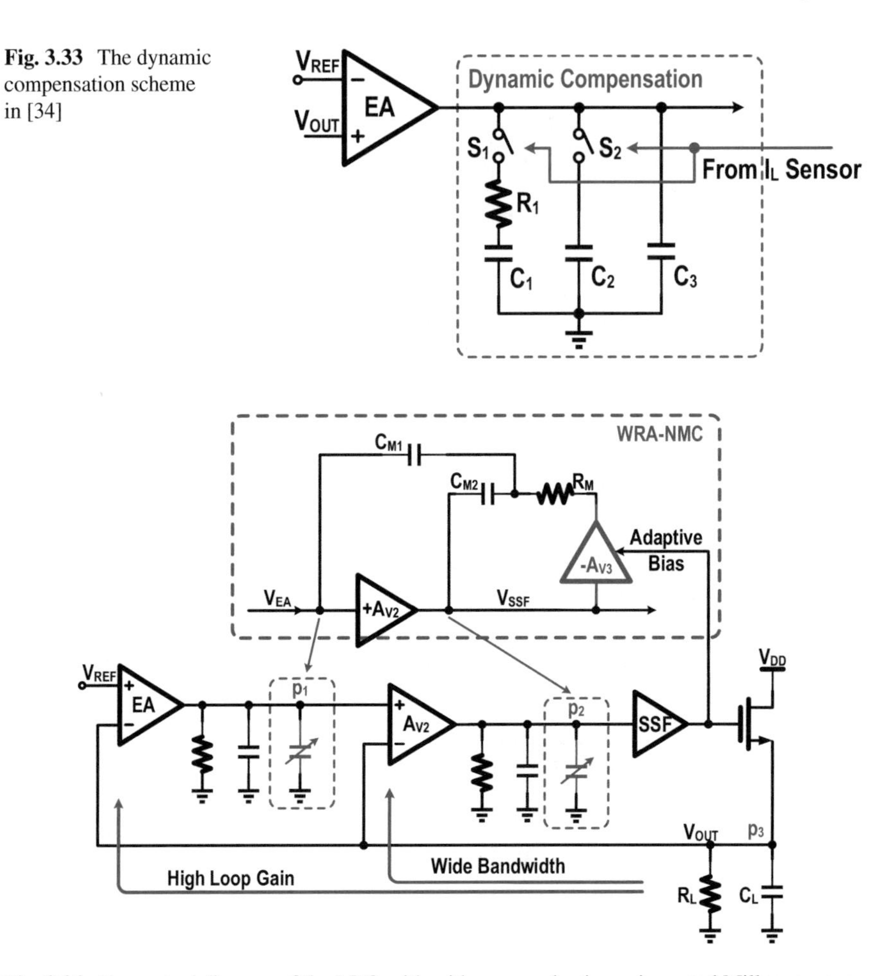

Fig. 3.33 The dynamic compensation scheme in [34]

Fig. 3.34 Conceptual diagram of the LDO with wide-range adaptive-gain nested Miller compensation [35]

load current: lower load currents result in higher A_{V3} gain. Under light-load conditions (small I_L), the output pole p_3 is at a low frequency, and the increased A_{V3} enlarges the equivalent Miller capacitances C_{M1} and C_{M2}, shifting poles p_1 and p_2 to lower frequencies, thus ensuring stability. Conversely, when I_L increases, the output pole p_3 shifts to a higher frequency, and the reduced A_{V3} moves the internal poles p_1 and p_2 to higher frequencies, which helps expanding the loop bandwidth and maintaining adequate phase margin.

In short, dynamic compensation involves adjusting the zero and pole positions in the loop according to load changes, enabling adaptation to varying load conditions. This approach is commonly used in LDOs to improve phase margin and expand bandwidth.

3.6 Biasing Circuits and Techniques

3.6.1 Adaptive Biasing

Quiescent current is a critical indicator for evaluating the performance of LDOs. In applications such as the Internet of Things (IoT), wearable devices, and implantable biomedical devices, low power consumption and efficient power management are essential to significantly extend the system's operating lifetime. To achieve power conservation, functional units typically operate in duty cycle mode, spending most of the time in standby or off states. Consequently, the power management unit operates under extremely light- or even no-load conditions for extended periods. In such scenarios, the quiescent power consumption of the power management units becomes the primary source of power dissipation within the system [36].

When functional units need to operate, the power management module must provide adequate output current and exhibit fast transient response performance. For LDOs, a lower quiescent current can restrict the loop bandwidth and the drive slew rate of the power transistor, thereby impacting transient response performance. Therefore, the challenge arises: how can we effectively balance static power consumption with transient performance.

Adaptive biasing can effectively mitigate this issue. As illustrated in Fig. 3.35, the bias current of the error amplifier and buffer in the LDO adjusts dynamically with changes in the load. During light-load conditions, the bias current decreases, resulting in a narrower loop bandwidth, but it remains sufficient to push certain parasitic poles to high-frequency regions significantly beyond the unity-gain bandwidth (UGB). In contrast, under heavy-load conditions, the loop bandwidth expands as the bias current increases, thereby pushing the parasitic poles to even higher frequencies and achieving an enhanced bandwidth. Adaptive biasing requires detecting changes in output current. However, when the load changes instantaneously, the LDO has not yet started to adjust, causing the bias current change to lag

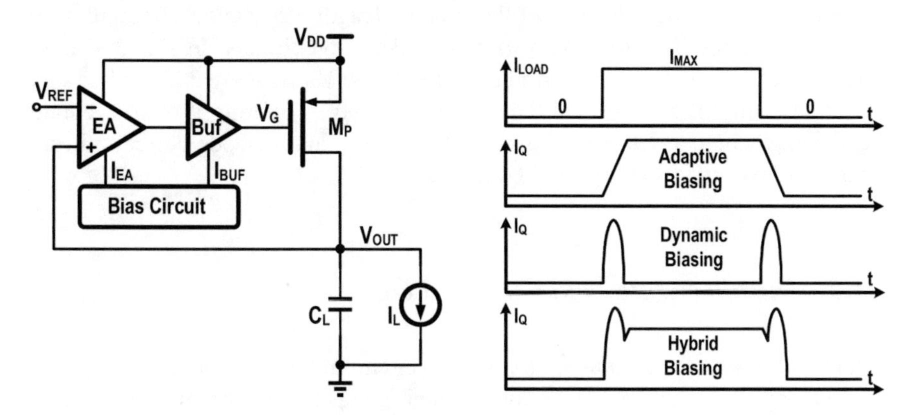

Fig. 3.35 Biasing schemes in LDO design

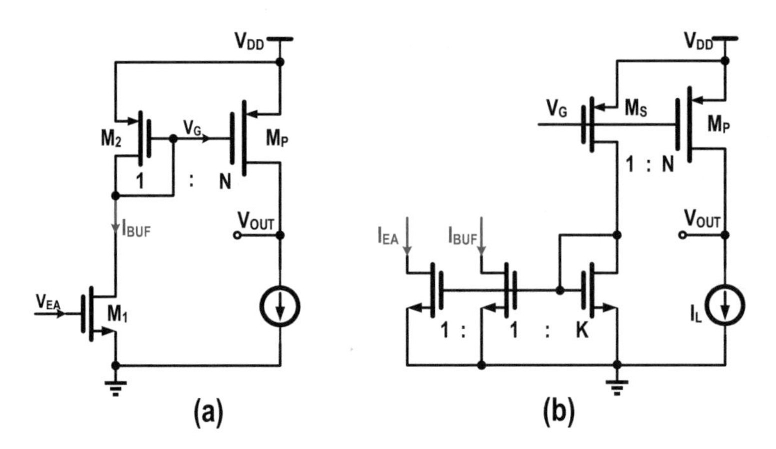

Fig. 3.36 (**a**) Common-source stage buffer and (**b**) adaptive biasing generation circuit

behind the load current change. Especially in the case of a load step-up, the transient response performance is still limited by the current bandwidth and slew rate.

Dynamic biasing increases the bias current solely during load transients, enhancing the LDO's bandwidth and slew rate precisely at these moments. This approach typically involves detecting the undershoot and overshoot of the output voltage. The hybrid biasing combines adaptive biasing and dynamic biasing techniques [10].

Figure 3.36 illustrates the application of adaptive biasing. In Fig. 3.36a, a diode-connected common-source stage buffer is depicted, where the buffer's bias current is proportional to the load current. As the load current increases, the gate pole of the power transistor shifts to a higher frequency. Figure 3.36b presents a typical adaptive biasing generation circuit. Here, M_S functions as a sensor transistor. The sensing current is scaled down by a factor of K and subsequently utilized for biasing the error amplifier and drive circuit.

Generally, the adaptive bias current only needs to follow the changes in the load current and does not require high accuracy. However, for more precise tracking of the output current, the influence of the channel length modulation effect must be minimized. Figure 3.37 presents two current detection circuits. In Fig. 3.37a, an open-loop clamp circuit is employed to make $V_{SEN} \approx V_{OUT}$. Figure 3.37b utilizes an operational amplifier for closed-loop feedback control, offering a more accurate V_{SEN} voltage compared to Fig. 3.34a. Additionally, using cascode current mirrors for M_3 and M_4 can further enhance current accuracy.

3.6.2 *Ultralow Quiescent Current LDO*

Figure 3.38 illustrates a 10 mA, 16 nA LDO circuit in [37]. The design comprises an error amplifier, adaptive bias circuit, dynamic compensation, buffer, power stage, and load current generator (LCG). Transistors M_1–M_{10} form the error amplifier,

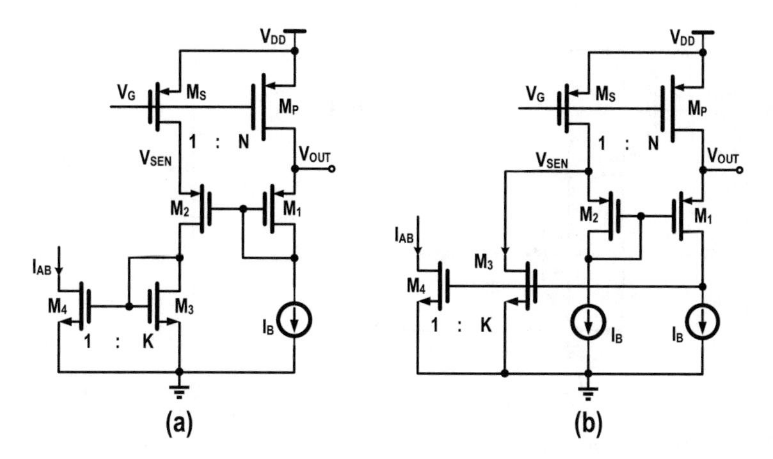

Fig. 3.37 (**a**) Open-loop voltage clamp circuit and (**b**) closed-loop voltage feedback

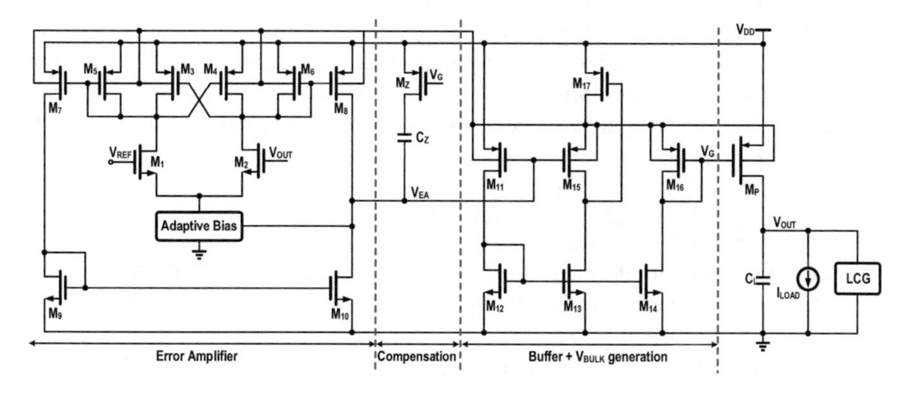

Fig. 3.38 The schematic of the low quiescent current LDO in [37]

where M_3 and M_4 are cross-coupled to increase the gain of the error amplifier. The LDO employs off-chip capacitors, and M_Z and C_Z are compensation circuits. The gate of M_Z is connected to V_G, and the impedance varies with the load, providing dynamic zero compensation.

The buffer circuit uses a differential flipped-voltage follower (DFVF) structure [38], ensuring that its input and output DC voltages are nearly identical. The drain output of M_{17} is used as the bulk voltage for the power transistor. Through bulk modulation [39], this configuration increases the output capacity of the power transistor, allowing for a smaller device size. Additionally, employing DFVF in the power stage offers another benefit: since V_{EA} equals V_G, the adaptive bias circuit can detect the load current using V_{EA} instead of V_G. This approach avoids the buffer's gate-driving delay effect, reducing the adaptive biasing response time to load changes.

Figure 3.39 depicts the adaptive biasing circuit in [37]. The gate voltage of M_{29}, V_{EA}, is used to detect the load current, which varies widely (over a range of >1000×).

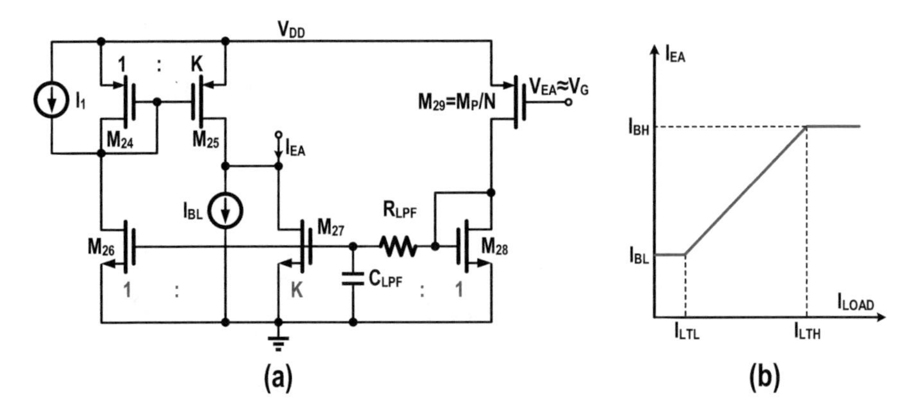

Fig. 3.39 (**a**) The schematic of adaptive bias and (**b**) I_{EA} variation versus load current

If the bias current changes across the same range, maintaining a suitable operating state for the error amplifier would be challenging. Therefore, it is essential to limit the bias current range. Under extremely light-load conditions, the current flowing through the current mirror transistors M_{28}, M_{27}, and M_{26} is zero. Consequently, the bias current I_1 raises the gate voltage of M_{24} and M_{25}, turning off M_{25}. At this point, the output bias current is

$$I_{EA} = I_{BL} \tag{3.76}$$

Through I_{BL}, the minimum value of the bias current is limited to ensure that the loop maintains sufficient gain and bandwidth for proper operation. As the load current increases, the current through M_{26}–M_{28} also increases. However, as long as the current through M_{26} remains less than I_1, the currents through M_{24} and M_{25} remain zero. At this point, the bias current I_{EA} is

$$I_{EA} = I_{BL} + \frac{K}{N} I_{LOAD} \tag{3.77}$$

As the load current continues to increase and the current through M_{26} exceeds I_1, the gate voltages of M_{24} and M_{25} decrease, allowing current to flow through M_{25}. At this point, I_{EA} will be clamped at

$$I_{EA} = I_{BL} + I_1 \times K \tag{3.78}$$

The low-pass filter, consisting of R_{LPF} and C_{LPF}, attenuates the kickback noise at the V_{EA}. Additionally, it enhances the transient performance during the load step-down but limits the slope of the bias current during the load step-up.

With advanced processes, power transistors may experience leakage, especially at high temperatures and FF corners. When leakage exceeds the load current, the output voltage rises above the set value. To prevent this, a leakage current generation (LCG) circuit can be applied to produce an appropriate amount of leakage

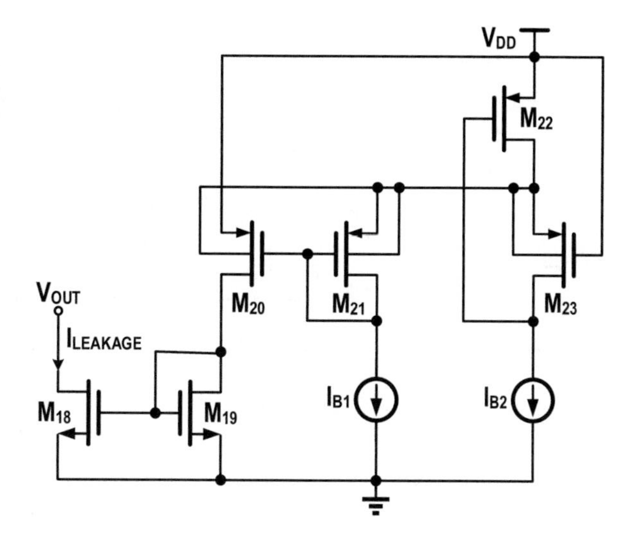

Fig. 3.40 The schematic of the leakage current generation (LCG)

current as the output load. Use the maximum leakage as the load will increase the quiescent current in typical scenarios. Figure 3.40 shows the LCG circuit, a scaled version of the DFVF buffer, and power transistor. M_{20} represents the power transistor, and its leakage current mirrors that of the power transistor. Current mirrors M_{18} and M_{19} amplify this leakage current and direct it to V_{OUT}. This solution adapts to process and temperature changes, always providing "just enough" discharge current for the LDO and ensuring minimal quiescent current under typical conditions.

Figure 3.41 illustrates a capacitorless LDO with a quiescent current of 54 nA [40]. The circuit includes an adaptive G_M OTA, an adaptive biasing circuit, a DFVF buffer, and a power stage. The compensation mechanism consists of a Miller capacitor (C_M) and a feedforward transistor (M_{24}). The transconductance of M_{24} (g_{mf}), which is related to the bias current, generates a dynamic zero. Under light-load conditions, the output pole shifts to a lower frequency. The g_{mf} decreases, and the zero also moves to a lower frequency, thereby enhancing the phase margin.

3.6.3 Dynamic Biasing

Dynamic biasing can increase the bias current during transient conditions. Figure 3.42 depicts the dynamic biasing circuit in [41]. The right side of the figure is the folded cascode FVF LDO. The output voltage is controlled by V_{SET}, generated by the one-stage control circuit. C_M is the Miller compensation capacitor of the FVF LDO, and I_1, I_2, and I_3 represent its bias currents.

For a load step-up, after V_{OUT} drops, capacitor C_1 couples this information to the gate of M_2, causing a decrease in the current through M_2 and subsequently through current mirrors M_3, M_4, and M_5. However, due to the isolation effect of R_1, the currents in current mirrors M_1 and M_{11} remain unchanged. This weakens the gate

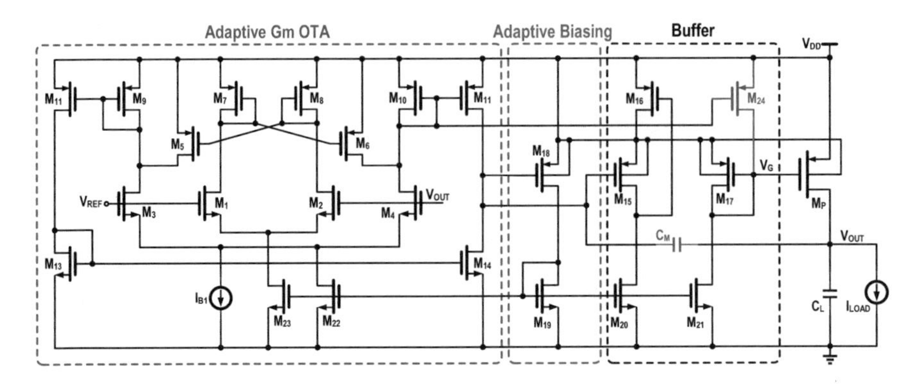

Fig. 3.41 The 54 nA quiescent current capacitorless LDO in [40]

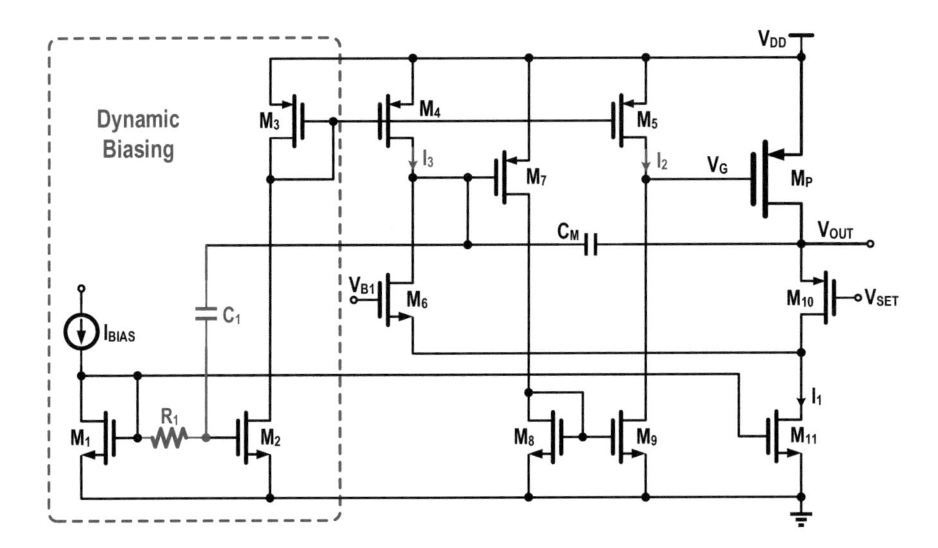

Fig. 3.42 The dynamic biasing circuit in [41]

voltage pull-up capability (I_3) of M_7 while keeping its pull-down capability constant. Consequently, the gate voltage of M_7 decreases, increasing the current through M_9. The pull-down current of V_G increases rapidly, while its pull-up current (I_2) decreases, causing V_G to drop rapidly.

Similarly, for a load step-down, following an overshoot in V_{OUT}, the pull-up capability of the FVF LDO is enhanced, while the pull-down capability is weakened. Dynamic biasing helps to minimize overshoot in load step-down conditions.

Figure 3.43 illustrates another dynamic biasing circuit from [42]. In this configuration, M_7, M_8, and M_P form the FVF LDO, while M_6 and M_{11} supply the upper bias current (I_2) and lower bias current (I_1), respectively. When V_{OUT} changes, capacitors C_{UP} and C_{DN} couple this change to the gates of M_4 and M_9, instantly adjusting the pull-up and pull-down bias currents of the FVF LDO.

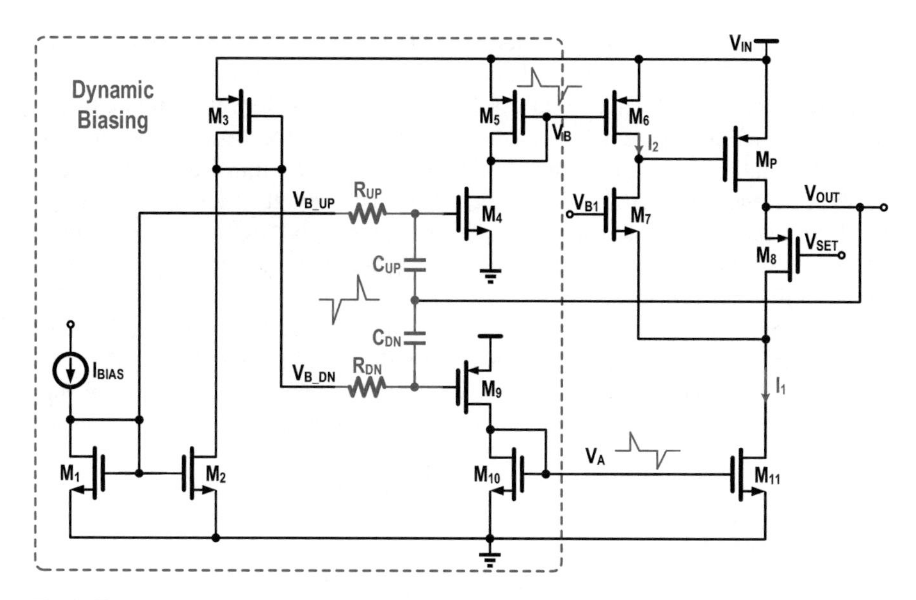

Fig. 3.43 The dynamic biasing circuit in [42]

3.7 PSRR Improvement Technology

When powering noise-sensitive modules such as ADCs, RF circuits, and PLL circuits, LDOs can provide a clean and stable power supply. In these applications, the power supply rejection ratio (PSRR) or power supply rejection (PSR) is a crucial metric for the LDO designs. This section begins by discussing the main limitation sources of PSRR in a typical LDO. It then discusses the key factors affecting PSRR for internal-pole-dominant (IPD) LDOs and output-pole-dominant (OPD) LDOs. Finally, the chapter explores several techniques for improving PSR.

3.7.1 PSRR Limitation Sources

The PSRR refers to an LDO's ability to suppress input power ripple, defined as

$$\mathrm{PSRR} = 20\log\left(\frac{V_{\mathrm{RIPPLE,OUT}}}{V_{\mathrm{RIPPLE,IN}}}\right) \tag{3.79}$$

$V_{\mathrm{RIPPLE,\,OUT}}$ and $V_{\mathrm{RIPPLE,\,IN}}$ represent the ripple amplitudes of the output voltage V_{OUT} and the input voltage V_{IN}, respectively. Generally, $V_{\mathrm{RIPPLE,\,OUT}}$ is less than $V_{\mathrm{RIPPLE,\,IN}}$, resulting in a generally negative PSRR. A smaller (more negative) PSRR value indicates better performance. In the industrial, the absolute value of PSRR is commonly used to quantify ripple rejection capability.

There are multiple ripple transmission paths between the input supply and the output of the LDO [43], as depicted in Fig. 3.44. Path 1 arises from the finite PSRR of the reference circuit. Path 2 is attributed to the finite PSRR of the error amplifier. Since power supply ripple will cause the change of V_{GS} and V_{DS} of the power transistor, path 3 and path 4 come through g_m and g_{ds} of the power transistor, respectively.

The reference voltage V_{REF} has a finite PSRR, and the reference ripple is amplified by $1/\beta$ and appears at the LDO output. Using the high PSRR bandgap and large RC low-pass filter can enhance the PSRR of V_{REF} and prevent it from degrading the overall PSRR performance.

The error amplifier PSRR influences the overall PSRR performance of the LDO [44]. Figure 3.45 illustrates two types of amplifiers and their PSRR small-signal models. Type A amplifier uses a PMOS current mirror connected to the power supply as a load, whereas type B amplifier uses an NMOS current mirror connected to ground.

For the type A amplifier in Fig. 3.45a, according to the low-frequency small-signal model, we can derive the following:

$$\begin{aligned} V_{EA} &= V_{DD}\left(\frac{r_{o2}}{r_{o2}+r_{o4}}\right)+I_1\left(r_{o2}\parallel r_{o4}\right) \\ &= V_{DD}\left(\frac{r_{o2}}{r_{o2}+r_{o4}}\right)+\frac{V_{DD}}{r_{o1}+{}^{1}\!/_{g_{m3}}}\left(r_{o2}\parallel r_{o4}\right) \end{aligned} \tag{3.80}$$

Assuming that $r_{o1} = r_{o2} = r_{o4} >> 1/g_{m3}$, and substituting into Eq. (3.60), we have

$$V_{EA} \approx V_{DD} \tag{3.81}$$

According to Eq. (3.60), it can be observed that at low frequencies, the ripple on the power supply passes through two paths in the amplifier and then superimposes

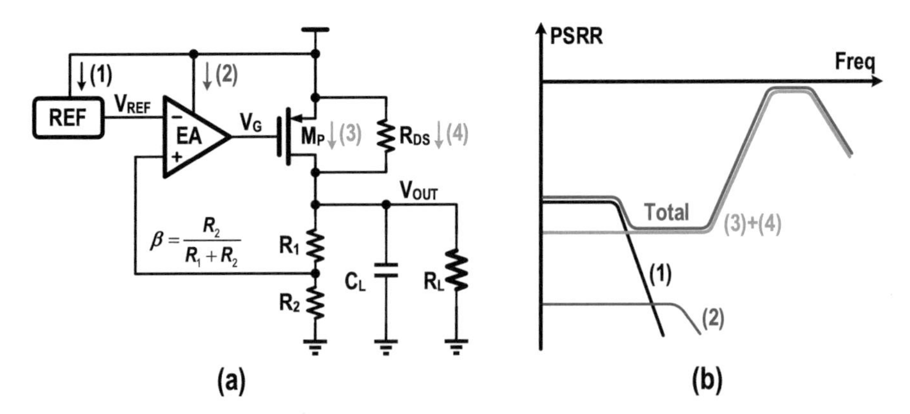

Fig. 3.44 (**a**) Four major supply ripple feed-through paths, and (**b**) respective contribution of the four paths in total PSRR performance

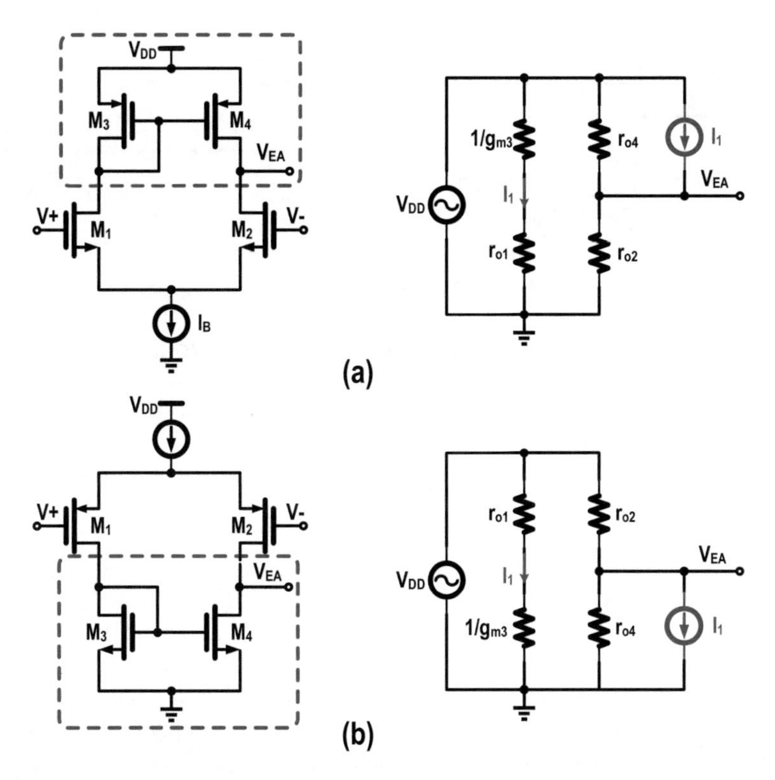

Fig. 3.45 (**a**) Type A amplifier and its small-signal model for PSRR; (**b**) type B amplifier and its small-signal model for PSRR

on V_{EA}, ultimately equal to V_{DD}, without attenuation. However, if a type A amplifier-driven PMOS power stage is used, the V_{GS} differential of the power transistor remains unchanged, eliminating the impact of path 3 on PSRR. This improves the overall PSRR performance of the LDO. Conversely, if a type A drives an NMOS power stage, the output voltage fluctuates with changes in V_G, thereby degrading the PSRR performance.

Figure 3.45b shows a type B amplifier. It consists of an NMOS current mirror load and a PMOS input pair. We can get

$$\begin{aligned} V_{EA} &= V_{DD}\left(\frac{r_{o4}}{r_{o2}+r_{o4}}\right) - I_1\left(r_{o2} \parallel r_{o4}\right) \\ &= V_{DD}\left(\frac{r_{o4}}{r_{o2}+r_{o4}}\right) - \frac{V_{DD}}{r_{o1}+1/g_{m3}}\left(r_{o2} \parallel r_{o4}\right) \approx 0 \end{aligned} \tag{3.82}$$

Based on Eq. (3.62), at low frequencies, the ripple on the power supply can be countered in V_{EA} after passing through the amplifier's two paths. Consequently, V_{EA} can be regarded as a clean signal source. Using type B amplifier to drive the NMOS

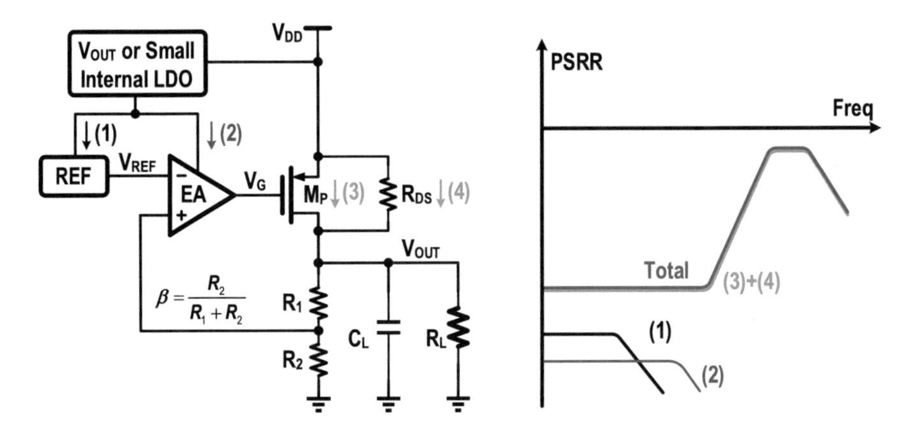

Fig. 3.46 In high PSRR LDO design, use "clean" power supply to eliminate the effect of path 1 and path 2

power stage will enhance the NMOS's PSRR performance. However, when type B amplifier drives a PMOS power stage, the impact of path 3 on PSRR remains.

Analyzing the PSRR pathways of type A and type B amplifiers helps us understand how power supply ripple propagates through operational amplifiers. In practical applications, due to parasitic capacitance, compensation circuits, and buffer circuits, it is often challenging to analyze how the power supply ripple reaches the gate of the power transistor after passing through the EA and various other circuits. Therefore, in high PSRR LDO designs, V_{OUT} or an additional clean power source is typically used as the power supply for the reference and error amplifiers to enhance their PSRR without affecting the overall PSRR performance of the LDO, as illustrated in Fig. 3.46. In this scenario, the overall PSRR performance of the LDO is mainly influenced by path 3 and path 4. Assuming that the gate voltage of the power transistor is ideal and clean, the ripple contributions from path 3 and path 4 are

$$V_{\text{OUT},3+4} = \left(g_{\text{m}} + g_{\text{ds}}\right) V_{\text{DD}} \times \left(r_{\text{ds}} \parallel R_{\text{L}} \parallel sC_{\text{L}}\right) \tag{3.83}$$

The analysis above does not account for the effect of the parasitic capacitance of the power transistor. Due to the large size of the power transistor, the parasitic capacitance can cause significant voltage division at the gate at mid-to-high frequencies. Moreover, there is a parasitic capacitance path from the gate to the output V_{OUT}. Therefore, the parasitic capacitance significantly impacts the PSRR at mid-to-high frequencies.

3.7.2 *PSRR Analysis: IPD LDO and OPD LDO*

To more clearly analyze the characteristics of PSRR across different frequency bands, Fig. 3.47a presents the complete small-signal model used for the derivation of the PSRR transfer function. In this model, we neglect the contribution of the voltage regulation circuits to the PSRR. Here, R_A represents the output impedance of the control circuit, and gm denotes the equivalent transconductance of the control circuit. C_{GS}, C_{DS}, and C_{GD} are the parasitic capacitances of the PMOS power transistor. The blue box highlights the LDO regulation loop, while the gray box represents the small-signal model for the power transistor. The transfer function for the blue box area is defined as follows:

$$T(s) = \frac{V_{OUT}(s)}{V_A(s)} \tag{3.84}$$

The work in [45] proposed an open-loop PSRR analysis method, in which the open-loop $PSRR_{OL}(s)$ refers to the transfer function of the ripple path from V_{DD} to V_{OUT} when the regulation loop of the LDO is cut. Figure 3.47b presents the small-signal model of the open-loop PSRR. In addition, [45] presents a simulation method for open-loop $PSRR_{OL}(s)$, as shown in Fig. 3.48.

The closed-loop $PSRR_{CL}(s)$ can be expressed as

$$\mathrm{PSRR}_{CL}(s) = \frac{\mathrm{PSRR}_{OL}(s)}{1+T(s)} \tag{3.85}$$

In practical simulations and measurements, PSRR typically refers to the closed-loop $PSRR_{CL}(s)$. Equation (3.60) explains why the PSRR curve is affected by the loop gain $T(s)$.

Based on the position of the dominant pole, LDOs are typically classified into internal-pole-dominant (IPD) LDOs and output-pole-dominant (OPD) LDOs. We will discuss the $PSRR_{OL}(s)$ and the transfer function $1 + T(s)$ for each type, to better understand the factors that influence PSRR performance and the mechanisms.

Internal-Pole-Dominant LDO

Based on Figs. 3.47b and 3.49 further illustrates the small-signal model for the transfer function of the open-loop PSRR. Here, R_O represents the output equivalent impedance, R_A denotes the gate equivalent impedance of the power transistor, and C_{LT} and C_G are the total equivalent capacitances at the output and the gate of the power transistor, respectively.

$$R_O = R_L \parallel r_{ds}, C_{LT} = C_L + C_{GD} + C_{DS},\ C_{LT} = C_L + C_{GD} + C_{DS}, \tag{3.86}$$

The open-loop PSRR transfer function can be obtained as derived in [26]:

$$\mathrm{PSRR}_{OL}(s) \approx (g_{mp} + g_{ds}) R_O \frac{\left(1+\frac{s}{z_1}\right)\left(1+\frac{s}{z_2}\right)}{\left(1+\frac{s}{p_1}\right)\left(1+\frac{s}{p_2}\right)} \tag{3.87}$$

$$p_1 = \frac{1}{R_A\left[C_{GS} + (1+g_{mp}R_O)C_{GD}\right]} \tag{3.88}$$

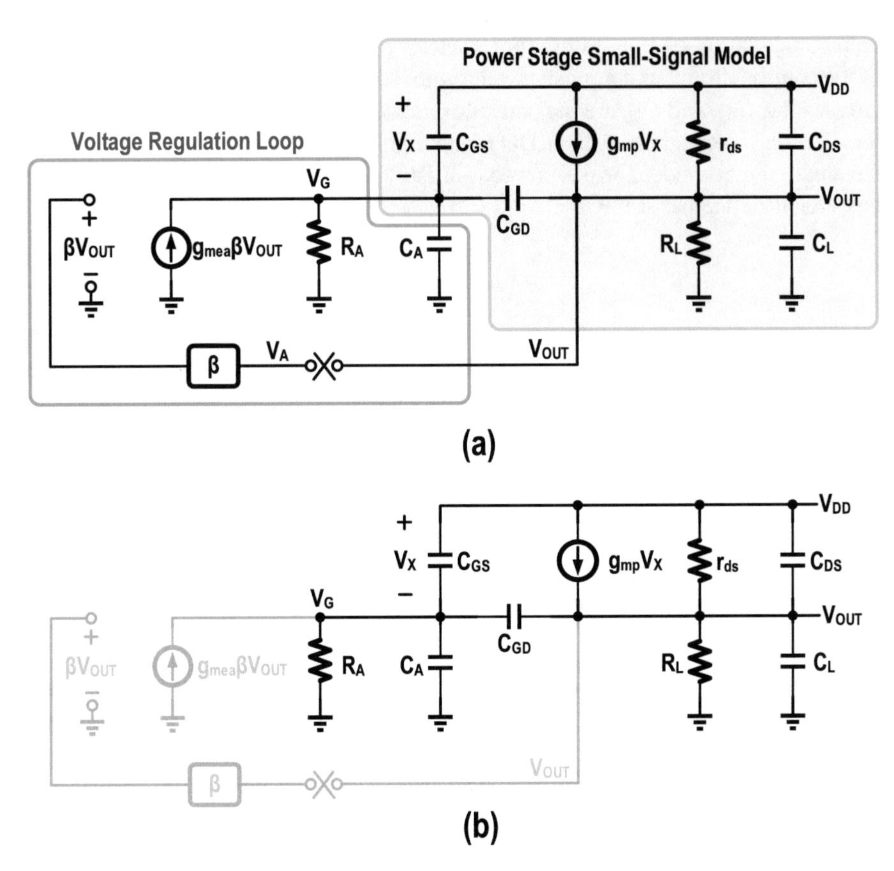

Fig. 3.47 (**a**) Complete small-signal model for the derivation of the transfer function of the PSRR, and (**b**) small-signal model for the transfer function of the open-loop PSRR

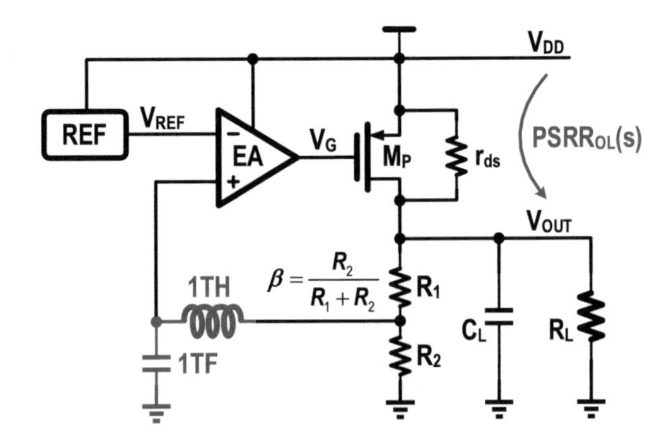

Fig. 3.48 Simulation setup for the open-loop PSRR

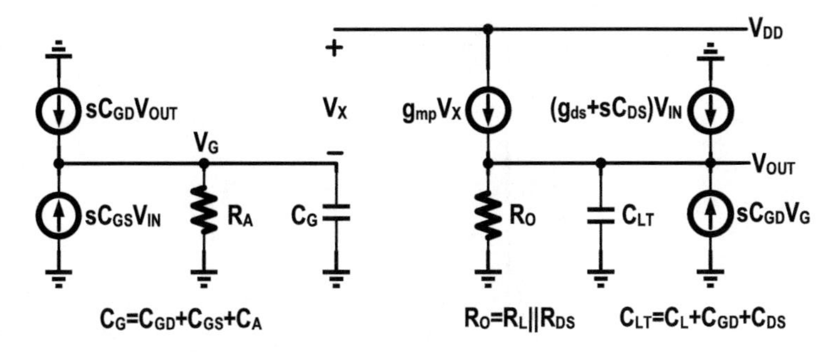

Fig. 3.49 Small-signal model for the open-loop PSRR

$$p_2 = \frac{C_{\mathrm{GS}} + \left(1 + g_E R_{\mathrm{O}}\right) C_{\mathrm{GD}}}{R_{\mathrm{O}}\left[\left(C_{\mathrm{GS}} + C_{\mathrm{GD}}\right) C_{\mathrm{LT}} - C_{\mathrm{GD}}^2\right]} \tag{3.89}$$

$$z_1 = \frac{g_{\mathrm{mp}} + g_{\mathrm{ds}}}{R_{\mathrm{A}}\left[\left(g_{\mathrm{mp}} + g_{\mathrm{ds}}\right) C_{\mathrm{GD}} + g_{\mathrm{ds}} C_{\mathrm{GS}}\right]} \approx \frac{1}{R_{\mathrm{A}} C_{\mathrm{GD}}} \tag{3.90}$$

$$z_2 = \frac{\left(g_{\mathrm{mp}} + g_{\mathrm{ds}}\right) C_{\mathrm{GD}} + g_{\mathrm{ds}} C_{\mathrm{GS}}}{C_{\mathrm{GS}} C_{\mathrm{GD}} + C_{\mathrm{GS}} C_{\mathrm{DS}} + C_{\mathrm{GD}} C_{\mathrm{DS}}} \tag{3.91}$$

The transfer function of the voltage regulation loop is

$$T(s) \approx \frac{A_0}{\left(1 + \frac{s}{p_{\mathrm{d}}}\right)\left(1 + \frac{s}{p_{\mathrm{nd}}}\right)} \tag{3.92}$$

where p_{d} and p_{nd} are the dominant pole and the nondominant pole of the regulation loop, respectively. And A_0 is the DC loop gain.

For IPD LDO, the p_{d} and p_{nd} are

$$p_d = \frac{1}{R_{\mathrm{A}}\left[C_{\mathrm{GS}} + \left(1 + g_{\mathrm{mp}} R_{\mathrm{O}}\right) C_{\mathrm{GD}}\right]},\ p_{\mathrm{nd}} = \frac{1}{R_{\mathrm{O}} C_{\mathrm{LT}}}, \tag{3.93}$$

$$\frac{1}{1 + T(s)} \approx \frac{\left(1 + \frac{s}{p_{\mathrm{d}}}\right)\left(1 + \frac{s}{p_{\mathrm{nd}}}\right)}{A_0\left(1 + \frac{s}{A_0}\frac{1}{p_{\mathrm{d}}} + \frac{s^2}{A_0}\frac{1}{p_{\mathrm{d}} p_{\mathrm{nd}}}\right)} = \frac{\left(1 + \frac{s}{p_{\mathrm{d}}}\right)\left(1 + \frac{s}{p_{\mathrm{nd}}}\right)}{A_0\left(1 + \frac{1}{Q_{\mathrm{x}}}\frac{s}{p_{\mathrm{x}}} + \frac{s^2}{p_{\mathrm{x}}}\right)} \tag{3.94}$$

$$p_x \approx \sqrt{A_0 p_{\mathrm{d}} p_{\mathrm{nd}}},\ Q_x \approx \sqrt{\frac{A_0 p_{\mathrm{d}}}{p_{\mathrm{nd}}}} \tag{3.95}$$

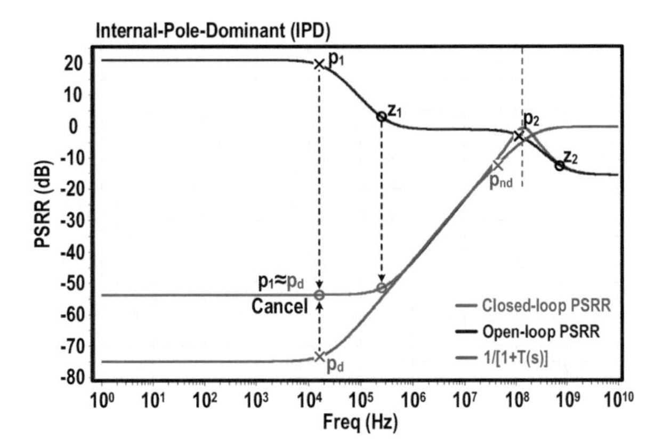

Fig. 3.50 PSRR simulation result of an IPD LDO

Near the unity-gain bandwidth (UGB) of the control loop, a pair of complex poles emerges at frequency p_x with a Q factor of Q_x.

Figure 3.50 shows the PSRR simulation results for the IPD LDO. The black curve represents the open-loop PSRR, the blue curve corresponds to $1/[1 + T(s)]$ related to the loop gain, and the red curve represents the closed-loop PSRR. According to Eq. (3.85), the red curve is the sum of the black and blue curves. For the IPD LDO, since $p_1 \approx p_d$, the closed-loop PSRR curve does not change at the loop's dominant pole. The first knee of the closed-loop PSRR curve corresponds to the z1 of the open-loop PSRR, while the second knee is near the unity gain of $T(s)$.

We will explore how the loop DC gain, load resistance R_L, drive impedance R_A, and output capacitance C_L affect closed-loop PSRR performance by adjusting these parameters.

Regarding Fig. 3.51a, increasing g_{mea} enhances the loop gain by 14 dB without significantly altering the positions of the poles and zeros. The absolute value of low-frequency closed-loop PSRR also increases by 14 dB. Therefore, improving the loop gain $T(s)$ effectively enhances low-frequency PSRR.

In Fig. 3.51b, for an IPD LDO, changing the load RL shifts both the open-loop PSRR and the position of the secondary dominant pole. The dominant pole (p_1, p_d) also shifts due to the Miller effect and variations in $g_{mp}R_O$. Generally, PSRR performance with light loads is better than that with heavy loads.

As illustrated in Fig. 3.51c, with the loop gain kept constant, increasing the driving impedance R_A of the power transistor will shift the dominant poles (p_1, p_d) and the open-loop PSRR zero (z_1) to lower frequencies. This significantly deteriorates the PSRR performance at mid-frequencies. Thus, to improve mid-frequency PSRR, R_A must be reduced. However, simply reducing R_A can affect the loop gain. Increasing the preceding-stage transconductance g_{mea} while reducing R_A will increase power consumption.

Lastly, as illustrated in Fig. 3.51d, by appropriately increasing the output capacitance C_L for an IPD LDO, both the secondary dominant pole (p_2, p_{nd}) and the peak frequency of the closed-loop PSRR will shift to lower frequencies, while the open-loop PSRR zero (z_2) remains unchanged, thereby improving the high-frequency closed-loop PSRR.

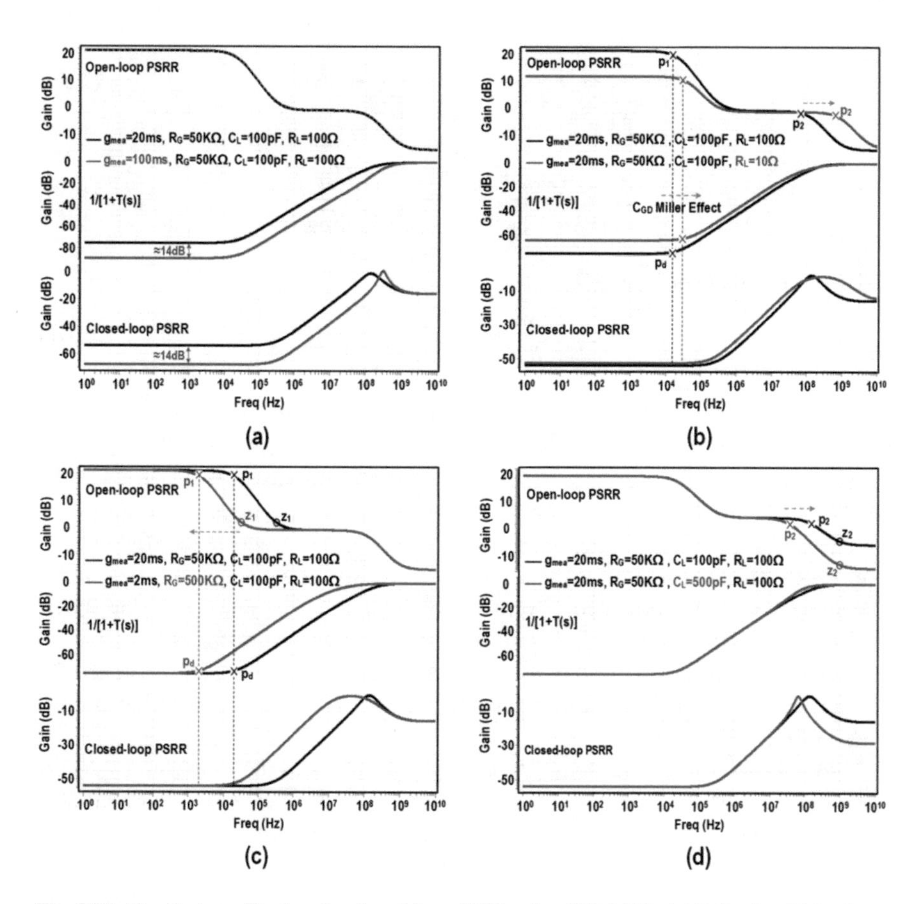

Fig. 3.51 Key factors affecting the closed-loop PSRR of an IPD LDO: (**a**) DC gain, (**b**) load current, (**c**) impedance R_A, and (**d**) output capacitor C_L

Output-Pole-Dominant LDO

The open-loop PSRR transfer function of OPD LDO can be obtained as derived in [45]:

$$\text{PSRR}_{\text{OL}}(s) \approx \left(g_{\text{mp}} + g_{\text{ds}}\right) R_{\text{O}} \frac{\left(1+\dfrac{s}{z_1}\right)\left(1+\dfrac{s}{z_2}\right)}{\left(1+\dfrac{s}{p_1}\right)\left(1+\dfrac{s}{p_2}\right)} \tag{3.96}$$

$$p_1 = \frac{1}{R_{\text{O}} C_{\text{LT}}} \tag{3.97}$$

$$p_2 = \frac{1}{R_{\text{A}}\left(C_{\text{GS}} + C_{\text{GD}}\right)} \tag{3.98}$$

$$z_1 = \frac{g_{\text{mp}} + g_{\text{ds}}}{R_{\text{A}}\left[\left(g_{\text{mp}} + g_{\text{ds}}\right)C_{\text{GD}} + g_{\text{ds}}C_{\text{GS}}\right]} \approx \frac{1}{R_{\text{A}}C_{\text{GD}}} \tag{3.99}$$

$$z_2 = \frac{\left(g_{\text{mp}} + g_{\text{ds}}\right)C_{\text{GD}} + g_{\text{ds}}C_{\text{GS}}}{C_{\text{GS}}C_{\text{GD}} + C_{\text{GS}}C_{\text{DS}} + C_{\text{GD}}C_{\text{DS}}} \tag{3.100}$$

For OPD LDO, the p_{d} and p_{nd} are

$$p_d = \frac{1}{R_{\text{O}}C_{\text{L}}},\ p_{\text{nd}} = \frac{1}{R_{\text{A}}C_{\text{G}}}, \tag{3.101}$$

Figure 3.52 shows the open-loop and closed-loop PSRR curves of the OPD LDO. The dominant pole p_1 of the open-loop PSRR is canceled out by the loop's dominant pole pd. Compared to the IPD LDO, the closed-loop PSRR curve of the OPD LDO has a flatter region in the mid-frequency range due to the reduction of R_{A}, which pushes the zero z_1 in the open-loop curve to higher frequencies. The closed-loop PSRR curve begins to decrease near the UGB of $T(s)$, obtaining good high-frequency PSRR performance. Overall, the OPD LDO exhibits superior PSRR performance in the mid- to high-frequency range compared to the IPD LDO. We will investigate the effects of parameters such as loop DC gain, load resistance R_{L}, driving impedance R_{A}, and output capacitor C_{L} on the closed-loop PSRR performance of the OPD LDO.

Figure 3.53a demonstrates that increasing g_{mea} enhances the loop gain by 14 dB while maintaining the pole-zero position. The magnitude of closed-loop PSRR also improves by 14 dB at mid-low frequencies, confirming that enhancing the loop gain $T(s)$ can effectively improve mid-low-frequency PSRR.

In the OPD LDO, the output pole serves as the dominant pole. Figure 3.53b illustrates that as load current increases, the dominant poles p_1 of open-loop PSRR and p_{d} of $T(s)$ shift to higher frequencies. At mid-low frequencies, substantial variations in open-loop PSRR and $T(s)$ gain occur due to changes in power stage $g_{\text{mp}}R_{\text{O}}$,

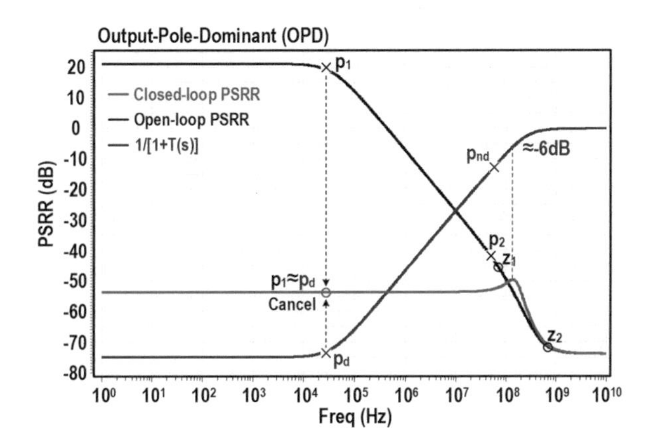

Fig. 3.52 PSRR simulation result of an OPD LDO

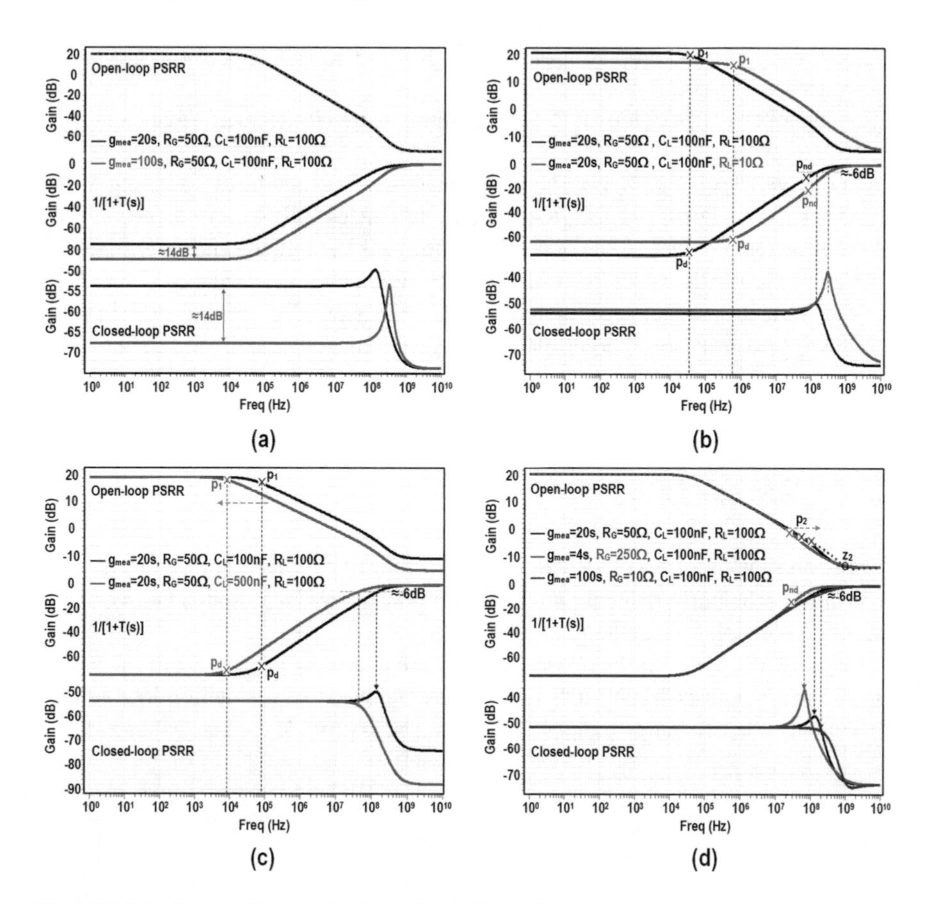

Fig. 3.53 Key factors affecting the closed-loop PSRR of an IPD LDO: (**a**) DC gain, (**b**) load current, (**c**) impedance R_A, and (**d**) output capacitor C_L

while closed-loop PSRR exhibits less variation. At high frequencies, light-load PSRR is better than heavy-load PSRR.

Figure 3.53c reveals that increasing output capacitor C_L shifts the dominant poles (p_1, p_d) to lower frequencies without affecting mid-low-frequency PSRR. The increased separation between p_d and p_{nd} results in a larger Q_X. Consequently, the closed-loop curve near the unity-gain bandwidth (UGB) declines smoothly and rapidly without any peaks.

As depicted in Fig. 3.53d, maintaining a constant loop gain $T(s)$ while adjusting the driving impedance R_A of the power transistor does not alter the mid-frequency closed-loop PSRR. However, the magnitude and frequency of the "peak" change. Increasing R_A brings p_d and p_{nd} closer together, reducing p_X and Q_X in Eq. (3.60), which in turn increases the peak amplitude.

Conclusions on PSRR

After discussing the PSRR simulation results of IPD LDO and OPD LDO, we can draw the following conclusions:

1. Increasing the DC loop gain of $T(s)$ helps improve PSRR performance at low frequencies.
2. Compared to IPD LDO, OPD LDO exhibits better PSRR performance at mid-high frequencies.
3. The drive impedance RA of the power transistor is a key factor affecting PSRR performance. For IPD LDO, RA directly influences PSRR performance at mid-frequencies. For OPD LDO, RA affects the "peak" amplitude and determines the worst-case PSRR value. Lower RA values yield better performance.
4. The parasitic capacitance of the power transistor also significantly impacts PSRR performance. Larger parasitic capacitance demands a smaller drive impedance RA.

3.7.3 Cascode LDO

A cascode LDO improves PSRR by cascading two LDO stages, acting as a double-layer filter, efficiently reducing the ripple on power supply. It is also known as the supply ripple isolation method.

Figure 3.54 illustrates five cascode LDO architectures proposed in [46–50]. In Fig. 3.54a, two cascaded NMOS LDOs are employed. To maintain a low voltage differential, a boost charge pump (CP) is required to provide the bias voltage for the

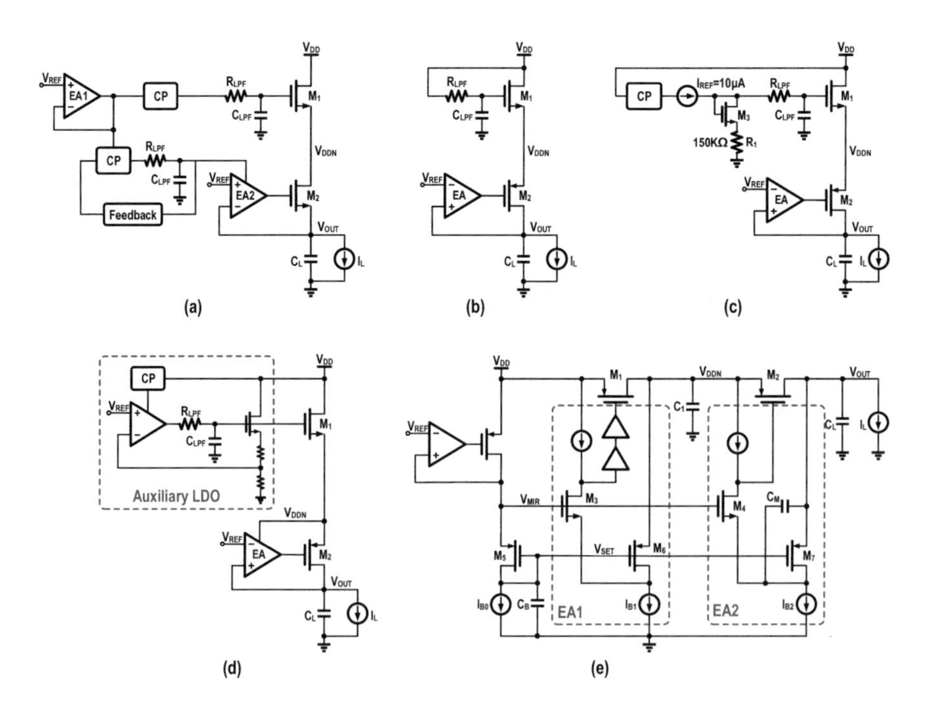

Fig. 3.54 Cascode LDOs presented in (**a**) [46], (**b**) [47], (**c**) [48], (**d**) [49], and (**e**) [50]

two-stage LDOs, and the error amplifier of the second-stage LDO also relies on the CP for power. In addition, a low-cutoff-frequency low-pass filter is necessary to suppress the switching noise generated by the CP. Due to the utilization of two CPs and filtering circuits, this configuration necessitates a large on-chip capacitor.

Figure 3.54b depicts a cascaded structure of an NMOS LDO and a PMOS LDO. The NMOS LDO operates under open-loop control. Since the bias voltage is derived from the input power supply, it requires a large voltage margin, making it unsuitable for low-voltage-differential applications.

In Fig. 3.54c, a cascaded NMOS and PMOS structure is used. To reduce the voltage margin, a boost CP is employed, and the bias voltage is generated by the bias current I_{REF}, M_3, and resistor R_1. Since both M_3 and M_1 are all NMOS transistors, they can offset some PVT variations, maintaining the stability of V_{DDN}. In Fig. 3.54d, the first-stage LDO uses a boost CP and a replica NMOS LDO structure [36], which allows for more accurate control of the DC voltage of V_{DDN}.

The structures presented in Fig. 3.54a–d consistently employ an NMOS LDO configuration for the first stage, and all are open-loop control. This is mainly due to the following reasons:

1. The NMOS source follower exhibits excellent PSRR performance, especially when a low-pass filter is connected to the gate.
2. The NMOS source follower inherently responds well to load transients; when VOUT drops, the NMOS source follower naturally provides additional output current.
3. Open-loop control simplifies the circuit architecture and reduces current consumption. It necessitates only a charge pump with very small load capacity, resulting in lower on-chip capacitance and quiescent current.

In Fig. 3.54e, the structure employs two cascaded PMOS LDOs. Two different folded cascode flipped-voltage follower circuits are used for EA1 and EA2. Both FVF structures utilize the same V_{SET} and bias voltage V_{MIR}. It is noteworthy that EA1 consumes more current and incorporates a buffer, making the first-stage LDO a fast OPD LDO. In contrast, the second-stage LDO is relatively slower and employs Miller compensation, classifying it as an IPD LDO. Compared to the previous NMOS+PMOS structure, the dual-PMOS structure eliminates the need for a boost charge pump and an additional clock circuit.

The cascode LDO can effectively improve PSRR performance, but it requires a higher voltage margin and a larger power transistor area. In addition, since the transient response is constrained by the two-stage LDO control, some load transient performance is sacrificed.

3.7.4 *Feedforward Ripple Cancellation*

Besides the OPD LDO and cascode LDO techniques we introduced above, in this section, we will discuss another technology for improving PSRR: feedforward ripple cancellation (FFRC). This is an active PSRR enhancement technique. The standard approach involves amplifying the power supply ripple to an appropriate ratio and injecting it into the gate of the power transistor (Fig. 3.55a) or the body of the power transistor (Fig. 3.55b). If the injected ripple V_{FF} is optimal, the power supply ripple component can be completely canceled at the output, thus enhancing the PSRR performance of the LDO.

At mid-high frequencies, the voltage component of the power ripple at the gate of the power transistor is

$$V_{RG} = V_R \times \frac{C_{GS}}{C_{GD} + C_{GS}} \tag{3.102}$$

where V_R represents the amplitude of the power supply ripple. Considering that the output ripple is much smaller than the supply ripple, the influence of the output ripple is ignored. C_{GS} and C_{GD} are the gate-source and gate-drain parasitic capacitances of the power transistor, respectively.

As illustrated in Fig. 3.55a, the power supply ripple mainly affects the output through transconductance gm and drain-source conductance gds. If the ripple V_{FF} is injected into the gate of the power transistor to precisely counteract the supply ripple, we can derive

$$\Delta i = \left(V_R - V_R \times \frac{C_{GS}}{C_{GD} + C_{GS}} - V_{FF} \right) g_m + V_R \times g_{ds} = 0 \tag{3.103}$$

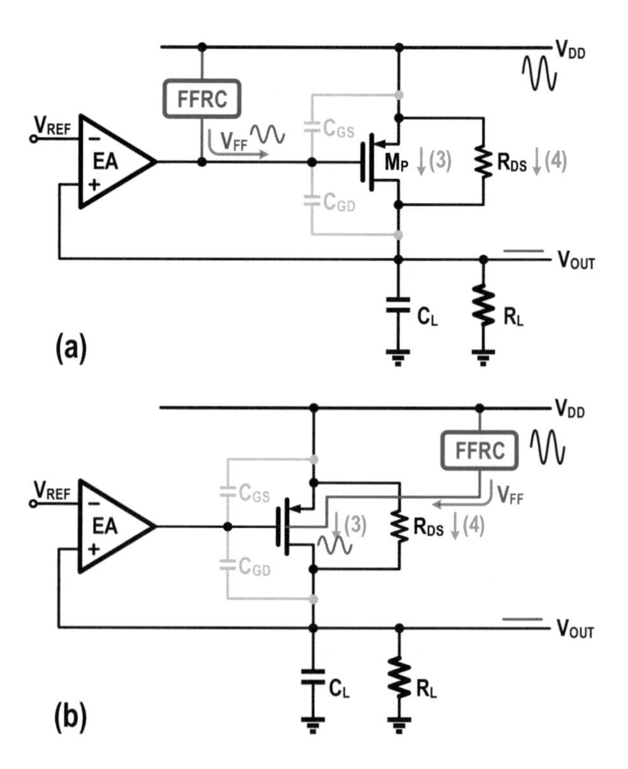

Fig. 3.55 Feedforward ripple cancellation technologies. (**a**) gate ripple injection [53–55] and (**b**) body ripple injection [51, 52]

The optimal gate injection ripple is

$$V_{FF}\big|_{OPT} = V_R\left(\frac{C_{GD}}{C_{GS}+C_{GD}}+\frac{g_{ds}}{g_m}\right) \tag{3.104}$$

Ignore the coupling effect of parasitic capacitance, or consider that the coupled power supply ripple is zero at low frequency; then Eq. (3.103) becomes

$$\Delta i = (V_R - V_{FF})g_m + V_R \times g_{ds} = 0 \tag{3.105}$$

$$V_{FF}\big|_{OPT} = V_R\left(\frac{g_m + g_{ds}}{g_m}\right) \tag{3.106}$$

As shown in Fig. 3.55b, if the ripple V_{FF} is injected into the body terminal of the power transistor to precisely counteract the power supply ripple effect, we can derive

$$\Delta i = \left(V_R - V_R \times \frac{C_{GS}}{C_{GD}+C_{GS}}\right)g_m + V_R \times g_{ds} + g_{mb}(V_R - V_{FF}) = 0 \tag{3.107}$$

The optimal body injection ripple is

$$V_{FF}\big|_{OPT} = V_R\left(\frac{C_{GD}}{C_{GS}+C_{GD}}\times\frac{g_m}{g_{mb}}+\frac{g_{ds}}{g_{mb}}+1\right) \tag{3.108}$$

Ignore the coupling effect of parasitic capacitance, or consider that the coupled power supply ripple is zero at low frequency; then Eq. (3.107) becomes

$$\Delta i = V_R \times g_m + V_R \times g_{ds} + g_{mb}(V_R - V_{FF}) = 0 \tag{3.109}$$

$$V_{FF}\big|_{OPT} = V_R\left(\frac{g_m}{g_{mb}}+\frac{g_{ds}}{g_{mb}}+1\right) \tag{3.110}$$

The implementation of ripple injection can be categorized into three methods: voltage-mode gate ripple injection [53, 54], current-mode gate ripple injection [6, 55], and body ripple injection [51, 52].

Voltage-Mode Gate Ripple Injection

Figure 3.56 presents an implementation of voltage-mode gate ripple injection. The operational amplifier A_{FF}, along with feedback resistors R_{FF1} and R_{FF2}, and capacitor C_F, forms an inverting proportional amplifier for the feedforward path. The feedforward power supply ripple can be expressed as

$$V_{FFA} = -V_R \times \frac{R_{F2}}{R_{F1}} \times (1 + SR_{F1}C_F) \tag{3.111}$$

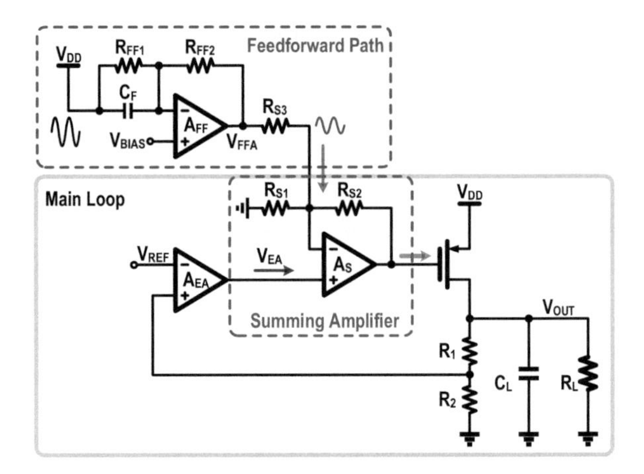

Fig. 3.56 Voltage-mode gate ripple injection in [53]

The operational amplifier AS, together with resistors R_{S1}, R_{S2}, and R_{S3}, forms a summing amplifier (SA). The SA inverts the feedforward power supply ripple V_{FFA} and sums it with the main loop control signal V_{EA}. This composite signal is then used to control the power transistor, ultimately injecting the optimal ripple into its gate:

$$V_{FF} = -V_R \times \frac{R_{F2}}{R_{F1}} \times \frac{R_{S2}}{R_{S3}} \times \left(1 + SR_{F1}C_F\right) \tag{3.112}$$

According to Eq. (3.106), we need to set

$$\frac{R_{F2}}{R_{F1}} \times \frac{R_{S2}}{R_{S3}} = \frac{g_m + g_{ds}}{g_m} \tag{3.113}$$

The optimal gate ripple signal injection can be achieved with this setting. In addition, it can be observed that Eq. (3.70) includes a zero, which is necessary for the system. Figure 3.57 illustrates the mathematical model of the voltage-mode gate ripple injection LDO.

Here, $H_{FF}(s)$ is the feedforward path transfer function, A_{S0} is the closed-loop DC gain of the summing amplifier, and ω_s is the gate pole of the power tube, set $R_{S1} = R_{S3}$:

$$A_{S0} = \frac{R_{S2}}{R_{S3}} \tag{3.114}$$

According to Fig. 3.57, the LDO PSRR transfer function can be derived as

$$\frac{V_{OUT}}{V_{DD}}(s) = \frac{1 + g_m \times r_{ds} \times \left[1 - H_{FF}(s)\dfrac{A_{S0}}{1 + s/\omega_S}\right]}{1 + \dfrac{r_{ds}}{Z_L(s)} + \dfrac{r_{ds}}{R_{F1} + R_{F2}} + \dfrac{g_m \times r_{ds} \times A_{E0} \times R_{F2}}{(R_{F1} + R_{F2})(1 + s/\omega_E)}} \tag{3.115}$$

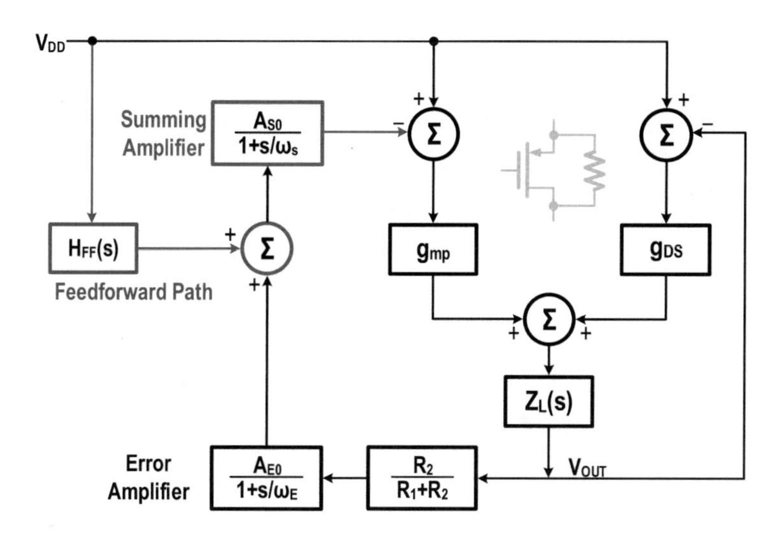

Fig. 3.57 Mathematical model of voltage-mode gate ripple injection LDO

To eliminate the ripple on V_{OUT}, Eq. (3.115) should be set to zero. The optimal feedforward path transfer function, HFF(s)$|_{OPT}$, is given by

$$H_{FF}(s)\big|_{OPT} = \frac{1+s/\omega_S}{A_{S0}} \times \left(\frac{g_m+g_{ds}}{g_m}\right) = \frac{R_{S3}}{R_{S2}} \times \left(\frac{g_m+g_{ds}}{g_m}\right) \times \left(1+s/\omega_S\right) \quad (3.116)$$

Equation (3.72) includes a zero generated by C_F and R_{FF1}, which can offset the influence of the power transistor's gate pole and extend the ripple suppression frequency range.

The bias current flowing through R_{FF1}, R_{FF2}, R_{S1}, and R_{S2} in [54] makes it challenging to use low-quiescent current amplifiers, resulting in higher overall LDO quiescent power consumption [54], shown in Fig. 3.58, which reduces this consumption by replacing these resistors with coupling capacitors. R_{B1} and R_{B2} provide a stable DC operating point for amplifiers, eliminating the need for additional voltage bias and allowing a wider gate voltage swing for the power transistor. However, [54] does not design a zero in the feedforward path for optimizing power consumption.

Current-Mode Gate Ripple Injection

Current-mode gate ripple injection involves converting the voltage ripple signal into current and then injecting an appropriately scaled current into the power transistor gate. Unlike voltage mode, the current method eliminates the need for a summing amplifier, simplifying the circuit design. Figure 3.59 illustrates the circuit implementation of current-mode FFRC, as discussed in [55].

The PMOS transistor M_{FF} is configured in a diode-connected arrangement, with its drain voltage V_{LPF} clamped to V_{B1}. The current through M_{FF} is mirrored by

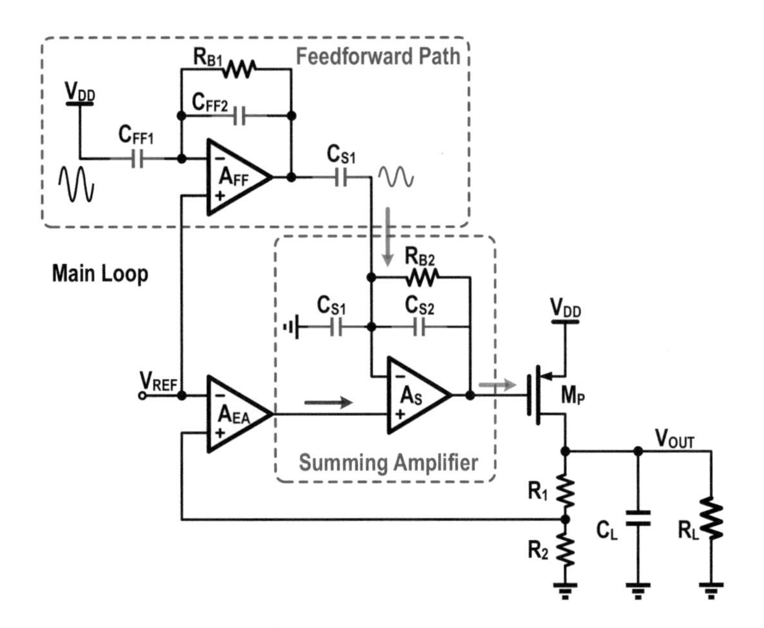

Fig. 3.58 Capacitive voltage-mode gate ripple injection in [54]

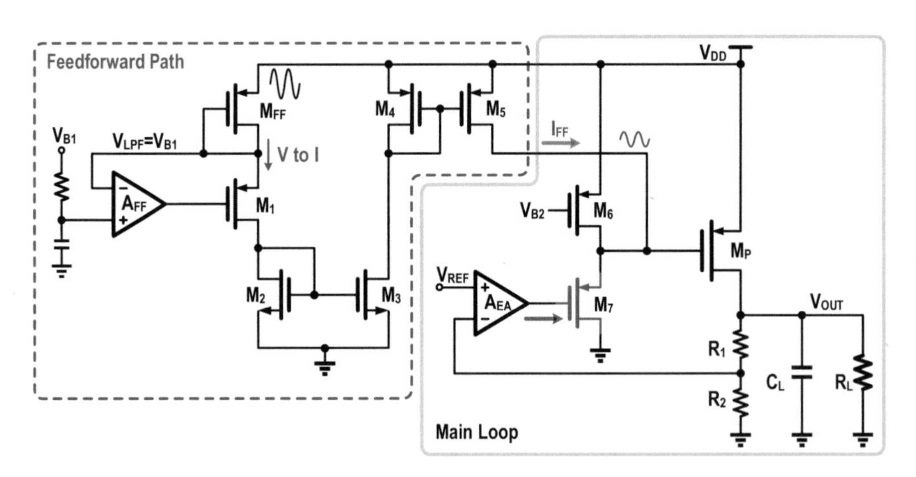

Fig. 3.59 Current-mode gate ripple injection in [55]

transistors M_2–M_5 and then being injected into the driver stage. The feedforward current I_{FF} is

$$\Delta I_{FF} = \left(g_{mff} + g_{dsff}\right)V_R \tag{3.117}$$

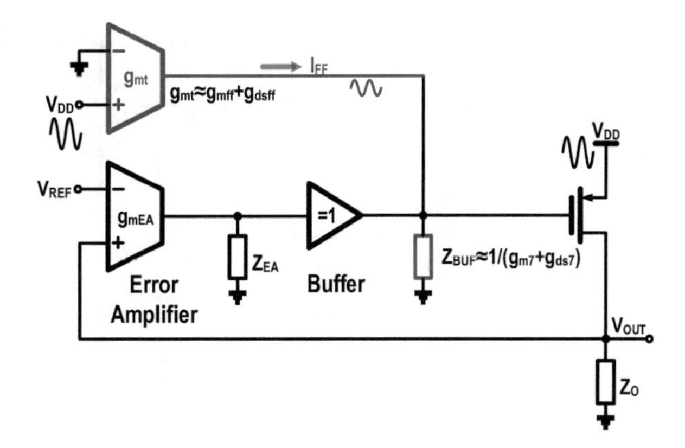

Fig. 3.60 Simplified model of current-mode FFRC

The study [40] uses a source follower buffer, and the output impedance of the buffer is

$$Z_{\mathrm{BUF}} \approx \frac{1}{g_{\mathrm{m7}} + g_{\mathrm{ds7}}} \,||\, sC_{\mathrm{G}} \approx \frac{1}{g_{\mathrm{m7}} + g_{\mathrm{ds7}}} \tag{3.118}$$

Figure 3.60 presents the simplified model of current-mode FFRC.

The injected ripple voltage is

$$V_{\mathrm{FF}} \approx Z_{\mathrm{BUF}} \times I_{\mathrm{FF}} = \frac{g_{\mathrm{mff}} + g_{\mathrm{dsff}}}{g_{\mathrm{m7}} + g_{\mathrm{ds7}}} \times V_{\mathrm{R}} \tag{3.119}$$

According to Eq. (3.65), set

$$\frac{g_{\mathrm{mff}} + g_{\mathrm{dsff}}}{g_{\mathrm{m7}} + g_{\mathrm{ds7}}} = \frac{g_{\mathrm{mp}} + g_{\mathrm{dsp}}}{g_{\mathrm{mp}}} \tag{3.120}$$

The optimal gate ripple injection can be achieved.

The work in [6] gave a simpler implementation of current-mode FFRC. As shown in Fig. 3.61, the LDO described in [6] also utilizes a source follower buffer. M_2 supplies the bias current for the drive circuit, with a low-pass filter placed between the current mirrors M_1 and M_2. It is essential that the cutoff frequency of this low-pass filter is low enough to stabilize the gate voltage of M_2. The power supply ripple is converted into a feedforward current through M_2 and injected into M_3. This simple yet effective structure is also referenced in literature [56]. In addition, to reduce the impact of EA on PSRR, the power supply for EA is derived from V_{OUT} and passes through a low-pass filter (LPF_1).

When the load current changes, the g_{m} and g_{ds} of the power transistor will also change. According to Eq. (3.106), the coefficient of the gate injection ripple should vary with the load current. The FFRC technologies described in [53–55] use fixed coefficient values, making it challenging to maintain the optimal coefficient.

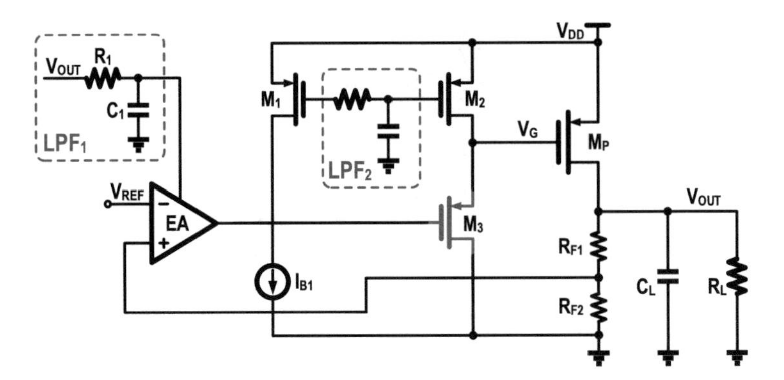

Fig. 3.61 Current-mode FFRC implementation in [6]

However, the measured results indicate that their PSRR is still effectively enhanced under varying loads.

Body Ripple Injection
According to Eq. (3.110), we have

$$\frac{V_{\mathrm{FF}}|_{\mathrm{OPT}}}{\mathrm{V_R}} = \frac{C_{\mathrm{GD}}}{C_{\mathrm{GS}} + C_{\mathrm{GD}}} \times \frac{g_{\mathrm{m}}}{g_{\mathrm{mb}}} + \frac{g_{\mathrm{ds}}}{g_{\mathrm{mb}}} + 1 \tag{3.121}$$

where $\mathrm{V_R}$ represents the power ripple. Since g_{mb} is much smaller than g_{m}, the ripple amplitude required for injection into the body is larger. This is one of the drawbacks of body ripple injection.

To address this issue, [51] introduces a large capacitor C_{C} between the source and gate of the power transistor. Figure 3.62 illustrates the implementation circuit for the body ripple injection described in [51].

Given that $C_{\mathrm{C}} >> C_{\mathrm{GD}}$, Eq. (3.121) can be simplified to

$$V_{\mathrm{FF}}|_{\mathrm{OPT}} = \left(1 + \frac{g_{\mathrm{ds}}}{g_{\mathrm{mb}}}\right) V_R \tag{3.122}$$

The circuit on the right is a feedforward circuit, functioning as a non-inverting amplifier composed of a mixing amplifier (MA) and feedback resistors R_1 and R_2. The reference voltage V_{REF} establishes the bias point, while the power ripple is coupled to the positive input of MA through capacitor C_{M}:

$$V_{\mathrm{FF}} = \left(1 + \frac{R_2}{R_1}\right) V_{\mathrm{R}} \tag{3.123}$$

Comparing Eqs. (3.122) and (3.123), we can obtain

$$\frac{R_2}{R_1} = \frac{g_{\mathrm{ds}}}{g_{\mathrm{mb}}} \tag{3.124}$$

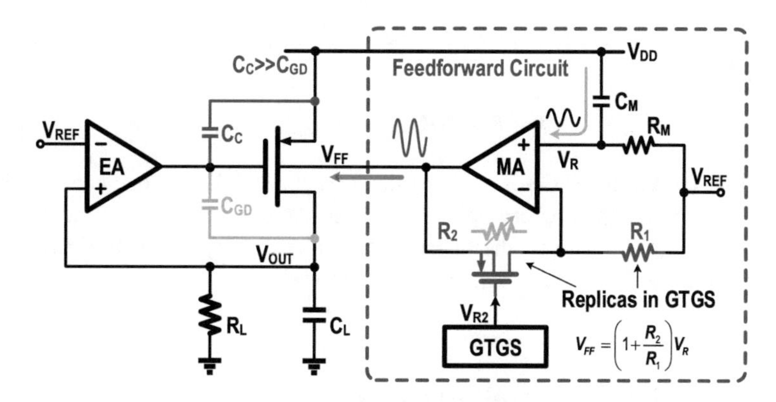

Fig. 3.62 Block diagram of the body ripple injection LDO in [51]

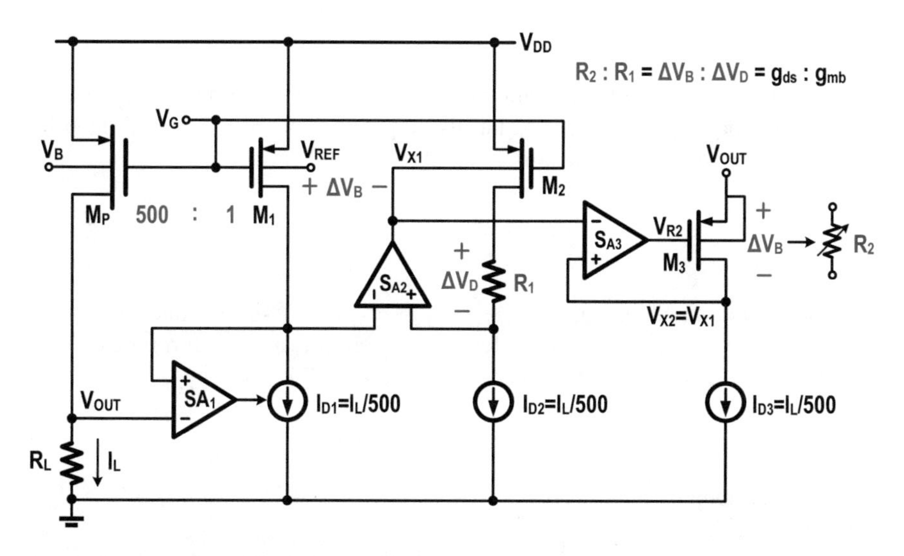

Fig. 3.63 Schematics of the GTGS in [51]

As the load varies, both g_{ds} and g_{mb} will also change. Hence, a g_{ds}-to-g_{mb} sensor (GTGS) is required to monitor the ratio of g_{ds} to g_{mb} and to determine the optimal injection coefficient.

Figure 3.63 presents the circuit diagram of the GTGS. Transistors M_1 and M_2 are scaled-down sensor transistors of the power transistor M_P, sharing the same gate voltage V_G. The size of M_1 and M_2 is 1/500 of that of M_P. M_1 and M_2 have identical sizes, currents, and gate voltages, but their body and drain voltages differ. Consequently, we can obtain

$$\Delta I_D = \Delta V_D \times g_{ds} - \Delta V_B \times g_{mb} = 0 \tag{3.125}$$

$$g_{ds} : g_{mb} = \Delta V_B : \Delta V_D \tag{3.126}$$

The PMOS transistor M_3 acts as resistor R_2. Since the same current flows through both R_1 and R_2:

$$R_2 : R_1 = \Delta V_B : \Delta V_D = g_{ds} : g_{mb} \tag{3.127}$$

The bias current in GTGS is 1/500 of the load current. As the load changes, both ΔV_B and ΔV_D will adjust accordingly. Consequently, the resistance of R_2 will also be dynamically adjusted. Hence, the structure described in [51] can theoretically maintain the optimal injection coefficient as the load varies.

3.8 Flipped-Voltage Follower LDOs

3.8.1 Introduction

Figure 3.64a depicts a common drain amplifier, often utilized as a voltage buffer. It can absorb a large current from V_O, but its output current capability is limited by I_B. Ignoring the body effect, its output impedance is

$$R_{O1} = \frac{1}{g_{m1}} \parallel r_{o1} \parallel R_B \tag{3.128}$$

where R_B is the impedance of the bias current. Figure 3.64b illustrates a flipped-voltage follower. M_2 provides shunt feedback, while M_1 and M_2 form a two-pole negative feedback loop [38]. The open-loop gain of FVF is given by

$$A_{OL} \approx g_{m2}\left(R_B \parallel g_{m1} r_{o1} r_{o2}\right) \tag{3.129}$$

The closed-loop output impedance of FVF is

$$R_{O2} \approx \frac{\frac{1}{g_{m1}}\left(1 + \frac{R_B}{r_{o1}}\right) \parallel r_{o2}}{g_{m2}\left(R_B \parallel g_{m1} r_{o1} r_{o2}\right)} \tag{3.130}$$

When the impedance of the bias current I_B is particularly large, $R_{O2} \approx g_{m1} r_{o1} g_{m2}$.

The flipped-voltage follower (FVF) can deliver a large current and features a fast local feedback loop, which ensures a high bandwidth, making it an ideal choice for LDO applications. Figure 3.65 illustrates the classic FVF LDO circuit [57]. The output voltage, V_{OUT}, is determined by the set voltage, V_{SET}, derived from the buffer formed by the first-stage two amplifiers. As a result, we can obtain

$$V_{OUT} = V_{SET} + V_{GS1} = V_{MIR} - V_{GS2} + V_{GS1} = V_{REF} - \left(V_{GS2} - V_{GS1}\right) \tag{3.131}$$

Fig. 3.64 (**a**) Voltage follower, and (**b**) flipped-voltage follower

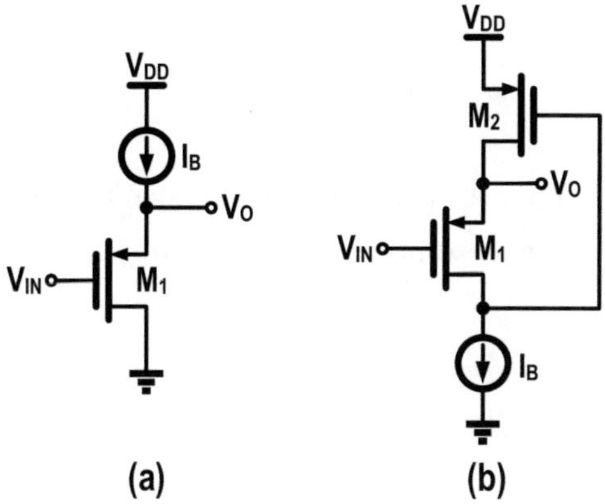

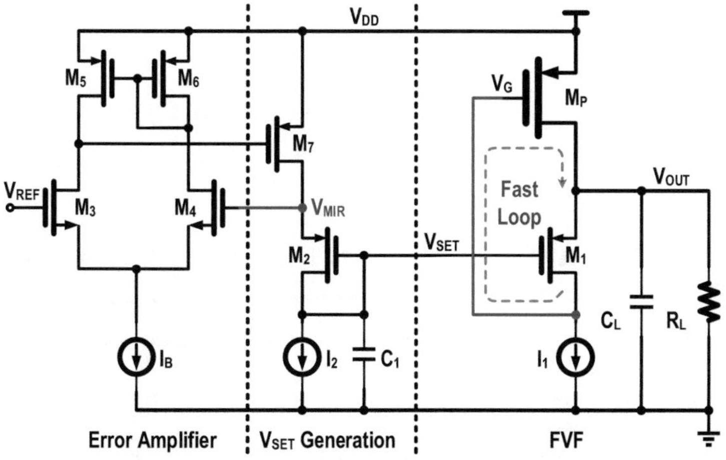

Fig. 3.65 A classic flipped-voltage follower-based LDO

By appropriately setting the bias currents I_1 and I_2 and the sizes of transistors M_1 and M_2, $V_{OUT} = V_{REF}$. Adding a ground capacitor C_{C1} at the V_{SET} node ensures the stability of the preamplifier and absorbs output kickback noise. The open-loop gain of the FVF stage is given by

$$A_{FVF} \approx -g_{mp}\left[R_{B1} \parallel \left(g_{m1} r_{o1} R_O\right)\right] \tag{3.132}$$

where R_{B2} is the impedance of bias current I_1, $R_O = R_L \parallel r_{OP}$.

There are two poles in the FVF: one located at the gate of the power transistor and the other at the output:

$$p_1 \approx \frac{-1}{\left[R_{B1} \parallel \left(g_{m1} r_{o1} R_O\right)\right] \times C_G} \tag{3.133}$$

$$p_2 \approx \frac{-1}{\left(\frac{1}{g_{m1}} \parallel R_O\right) \times C_L} \tag{3.134}$$

C_G is the parasitic capacitance at the gate of the power transistor. At light-load currents, assuming $g_{m1}r_{o1}R_O >> R_{B1}, R_O >> \frac{1}{g_{m1}}$:

$$p_1 \approx \frac{-1}{R_{B1} \times C_G} \tag{3.135}$$

$$p_2 \approx \frac{-g_{m1}}{C_L} \tag{3.136}$$

At heavy-load currents, assuming $g_{m1}r_{o1}R_O << R_{B1}, R_O << \frac{1}{g_{m1}}$:

$$p_1 \approx \frac{-1}{g_{m1}r_{o1}R_O \times C_G} \tag{3.137}$$

$$p_2 \approx \frac{-1}{R_O \times C_L} \tag{3.138}$$

The basic FVF LDO described in [57] faces several challenges in specific applications:

1. The voltage swing amplitude of the power transistor gate is limited, with the maximum gate voltage being restricted by V_{OUT}.
2. The swing rate of the power transistor gate is constrained by the bias current I_B, which impacts transient performance.
3. The gain of the FVF stage is low, leading to relatively poor load regulation.

To mitigate the abovementioned problems, it is necessary to insert a buffer before the gate of the power transistor, as shown in Fig. 3.66. The next section will introduce various types of buffered FVF LDOs.

3.8.2 *Buffered FVF LDO*

To overcome the loop gain limitation in the basic FVF LDO and enhance the swing of V_G, a folded cascode FVF structure (CA-FVF) can be utilized, as shown in Fig. 3.67a. The CA-FVF incorporates a common gate amplifier consisting of M_2 and the bias current I_2. This allows the maximum voltage of V_G to reach V_{DD}, thereby eliminating the limitation imposed by the output voltage V_{OUT} in the basic FVF structure. By selecting an appropriate bias voltage V_{B1}, the bias current I_1 is ensured to be in the saturation region.

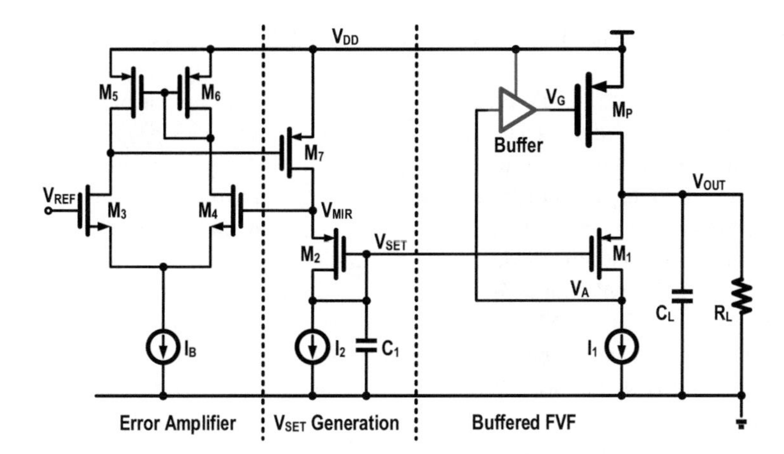

Fig. 3.66 Buffered FVF LDO

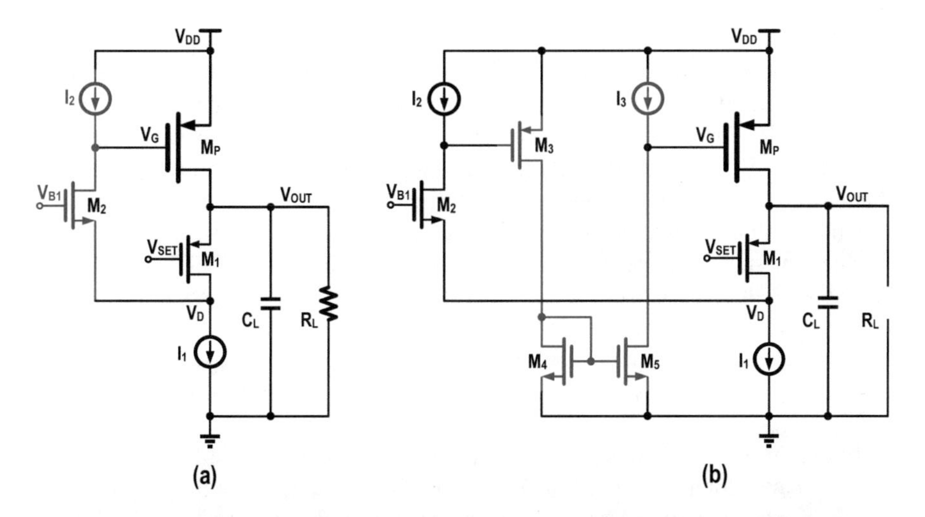

Fig. 3.67 (**a**) A folded cascode FVF structure, and (**b**) an FVF LDO with extra amplifier stages

Based on Fig. 3.67a, an extra common-source amplifier can be introduced to further enhance the FVF gain and improve the lower swing of the output V_G, as described in [41], shown in Fig. 3.67b. However, it is crucial to note that the use of multistage amplification introduces multiple parasitic poles, necessitating careful consideration of stability issues.

Since the parasitic capacitance of the power transistor is large and the drain of M_1 is a high-resistance node, the pole frequency at the V_D node is low in the basic FVF LDO structure. M_2 is introduced as a source follower buffer (SF) to mitigate this issue and reduce the gate drive impedance, as shown in Fig. 3.68a. This configuration can split the original low-frequency pole into two relatively high-frequency poles and improve the slew rate of V_G [58]. Building on the SF-FVF, a common-gate amplifier (M_3 and I_3) is added, as depicted in Fig. 3.68b. This modification enhances the loop gain while further reducing the drive impedance [59].

Figure 3.69a illustrates an FVF LDO utilizing a current mirror buffer (CB). M_2 is connected as a diode, and the current flowing through M_2 is proportional to the output current. This diode connection of M_2 reduces the gate drive impedance of V_G. In Fig. 3.69b, the resistor R_1 and capacitor C_1 add a zero, thereby extending the loop bandwidth [60]. However, since the drive current flows into I_1, the current through M_1 varies with the load, causing significant changes in loop gain and resulting in poor load regulation [60]. To address this issue, an adaptive biasing circuit (comprising M_3, M_4, and M_5) is added to absorb the varying drive current, thereby improving load regulation performance (Fig. 3.69c).

To further reduce the driving impedance, a super source follower (SSF) can be employed as a buffer [61], as depicted in Fig. 3.70a. The study [61] incorporates a diode-connected M_4 to generate an adaptive bias current that tracks load variations, thereby reducing the quiescent current and improving transient performance. [34] adopts an enhanced SSF structure to further minimize the driving impedance (Fig. 3.70b).

The basic FVF structure is a variation type of replica LDO. The output voltage accuracy, particularly the load regulation performance, is mainly determined by the gain of the FVF stage. To achieve higher bandwidth and transient performance, the gain of the FVF stage is relatively low, leading to poor load regulation. To address this issue, some works introduce V_{OUT} information into the first-stage amplifier to enhance the load regulation, as shown in Fig. 3.71.

The first-stage V_{SET} generation circuit and the second-stage FVF circuit of the basic FVF LDO are controlled independently, allowing for separate stability considerations. Introducing V_{OUT} feedback in the first stage creates two nested loops, requiring careful attention to stability. In Fig. 3.71a, the reverse input pair of the EA is divided into two parts with a 1:3 ratio: the gate of M_3 connects to V_{MIR} and the gate of M_2 to V_{OUT}.

In Fig. 3.71b, the input signal of the EA is V_{OUT}; for stability, the study [34] utilized dynamic compensation. In [27], V_{SET} is directly taken from the output of the EA. To ensure stability, a larger C_B capacitor is required to reduce the bandwidth of the first stage.

3.8.3 *OPD and IPD FVF LDOs*

As mentioned above, based on the location of the dominant pole, LDOs can be classified into output-pole-dominant (OPD) and internal-pole-dominant (IPD) types. In OPD LDOs, the output pole is the dominant pole, with the internal pole shifted as far as possible outside the unity-gain bandwidth. Consequently, OPD LDOs function like single-pole systems and do not require additional compensation circuits. In addition, compared to IPD LDOs, OPD LDOs exhibit better mid-high-frequency PSRR performance. However, to ensure stability and constrain the dominant pole

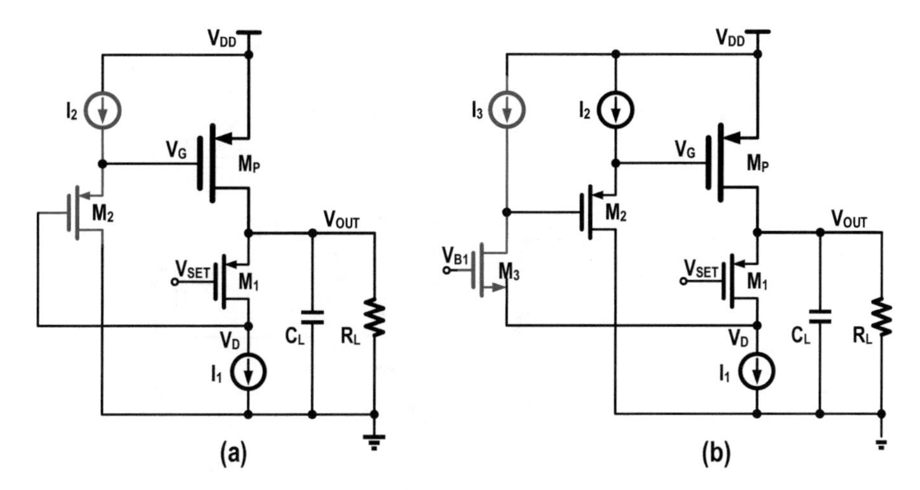

Fig. 3.68 (**a**) Source follower buffered FVF LDO, and (**b**) folded SF-FVF LDO

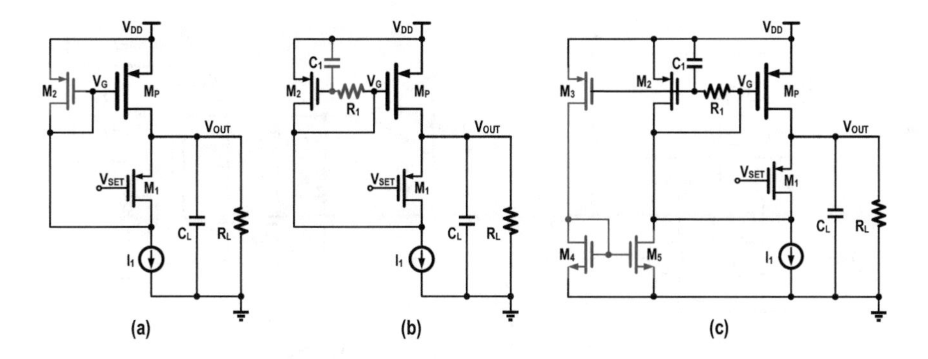

Fig. 3.69 (*a*) Current mirror buffer (CB) FVF. (**b**) Enhanced CB FVF. (**c**) Enhanced CB FVF with self-adaptive bias [60]

frequency, OPD LDOs typically have lower load capacity and often require a larger output capacitor. A large power transistor and drive circuit can significantly increase the difficulty of pushing the internal pole to a high frequency.

Consequently, OPD LDOs are generally suited for low-current, high-bandwidth applications. For applications demanding PSRR performance across a broad frequency spectrum, the OPD FVF LDO is the preferred choice. With advancements in technology and benefiting from the scaling device size, the high-bandwidth OPD FVF-LDOs are becoming more and more attractive [34], [60, 61]. For example, the OPD FVF LDO described in [60] utilizes a 10 nm process and achieves an impressive bandwidth of 1 GHz.

On the other hand, when the output capacitor is small, and the load current is large, the IPD LDO structure is more suitable but requires a compensation circuit.

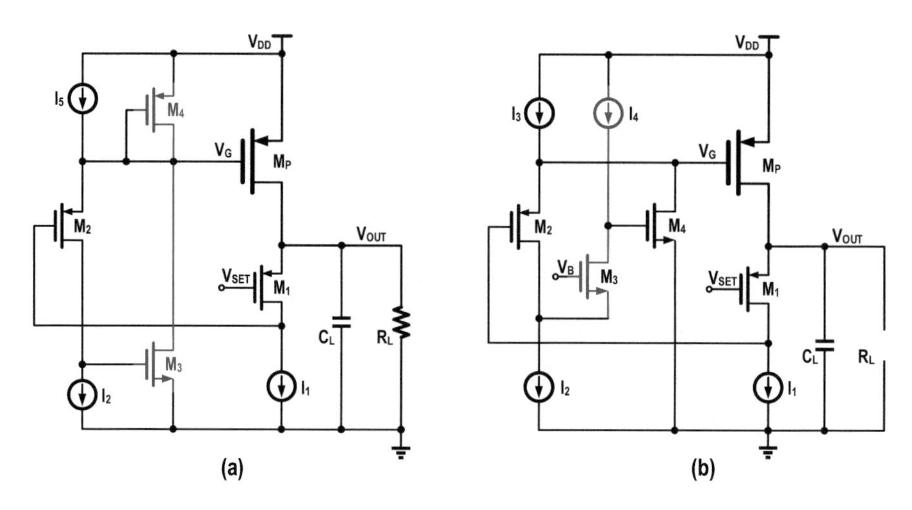

Fig. 3.70 (**a**) Super source follower (SSF) FVF and (**b**) enhanced SSF FVF

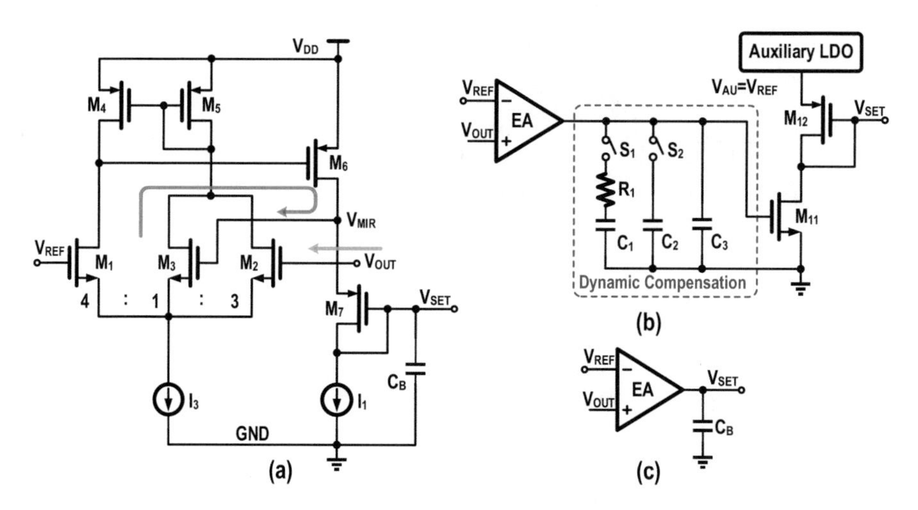

Fig. 3.71 Load regulation improvement techniques. In (**a**) [61], (**b**) [34], and (**c**) [27]

Figure 3.72 illustrates the compensation circuits employed by several IPD FVF LDOs.

In Fig. 3.72, all three structures have capacitor C_B connected to the V_{SET} terminal to reduce the first-stage bandwidth. In the FVF stage, Miller compensation is utilized. In the work [41], it is connected across the gate of M_3, while in [27, 62], it is connected to the drain of M_1. In addition, [27] introduces damping-factor-control frequency compensation at the power transistor gate to improve the phase margin. [62] also presents a similar structure at the drain of M_1, which is referred to as dual-summed Miller frequency compensation in [62].

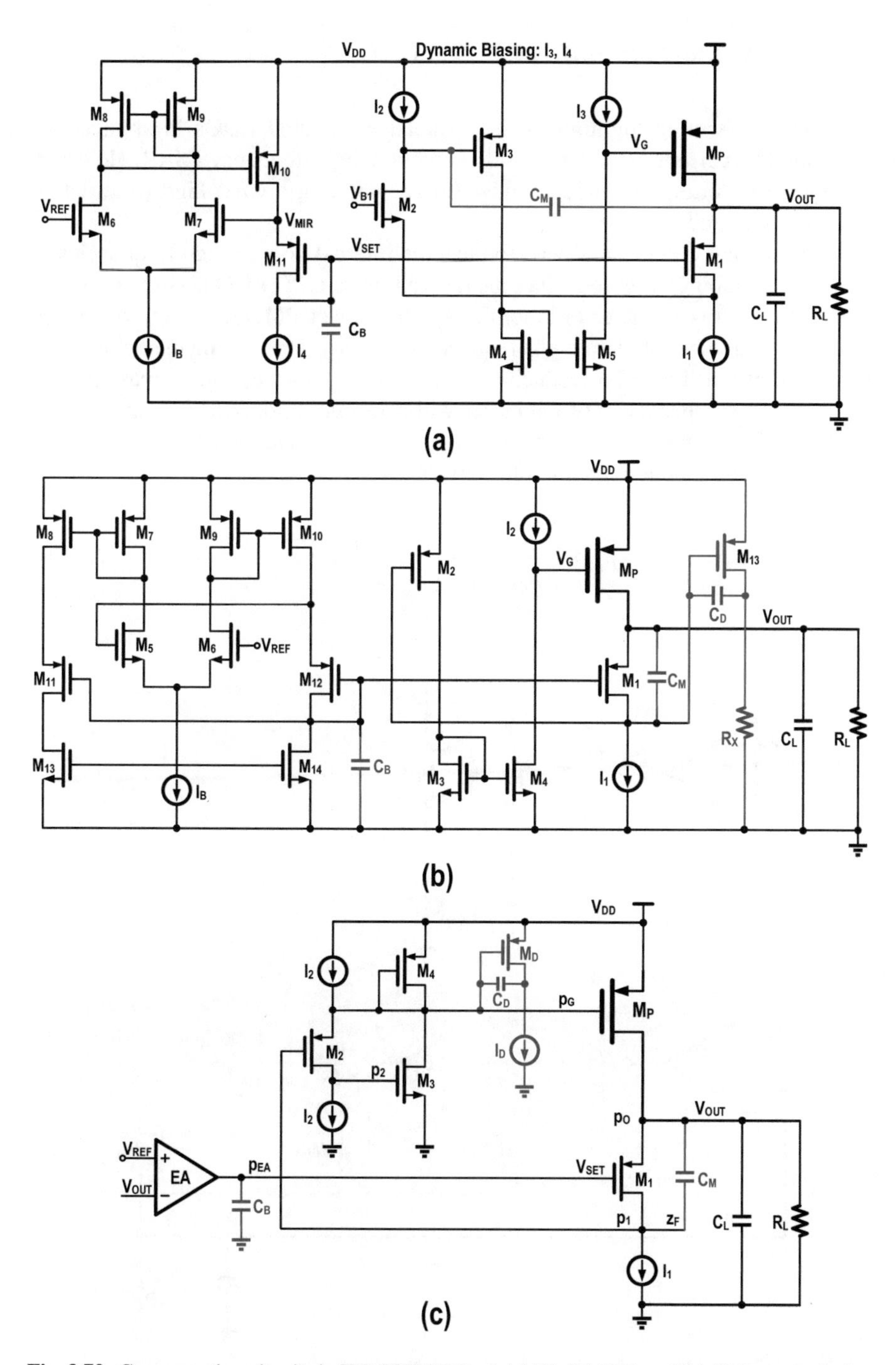

Fig. 3.72 Compensation circuits in IPD FVF LDOs. (**a**) [41], (**b**) [62], and (**c**) [27], respectively

3.8.4 Domino-Like Buffered FVF LDO

The previous section highlighted the advantages of OPD LDOs in applications requiring low current, wide bandwidth, and good high-frequency PSRR. However, if large load current and wide bandwidth are necessary, what kind of structure should be chosen?

Conventional OPD LDOs need to push the internal pole to high frequencies to ensure loop stability. Figure 3.73a illustrates a typical OPD LDO design. By inserting a buffer between the error amplifier and the power PMOS, the low-frequency gate pole is divided into two high-frequency poles, p_{EA} and p_G. Enhancing the driving capability of the buffer (reducing the driving impedance) can increase p_G, but the large input capacitance of the buffer will cause p_{EA} to decrease. Consequently, there is an inherent upper limit to elevating the nondominant pole.

The equivalent nondominant pole is given by

$$p_{nd} \approx \left[\frac{1}{p_{EA}} + \frac{1}{p_G}\right]^{-1} \leq \frac{1}{2}\sqrt{p_{EA}\, p_G} \tag{3.139}$$

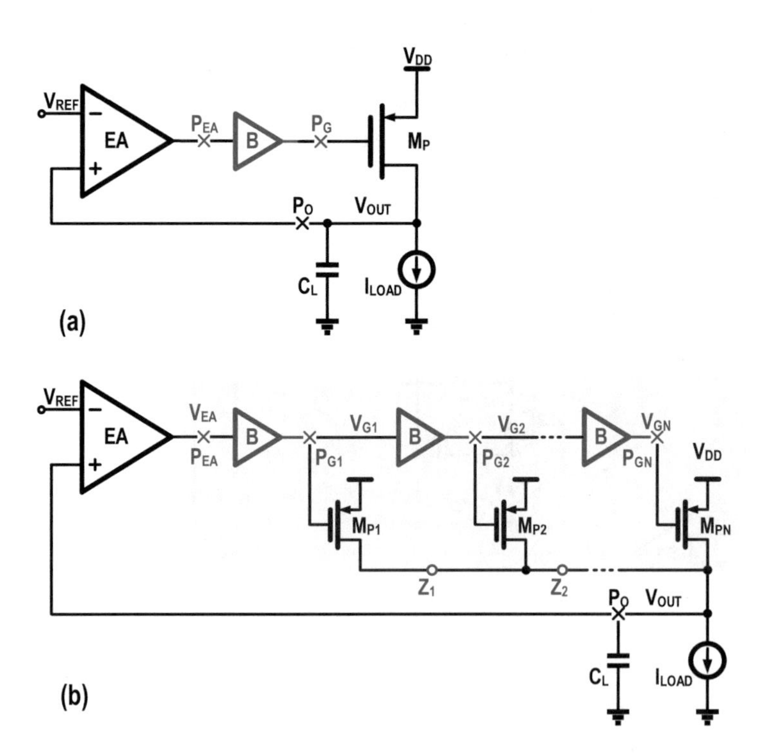

Fig. 3.73 (**a**) A conventional OPD LDO, and (**b**) a domino-like buffered LDO

Due to the limitation imposed by the nondominant pole, the OPD LDO requires a relatively large-output capacitor C_L or can only support a small load to ensure that the output pole remains at a lower frequency.

To enhance the load capacity of OPD LDOs and reduce the C_L requirement, [60] proposes a domino-like buffered (DLB) LDO architecture, as illustrated in Fig. 3.73b. In this design, the power transistor M_P is divided into N smaller power transistors, each with its independent buffer. Each buffer's output is connected to the gate of its respective power transistor and to the input of the next buffer stage. This in-series arrangement and sequential activation resemble the action of falling dominoes.

Dividing the power transistors can effectively reduce the gate parasitic capacitance of each power transistor. From the standpoint of power transistor size alone, the parasitic capacitance of each smaller transistor is 1/N of the original. Considering the Miller effect, the equivalent parasitic capacitance can be further diminished. As depicted in Fig. 3.74a, the equivalent gate parasitic capacitance of a conventional single power transistor is

$$C_G = C_{GS} + \left[1 + g_{mp} R_O\right] C_{GD} \tag{3.140}$$

For segmented power transistor in Fig. 3.74b,

$$C_{G1} = \frac{C_{GS}}{N} + \left[1 + \frac{g_{mp}}{N} R_{O1}\right] \frac{C_{GD}}{N} = \frac{1}{N}\left[C_{GS} + \left(1 + \frac{g_{mp}}{N} R_{O1}\right) C_{GD}\right] \tag{3.141}$$

$$R_{O1} = R_L \,||\, (N \times r_{ds}) \,||\, (N \times r_{ds}) \,||\, \ldots \,||\, (N \times r_{ds}) = R_O \tag{3.142}$$

$$C_{G1} = \frac{C_{GS}}{N} + \left[1 + \frac{g_{mp}}{N} R_{O1}\right] \frac{C_{GD}}{N} = \frac{1}{N}\left[C_{GS} + \left(1 + \frac{g_{mp}}{N} R_{O}\right) C_{GD}\right] \tag{3.143}$$

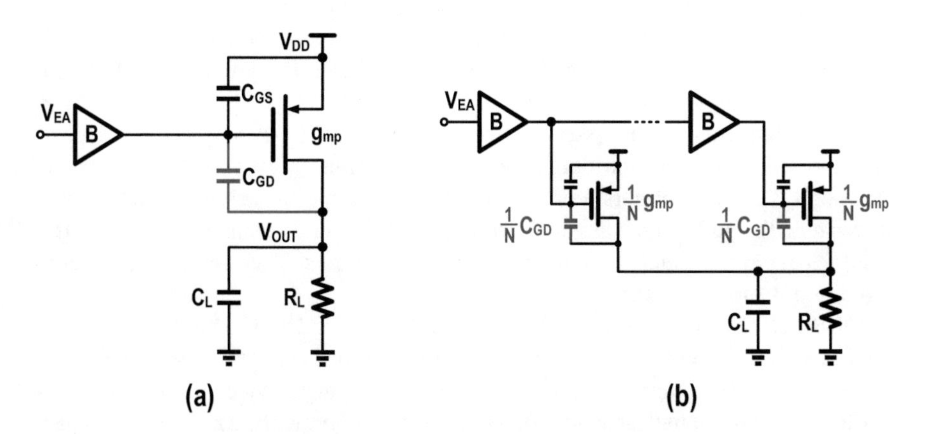

Fig. 3.74 (**a**). A single power transistor, and (**b**) segmented power transistors

By comparing Eqs. (3.65) and (3.67), it is evident that segmented power transistors can significantly mitigate the additive impact of the Miller effect on C_{GD}:

$$C_{G1} < \frac{1}{N} \times C_G \tag{3.144}$$

In the conventional OPD LDO, the gate pole p_G is

$$p_G = \frac{-1}{R_B \times C_G} \tag{3.145}$$

For the segmented power transistors, the gate pole p_{G1} is

$$p_{G1} = \frac{-1}{R_B \times C_{G1}} > N \times p_G \tag{3.146}$$

Therefore, the gate pole p_G in the traditional scheme is divided into N poles, each with frequencies higher than $N \times p_G$. In addition, with respect to the last buffer B_N and the power stage M_{PN}, the preceding $N - 1$ stages form a feedforward path, introducing $N - 1$ beneficial zeros, which offset the influence of $N - 1$ poles [59], as illustrated in Fig. 3.75.

When N is large, the gate pole can be pushed to a higher frequency, but this requires more buffers, thereby increasing static power consumption. Moreover, a larger N results in complex zeros, potentially diminishing the pole-zero cancellation effect. In [60], with the design trade-off, $N = 4$ was selected.

Figure 3.76 shows the circuit diagram of the DLB LDO, which adopts the FVF LDO structure. DLB technology pushes the nondominant pole in the FVF stage to a higher frequency. The first-stage error amplifier enhances the overall loop gain and improves load regulation. The C_C capacitor is used to compensate for the first stage. To further enhance PSRR performance, the current-mode ripple injection technique introduced in [6] is utilized in [60]. By adding resistor R_1 and capacitor C_1 at the gate stage of M_2, the supply ripple information is injected into the V_B node.

Figure 3.77 shows the domino buffer circuit. A bilaterally symmetrical OTA connected in a unity-gain configuration ensures that the buffer's input and output voltages are identical. As discussed in Sect. 3.7.1, this OTA belongs to the type B amplifier category, and the output node OUT exhibits high PSRR performance. Although type A amplifiers are generally more suitable for driving PMOS power transistors, the buffer input already employs ripple injection technology, so the OTA needs to avoid introducing additional ripple components. Therefore, a type B configuration is more appropriate.

DLB LDO is an excellent structure for implementing OPD LDO. With the same load capacitor and power transistor size (equivalent load capacity), the internal pole can be pushed to a higher frequency, thereby achieving a wider unity-gain bandwidth. If with the same capacitor and bandwidth requirements, a higher load capacity can be obtained by connecting more power unit stages in series. On the other hand, with the same load capacity and bandwidth requirements, the output capacitance requirements can be reduced.

3.9 NMOS LDO

3.9.1 NMOS LDO and PMOS LDO Comparison

Figure 3.78 illustrates two types of power transistor LDOs: the PMOS power transistor LDO and the NMOS power transistor LDO. For the NMOS LDO, the output voltage must be one V_{GS} lower than the gate voltage of the power transistor. To design a low-dropout voltage LDO with NMOS transistors, an additional power supply with higher voltages is needed to power the EA and buffer, or a boost charge pump must be designed to provide the necessary power. In contrast, PMOS transistors usually operate with a single power supply.

Aside from the difference in power supply requirements, what are the other distinctions between NMOS and PMOS LDOs?

Table 3.3 highlights some characteristic differences between NMOS and PMOS LDOs. Given the same size and operating conditions, NMOS transconductance is generally greater than that of PMOS. Consequently, the NMOS power transistor is typically smaller than the PMOS power transistor.

Regarding the output stage, the PMOS power transistor functions as a common-source amplifier, whereas the NMOS LDO operates as a common-drain amplifier (source follower). Due to this configuration, the NMOS LDO exhibits lower output impedance and a relatively higher frequency output pole. The PMOS power stage can provide a higher DC gain, whereas the gain of the NMOS power stage is approximately 1. Therefore, the NMOS LDO's error EA often needs to achieve higher gain.

The most used compensation for PMOS capless LDOs is Miller compensation, whereas NMOS LDOs can utilize pole-zero pair compensation, as shown in Fig. 3.78b. While lowering the P_{EA} pole, a left-half-plane zero can be generated to

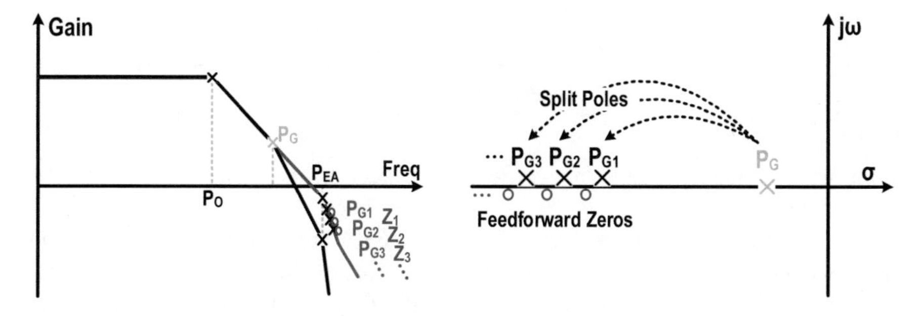

Fig. 3.75 Pole separation and feedforward zeros

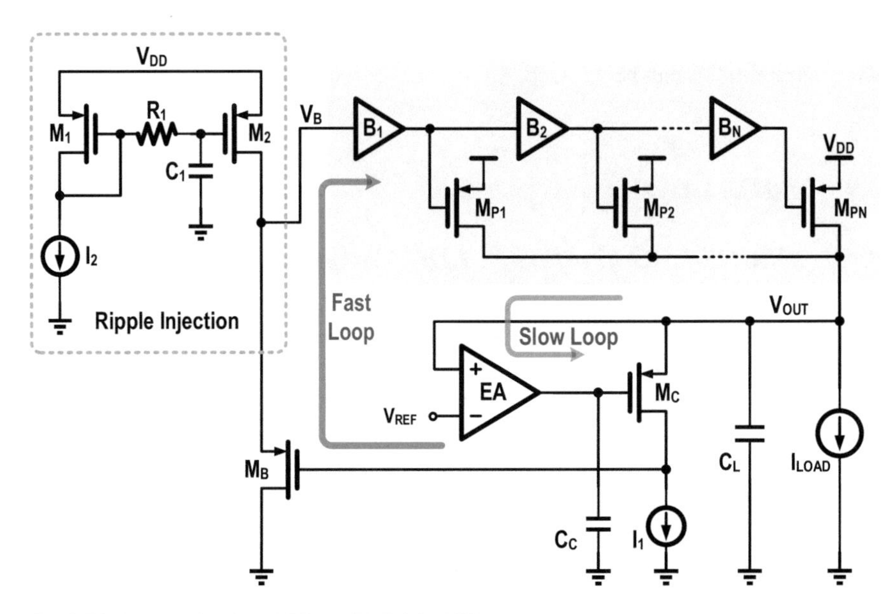

Fig. 3.76 Schematic of the DLB FVF LDO in [59]

Fig. 3.77 Schematic of domino buffer

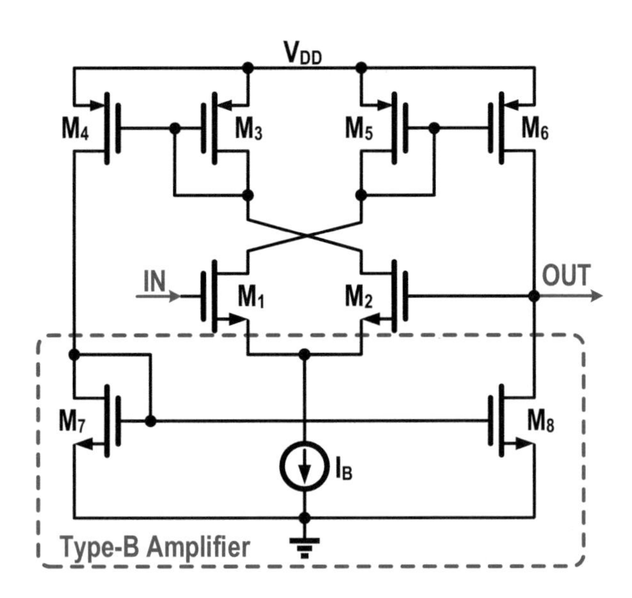

improve the phase margin. In addition, C_C capacitor also helps enhance PSRR performance.

For PMOS devices, the factors influencing PSRR are transconductance g_m and drain-source conductance g_{ds}. Given that $C_C >> C_{GD}$, the input power supply ripple coupled to the gate by the parasitic capacitor C_{GD} can be effectively reduced.

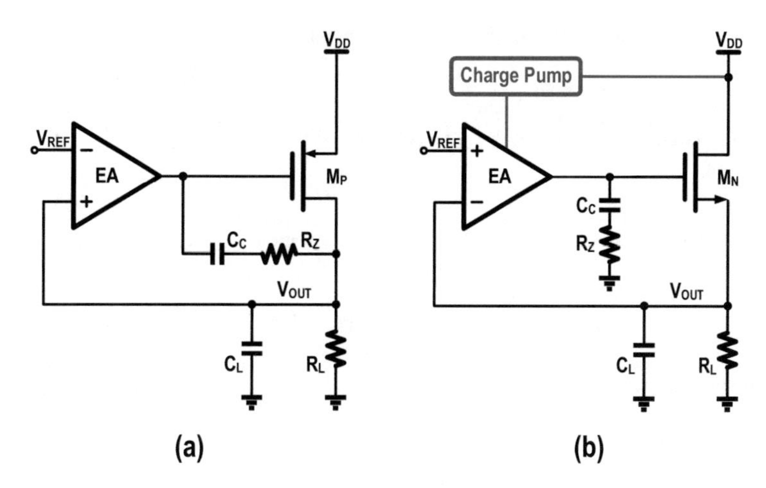

Fig. 3.78 (**a**) PMOS LDO and (**b**) NMOS LDO

Therefore, for NMOS devices, the primary influencing factor is the drain-source conductance g_{ds}, making it easier to achieve good PSRR performance.

The capacitor C_C impacts the slew rate of the power transistor gate, thereby limiting the response speed of the loop adjustment. However, NMOS transistors have an inherently fast V_{GS} response: when the output voltage drops, the increased V_{GS} voltage enables a timely provision of a larger output current.

NMOS LDOs require a higher power supply for the control circuit. Figure 3.79 illustrates a dual-branch voltage multiplier with a voltage conversion ratio of 2. Since the current demand for the error amplifier and buffer is relatively small, the flying capacitor C_H and the filtering capacitor typically range from a few pF to more than 10 pF. The charge pump also necessitates a clock circuit, and both the charge pump and the clock circuit contribute additional power consumption. Consequently, NMOS LDOs are generally not suitable for applications requiring low-dropout voltage and ultralow quiescent current.

3.9.2 Replica NMOS LDO

Figure 3.80 illustrates the replica NMOS LDO [63]. The error amplifier, M_{N1}, and the bias current I_B form a feedback network. The gate and drain voltages of M_{N2} and M_{N1} are connected, resulting in the following output voltage:

$$V_{OUT} = V_{MIR} + V_{GS1} - V_{GS2} = V_{REF} + \left(V_{GS1} - V_{GS2}\right) \tag{3.147}$$

M_{N1} and M_{N2} are both NMOS transistors, which can compensate for PVT variations. However, as the load current increases, V_{OUT} will gradually decrease (V_{GS} voltage increase), resulting in poor load regulation for the replica NMOS LDO.

Table 3.3 The comparison of NMOS LDOs and PMOS LDOs

	PMOS LDO	NMOS LDO
Power transistor area	Relatively large	Relatively small
Output impedance R_O	$R_L \parallel r_{op}$, no compensation	$R_L \parallel r_{on} \parallel \frac{1}{g_{mn}}$
Output pole p_O	$\frac{-1}{(R_L \parallel r_{op})C_L}$, no compensation	$\approx \frac{-g_{mn}}{C_L}$
	$\approx \frac{-g_{mp}}{C_L}$, Miller compensation	
Output stage gain A_V	$A_V = -g_{mp} \times (R_L \parallel r_{op})$	$A_V = \frac{g_{mn}}{\frac{1}{R_L} + \frac{1}{r_{on}} + g_{mn}} \approx 1$
Common compensation	Miller compensation	Pole-zero pair
	$p_{EA} = \frac{-1}{R_1 g_{mp}(R_L \parallel r_{op})C_C}$ $z_1 = \frac{-1}{R_Z C_C}$	$p_{EA} = \frac{-1}{(R_1 + R_Z)C_C}$ $z_1 = \frac{-1}{R_Z C_C}$
PSRR	$\Delta I_R \approx (g_{mp} + g_{dsp})V_R$	$\Delta I_R \approx g_{dsn} V_R$
Transient response	Loop regulation	Loop regulation+ Intrinsic V_{GS} response

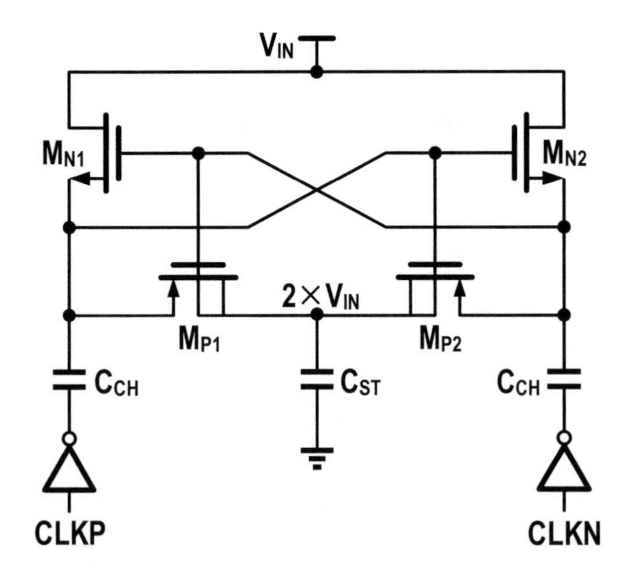

Fig. 3.79 Schematic of the charge pump

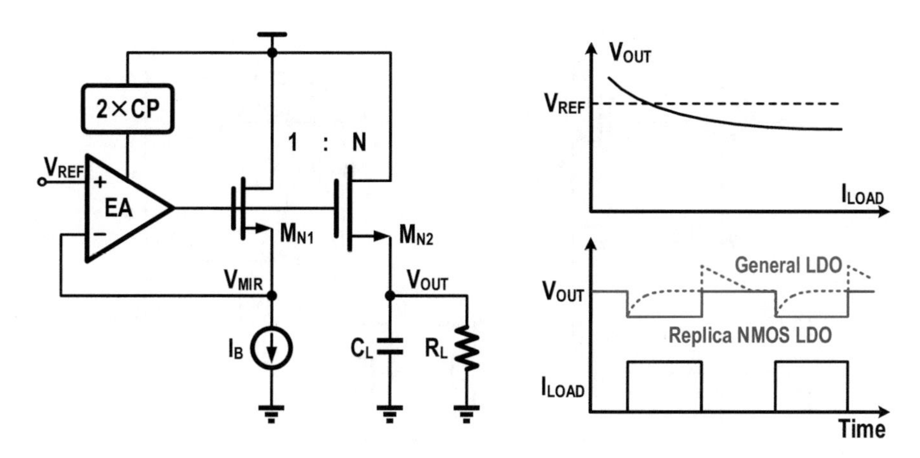

Fig. 3.80 Replica NMOS LDO

There have also been attempts to optimize the load transient performance of replica NMOS LDO. In [61], by detecting the load current, the bias current I_B is dynamically adjusted, causing the gate voltage of the power transistor to change with the load, thereby improving the load regulation.

During load transients, a closed-loop controlled LDO typically exhibits output voltage overshoot and undershoot; however, it eventually stabilizes at the target value over time. In contrast, the replica NMOS LDO is open loop controlled, with output voltage exceeding V_{REF} under light load and falling below V_{REF} under heavy load. A significant step change is observed during transitions between light and heavy loads, as illustrated in the lower right corner of Fig. 3.80.

The replica NMOS LDO is well suited for powering specific digital circuits, as it can rapidly respond to high-frequency current changes and maintain the output voltage stably around V_{REF}. In addition, the lower DC output voltage value under heavy-load conditions can help reduce the power consumption of digital circuits during such periods.

3.9.3 Adaptive Biasing in NMOS LDO

Adaptive biasing is commonly used in LDOs, as described in Sect. 3.6. For PMOS LDOs, the output current detection circuit is straightforward. By connecting the gate of the sensing PMOS transistor to the gate of the power transistors, the same V_{GS} voltage is ensured, resulting in a sensing current proportional to the output current. Figure 3.81 shows two current detection circuits for NMOS LDOs. In Fig. 3.81a, the clamping circuit composed of M_{P1}, M_{P2}, M_{N3}, and M_{N4} ensures $V_{SEN} \approx V_{OUT}$, making the source, drain, and gate voltages of the sensing transistor M_{N1} and the power transistor M_{N2} nearly identical. Thus, the current through M_{N1} is proportional to the output current.

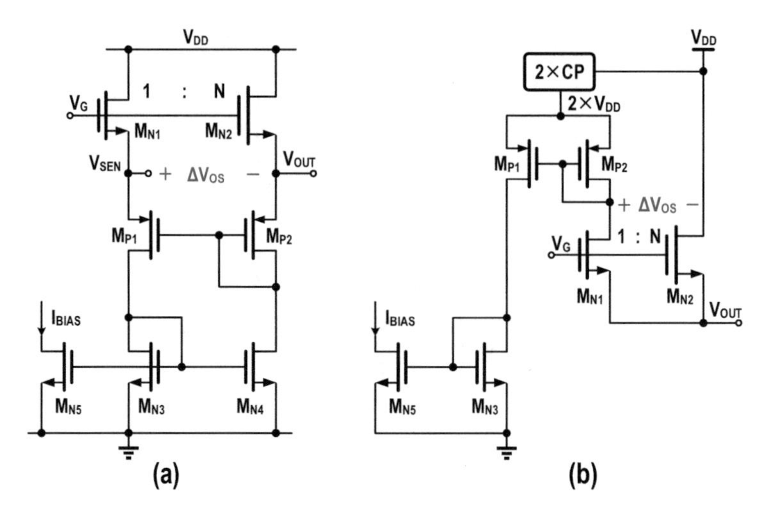

Fig. 3.81 NMOS current sensing circuits. (**a**) Source sensing and (**b**) drain sensing

The offset voltage due to MOS manufacturing mismatch can impact the accuracy of the sampling current. The clamp circuit shown in Fig. 3.81a is commonly used in PMOS LDOs. The error caused by the offset voltage ΔV_{OS} in PMOS LDOs is

$$\Delta I = g_{ds1} \times \Delta V_{OS} \tag{3.148}$$

The error caused by the offset voltage in the NMOS LDO is larger and is given by

$$\Delta I = g_{m1} \times \Delta V_{OS} \tag{3.149}$$

where g_m and g_{ds} are the transconductance and the drain-source conductance of the sensing transistor, respectively.

Figure 3.81b illustrates the method of sampling from the drain of the sampling transistor [64]. The gate of the sensing transistor, M_{N1}, is connected to V_G, and the source is connected to V_{OUT}, ensuring that its V_{GS} voltage matches that of the power transistor. A diode-connected PMOS transistor, M_{P2}, is connected in series with the drain of M_{N1}. The sampling current is mirrored through the current mirrors M_{P2} and M_{P1}. Compared to the source sampling method [35] shown in Fig. 3.81a, drain sampling ensures that the V_{GS} voltage remains consistent. When M_{N1} and M_{N2} are operating in the saturation region, the sampling accuracy is higher. The disadvantage of this method is that the charge pump needs to output a larger current, necessitating a larger output capacitor or a higher operating frequency.

Figure 3.81b illustrates drain sampling from the sensing transistor M_{N1}. The gate of M_{N1} is connected to V_G, and the source to V_{OUT}, ensuring that its V_{GS} voltage matches that of the power transistor. A diode-connected PMOS transistor, M_{P2}, is in series with M_{N1}'s drain. The sampling current is mirrored through M_{P2} and M_{P1}. Compared to the source sampling in Fig. 3.81a, drain sampling maintains consistent V_{GS} voltage. When M_{N1} and M_{N2} operate in the saturation region, sampling accuracy is higher. However, this requires the charge pump to deliver more current, needing a larger output capacitor or a higher operating frequency.

3.9.4 Multistage NMOS LDOs

If the load capability is high or the dropout voltage is low, the power transistor's size and gate parasitic capacitance are substantial. Hence, a buffer circuit must be inserted between the error amplifier and power stage. Insufficient buffer driving capability can affect load transient performance, causing excessive overshoot or undershoot and slow recovery.

Besides increasing drive current, the buffer circuits discussed in Sect. 3.4 can also enhance the driving slew rate. The studies [7, 29] employ a super source follower (SSF) buffer. Specifically, the work [29] describes a capacitorless NMOS LDO featuring nested Miller compensation and adaptive gain control, supporting load currents from 0.1 to 300 mA and maintaining a phase margin greater than 60° in all load conditions.

The study [29] describes an NMOS LDO with off-chip capacitors, as shown in Fig. 3.82. This LDO uses two power supplies: V_{BAT} for control and buffer, and V_{DD} for the power stage. Compensation is achieved through cascode Miller compensation, with one terminal of the capacitor C_C connected to V_{OUT} and the other end to the source of M_6, opposite in phase to the V_{EA}. To enhance PSRR performance, voltage-mode ripple injection technology is employed. The AUX circuit sets the optimal ripple injection factor, and the output V_X is connected to the summing stage M_{14}. This setup injects a signal with an opposite ripple phase into the gate of the power transistor, thereby offsetting the effect of the V_{DD} ripple on V_{OUT}.

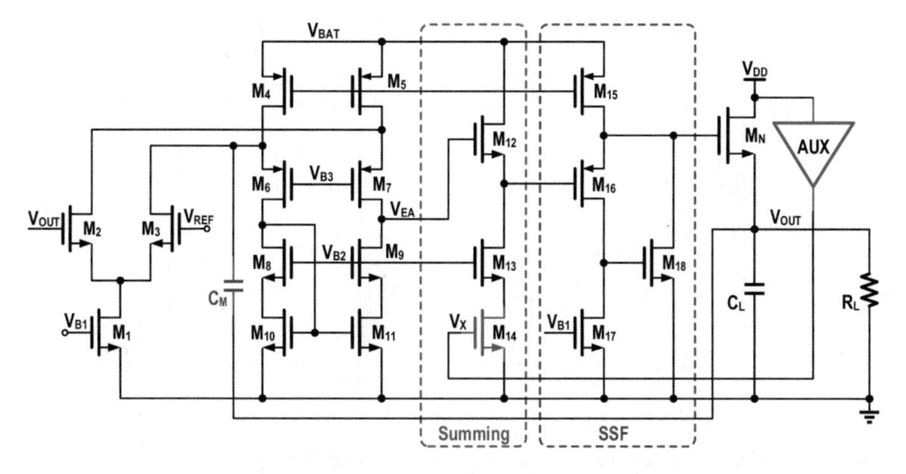

Fig. 3.82 Schematic of the NMOS LDO in [29]

The study [10] also describes an NMOS LDO that uses off-chip capacitors. It employs hybrid adaptive bias current technology to improve transient response and reduce quiescent current. The load current can reach 150 mA, while the quiescent current remains as low as 1.24 μA. In addition, this LDO is equipped with pull-up and pull-down buffers to ensure a high slew rate for both pull-up and pull-down actions. The compensation scheme uses dynamic zero-pole compensation technology, which can adjust the zero position according to load changes.

The structures and design concepts of the NMOS LDOs with external capacitors introduced earlier can also be applied to capless LDOs. In addition, some buffers used in PMOS LDOs are applicable to NMOS LDOs. However, there are also buffers specifically designed for NMOS LDOs.

It should be noticed that the NMOS LDO driver has a driving dead zone. When the gate voltage V_G is lower than the output voltage V_{OUT}, the power NMOS transistor stops providing current. In this situation, loop regulation is ineffective. During a load transient from high to low (H-L), after an overshoot, the drop in the output voltage depends entirely on the load current. Suppose the load capacitance C_L is large, and the light-load current is minimal. In that case, the output voltage may remain in an overshoot state for a prolonged period, causing the gate voltage V_G of the power transistor to enter the driving dead zone (i.e., V_G approaches 0 V). If a load transient occurs from low to high (L-H) before the output voltage and V_G have recovered, the delay in V_G exiting the dead zone and entering the adjustment state can cause a significant output voltage drop. This issue is particularly noticeable in flash memory applications such as universal flash storage (UFS) and embedded multimedia cards (eMMCs), where the load cycle from heavy to light to heavy (H-L-H) is very short, posing additional challenges for LDO design.

Figure 3.83 presents two buffer solutions [65, 66] for managing such short-period transients, which can clamp the driving voltage and prevent entering the driving dead zone. Figure 3.83a illustrates a virtual ground-based dynamic power recycling (VGDPR) buffer [65]. This buffer sets V_{OUT} to virtual ground, and the diode-connected M_S provides low impedance at V_G, resulting in a high-frequency

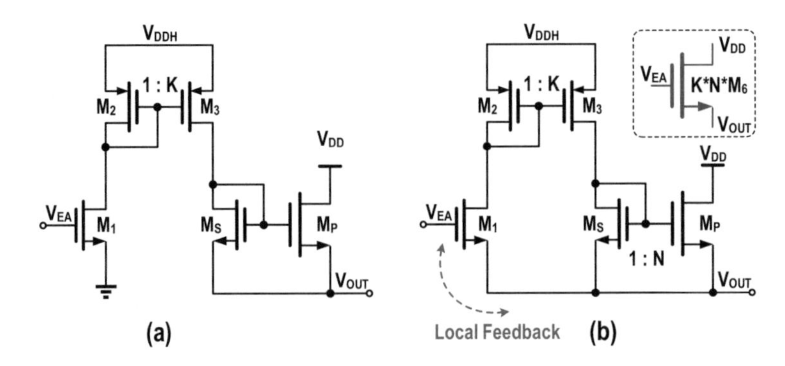

Fig. 3.83 (**a**) Virtual ground-based dynamic power recycling (VGDPR) buffer in [65], and (**b**) transconductance magnified (TM) MOS in [66]

pole P_G. In addition, the drive current varies with the load, enabling a high slew rate. More importantly, in load step-down scenarios, the output V_G of the VGDPR buffer does not enter the drive dead zone deeply but is clamped above V_{OUT}. For H-L-H load transients, this design can immediately supply drive current, effectively reducing voltage drop.

Figure 3.83b illustrates an enhanced version. The source of M_1 is also connected to V_{OUT}, creating a local feedback loop. The entire power stage can be considered equivalent to a super MOS with a gain of $K \times N$ times that of M_1. Therefore, it is called transconductance magnified (TM) MOS. The open-loop output impedance of the TM-MOS is

$$R_{O,OL} \approx \frac{1}{g_{m1}} \,||\, r_{dsp} \,||\left(\frac{1}{g_{mp}}\frac{g_{ms}}{Kg_{m1}}\right) \tag{3.150}$$

The open-loop output impedance of the VGDPR structure in Fig. 3.31a is

$$R_{O,OL} \approx r_{dsp} \tag{3.151}$$

Compared to the VGDPR structure, TM-MOS exhibits a lower open-loop output impedance and saves one channel of static current. In addition, transient simulation results indicate that overshoot and droop can be reduced by at least 10%. This is a very effective improvement.

References

1. G.A. Rincon-Mora, *Analog IC Design with Low-Dropout Regulators* (McGraw-Hill, 2009)
2. B. Razavi, *Design of Analog CMOS Integrated Circuits* (McGraw-Hill, 2017)
3. A. Sheikholeslami, Looking into a Node. IEEE Solid-State Circuit Magazine **6**(2), 8–10 (2014)
4. M. Ho, K.N. Leung, K.L. Mak, A low-power fast-transient 90-nm low dropout regulator with multiple small-gain stages. IEEE J. Solid-State Circuits **45**(11), 2466–2475 (2010)
5. Y.-H. Lam, W.-H. Ki, A 0.9V 0.35μm adaptively biased CMOS LDO regulator with fast transient response, in *IEEE International Solid-State Circuits Conference (ISSCC) Digest of Technical Papers*, (2008), pp. 442–443
6. J. Guo, K.N. Leung, A 25mA CMOS LDO with -86dB PSRR at 2.5MHz, in *IEEE Asian Solid-State Circuits Conference (ASSCC)*, (2013), pp. 381–384
7. K. Li, C. Yang, T. Guo, Y. Zheng, A multi-loop slew-rate-enhanced NMOS LDO handling 1-A-load-current step with fast transient for 5G applications. IEEE J. Solid-State Circuits **55**(11), 3076–3086 (2020)
8. M.A. Shyoukh, H. Lee, R. Perez, A transient-enhanced low-quiescent current low-dropout regulator with buffer impedance attenuation. IEEE J. Solid-State Circuits **42**(8), 1732–1742 (2007)
9. Y. Lu, C. Li, Y. Zhu, M. Huang, S.-P. U, R. Martins, A 312 ps response-time LDO with enhanced super source follower in 28 nm CMOS. Electron. Lett. **52**(16), 1368–1370 (2016)
10. R. Magod, B. Bakkaloglu, S. Manandhar, A 1.24 μA quiescent current NMOS low dropout regulator with integrated low-power oscillator-driven charge-pump and switched-capacitor pole tracking compensation. IEEE J. Solid-State Circuits **53**(8), 2356–2367 (2018)

11. K.N. Leung, P.K.T. Mok, Analysis of multistage amplifier-frequency compensation. IEEE Trans. Circuits Syst. I: Regul. Pap. **48**(9), 1041–1056 (2001)
12. X. Fan, C. Mishra, E.S. Sinencio, Single Miller capacitor frequency compensation technique for low-power multistage amplifiers. IEEE J. Solid-State Circuits **40**(3), 584–592 (2005)
13. P.E. Allen, D.R. Holberg, *CMOS Analog Circuit Design* (Oxford University Press, 2011)
14. A.D. Grasso, G. Palumbo, S. Pennisi, Comparison of the frequency compensation techniques for CMOS two-stage Miller OTAs. IEEE Trans. Circuits Syst. II Express Briefs **55**(11), 1099–1103 (2008)
15. W. Qu, S. Singh, Y. Lee, Y.-S. Son, G.-H. Cho, Design-oriented analysis for Miller compensation and its application to multistage amplifier design. IEEE J. Solid-State Circuits **52**(2), 517–527 (2017)
16. G.A. Rincon-Mora, Active capacitor multiplier in Miller-compensated circuits. IEEE J. Solid-State Circuits **35**(1), 26–32 (2000)
17. R.J. Milliken, J.S. Martinez, Full On-chip CMOS low-dropout voltage regulator. IEEE Trans. Circuits Syst. I: Regul. Pap. **54**(9), 1879–1890 (2007)
18. X. Peng, W. Sansen, L. Hou, J. Wang, W. Wu, Impedance adapting compensation for low-power multistage amplifiers. IEEE J. Solid-State Circuits **46**(2), 445–451 (2011)
19. D.B. Ribner, M.A. Copeland, Design techniques for cascaded CMOS OP amps with improved PSRR and common-mode input range. IEEE J. Solid-State Circuits **19**(6), 919–925 (1984)
20. M. Tan, W.-H. Ki, A cascode Miller-compensated three-stage amplifier with local impedance attenuation for optimized complex-pole control. IEEE J. Solid-State Circuits **50**(2), 440–449 (2015)
21. K.N. Leung, P.K.T. Mok, W.H. Ki, A novel frequency compensation technique for low-voltage low-dropout regulator, in *IEEE International Symposium on Circuits and Systems (ISCAS)*, (1999), pp. 102–105
22. S. Bu, J. Guo, K.N. Leung, A 200-ps response-time output-capacitorless low-dropout regulator with unity-gain bandwidth>100MHz in 130-nm CMOS. IEEE Trans. Power Electron. **33**(4), 3232–3246 (2018)
23. A. Maity, A. Patra, Tradeoffs aware design procedure for an adaptively biased capacitorless low dropout regulator using nested Miller compensation. IEEE Trans. Power Electron. **31**(1), 369–380 (2018)
24. A. Garimella, M.W. Rashid, P. Furth, Reverse nested Miller compensation using current buffers in a three-stage LDO. IEEE Trans. Circuits Syst. II, Exp. Briefs **57**(4), 250–254 (2010)
25. K.N. Leung, P.K.T. Mok, A capacitor-free CMOS low-dropout regulator with damping-factor-control frequency compensation. IEEE J. Solid-State Circuits **38**(10), 1691–1702 (2003)
26. K.N. Leung, P.K.T. Mok, W.-H. Ki, J.K.O. Sin, Three-stage large capacitive load amplifier with damping-factor-control frequency compensation. IEEE J. Solid-State Circuits **35**(2), 221–230 (2000)
27. M. Huang, H. Feng, Y. Lu, A fully integrated FVF-based low-dropout regulator with wide load capacitance and current ranges. IEEE Trans. Power Electron. **34**(12), 11880–11888 (2019)
28. S.K. Lau, P.K.T. Mok, K.N. Leung, A low-dropout regulator for SoC with Q-reduction. IEEE J. Solid-State Circuits **42**(3), 658–664 (2007)
29. J. Jiang, W. Shu, J.S. Chang, A 65-nm CMOS low dropout regulator featuring >60-dB PSRR over 10-MHz frequency range and 100-mA load current range. IEEE J. Solid-State Circuits **53**(8), 2331–2342 (2018)
30. X. Han, L. Wu, Y. Gao, W.-H. Ki, An adaptively biased output-capacitor-free low-dropout regulator with supply ripple subtraction and pole-tracking-compensation. IEEE Trans. Power Electron. **36**(11), 12795–12804 (2021)
31. C. Zhan, G. Cai, W.-H. Ki, A transient-enhanced output-capacitor-free low-dropout regulator with dynamic Miller compensation. IEEE Trans. Very Large Scale Integr. VLSI Syst. **27**(1), 243–247 (2019)
32. S. Bu, K.N. Leung, Y. Lu, J. Guo, Y. Zheng, A fully integrated low-dropout regulator with differentiator-based active zero compensation. IEEE Trans. Circuits Syst. I: Regul. Pap. **65**(10), 3578–3591 (2018)

33. K.C. Kwok, P.K.T. Mok, Pole-zero tracking frequency compensation for low dropout regulator, in *IEEE International Symposium on Circuits and Systems (ISCAS)*, (2002), pp. 735–738
34. G. Cai, Y. Lu, C. Zhan, R.P. Martins, A fully integrated FVF LDO with enhanced full-spectrum power supply rejection. IEEE Trans. Power Electron. **36**(4), 4326–4337 (2021)
35. H. Park, W. Jung, M. Kim, H.-M. Lee, A wide-load-range and high-slew capacitor less NMOS LDO with adaptive-gain nested Miller compensation and pre-emphasis inverse biasing. IEEE J. Solid-State Circuits **58**(10), 2696–2708 (2023)
36. J.S. Kim, K. Javed, K.H. Min, J. Roh, A 13.5-nA quiescent current LDO with adaptive ultra-low-power mode for low-power IoT applications. IEEE Trans. Circuits Syst. II Express Briefs **70**(9), 3278–3282 (2023)
37. N. Adorni, S. Stanzione, A. Boni, A 10-mA LDO with 16-nA IQ and operating from 800-mV supply. IEEE J. Solid-State Circuits **55**(2), 404–413 (2020)
38. R.G. Carvajal, J.R. Angulo, A.J. Lopez-Martin, A. Torralba, J.A.G. Galan, A. Carlosena, The flipped voltage follower: a useful cell for low-voltage low-power circuit design. IEEE Trans. Circuits Syst. I: Regul. Pap. **52**(7), 1276–1291 (2014)
39. K. Keikhosravy, S. Mirabbasi, A 0.13-μm CMOS low-power capacitor-less LDO regulator using bulk-modulation technique. IEEE Trans. Circuits Syst. I: Regul. Pap. **61**(11), 3105–3114 (2014)
40. T. Kim, B. Kim, J. Roh, A 54-nA quiescent current capless LDO with −39-dB PSRR at 1 MHz using a load-tracking bandwidth extension technique. IEEE Trans. Circuits Syst. II Express Briefs **71**(3), 1556–1560 (2024)
41. J. Guo, K.N. Leung, A 6-μW chip-area-efficient output-capacitorless ldo in 90-nm cmos technology. IEEE J. Solid-State Circuits **45**(9), 1896–1905 (2010)
42. P. Ying, K.N. Leung, An output-capacitorless low-dropout regulator with direct voltage-spike detection. IEEE J. Solid-State Circuits **45**(2), 458–466 (2010)
43. Y. Lee, J.-E. Park, Analysis of power-supply-rejection enhancement techniques for low-dropout regulators. IEEE Access **12**, 59976–59995 (2024)
44. V. Gupta, G.A. Rincon-Mora, P. Raha, Analysis and design of monolithic, high PSR, linear regulators for SoC applications, in *Proceedings of IEEE International SOC Conference*, (2004), pp. 311–315
45. F. Chen, Y. Lu, P.K.T. Mok, Transfer function analysis of the power supply rejection ratio of low-dropout regulators and the feed-forward ripple cancellation scheme. IEEE Trans. Circuits Syst. II Express Briefs **69**(8), 3061–3073 (2022)
46. C. Lee, K. McClellan, J. Choma Jr., A supply-noise-insensitive CMOS PLL with a voltage regulator using DC-DC capacitive converter. IEEE J. Solid-State Circuits **36**(10), 1453–1463 (2001)
47. J.M. Ingino, V.R. von Kaenel, A 4-GHz clock system for a high-performance system-on-a-chip design. IEEE J. Solid-State Circuits **36**(11), 1693–1698 (2001)
48. V. Gupta, G.A. Rincon-Mora, A 5 mA 0.6 μm CMOS Miller Compensated LDO Regulator with 27 dB Worst-Case Power-Supply Rejection Using 60 pF of On-Chip Capacitance, in *IEEE International Solid-State Circuits Conference (ISSCC) Digest of Technical Papers*, (2007), pp. 520–521
49. C. Zhan, W.-H. Ki, Analysis and design of output-capacitor-free low-dropout regulators with low quiescent current and high-power supply rejection. IEEE Trans. Circuits Syst. I: Regul. Pap. **61**(2), 625–636 (2014)
50. Y. Lu, A reconfigurable switched-capacitor DC-DC converter and cascode LDO for dynamic voltage scaling and high PSR, in *IEEE Asia Pacific Conference on Circuits and Systems (APCCAS)*, (2018), pp. 509–511
51. Y. Lim, L. Lee, S. Park, Y. Jo, J. Choi, An external capacitorless low-dropout regulator with high PSR at all frequencies from 10 kHz to 1 GHz using an adaptive supply-ripple cancellation technique. IEEE J. Solid-State Circuits **53**(9), 2675–2685 (2018)
52. F. Lavalle-Aviles, J. Torres, E. Sanchez-Sinencio, A high-power supply rejection and fast settling time capacitor-less LDO. IEEE Trans. Power Electron **34**(1), 474–484 (2019)

53. M. El-Nozahi, A. Amer, J. Torres, K. Entesari, E. Sanchez-Sinencio, High PSR low drop-out regulator with feed-forward ripple cancellation technique. IEEE J. Solid-State Circuits **45**(3), 565–577 (2010)
54. T. Guo, W. Kang, J. Roh, A 0.9-μA quiescent current high psrr low dropout regulator using a capacitive feed-forward ripple cancellation technique. IEEE J. Solid-State Circuits **57**(10), 3139–3149 (2022)
55. K. Joshi, S. Manandhar, B. Bakkaloglu, A 5.6μA a wide bandwidth, high power supply rejection linear low-dropout regulator with 68 dB of PSR up to 2 MHz. IEEE J. Solid-State Circuits **55**(8), 2151–2160 (2020)
56. J.-G. Lee, H.-H. Bae, S. Jang, H.-S. Kim, A Fully Integrated, Domino-Like-Buffered Analog LDO Achieving –28dB Worst-Case Power-Supply Rejection Across the Frequency Spectrum from 10Hz to 1GHz with 50pF On-Chip Capacitance, in *IEEE Solid-State Circuits Conference (ISSCC) Digest of Technical Papers*, (2024), pp. 456–457
57. T.Y. Man, K.N. Leung, C.Y. Leung, P.K.T. Mok, M. Chan, Development of single-transistor-control LDO based on flipped voltage follower for SoC. IEEE Trans. Circuits Syst. I: Regul. Pap. **55**(5), 1392–1401 (2008)
58. P. Hazucha, T. Kamik, B.A. Bloechel, C. Parsons, D. Finan, S. Borkar, Area-efficient linear regulator with ultra-fast load regulation. IEEE J. Solid-State Circuits **40**, 933–940 (2005)
59. H. Chen, K.N. Leung, A fast-transient LDO based on buffered flipped voltage follower, in *2010 IEEE International Conference of Electron Devices and Solid-State Circuits (EDSSC)*, (2010), pp. 1–4
60. J. Jung, J.-H. Choi, K. Jun Roh, J. Park, W.-M. Lim, T.-S. Kim, H.-K. Jeong, M. Kwak, J.-Y. Youn, J.-D. Ihm, C. Yoo, Y. Choi, J.-H. Choi, H. Ko, A 4 ns settling time FVF-based fast LDO using bandwidth extension techniques for HBM3. IEEE J. Solid-State Circuits **59**(10), 3307–3316 (2024)
61. Y. Lu, Y. Wang, Q. Pan, W.-H. Ki, C.P. Yue, A fully integrated low dropout regulator with full-spectrum power supply rejection. IEEE Trans. Circuits Syst. I: Regul. Pap. **62**(3), 707–716 (2015)
62. X.L. Tan, K.C. Koay, S.S. Chong, P.K. Chan, A FVF LDO regulator with dual-summed Miller frequency compensation for wide load capacitance range applications. IEEE Trans. Circuits Syst. I: Regul. Pap. **61**(5), 1304–1312 (2014)
63. G.W. den Besten, B. Nata, Embedded 5 V-to-3.3 V voltage regulator for supplying digital IC's in 3.3 V CMOS technology. IEEE J. Solid-State Circuits **33**(7), 956–962 (1998)
64. D. Mandal, C. Desai, B. Bakkaloglu, S. Kiaei, Adaptively biased output Cap-less NMOS LDO with 19 ns settling time. IEEE Trans. Circuits Syst. II Express Briefs **66**(2), 167–171 (2019)
65. W.-C. Chen, T.-C. Huang, C.-C. Chiu, C.-W. Chang, K.-C. Hsu, 94% power-recycle and near-zero driving-dead-zone N-type low dropout regulator with 20 mV undershoot at short-period load transient of flash memory in smart phone, in *IEEE International Solid-State Circuits Conference (ISSCC) Digest of Technical Papers*, (2018), pp. 436–438
66. X. Ming, J.J. Kuang, X.C. Gong, J. Zhang, Z. Wang, B. Zhang, An NMOS LDO with TM-MOS and dynamic clamp technique handling up to Sub-10-μs short-period load transient. IEEE J. Solid-State Circuits **59**(2), 583–594 (2024)

Chapter 4
Digital LDO

With the advancement of CMOS technology, digitally controlled voltage regulators have become a popular choice in recent years, especially for digital systems. Figure 4.1 presents an analog power stage and the digital power stage. The power transistor in an analog LDO is a single entity, where the output voltage is regulated by adjusting the V_{GS} voltage of the power transistor. In a digital LDO, the power transistor is divided into multiple unit cells, and the output voltage is stabilized by controlling the number of "on" power cells. Since the race hazards in the binary control signals may cause spike in output voltage, people usually employ a thermometer code as the control signals of the power cells.

Obviously, analog regulation is continuous, and the output voltage has no ripple. On the other hand, digital regulation is discrete, while the load current is continuous. Due to the finite resolution of the digital power stage, to ensure output voltage accuracy, the output ripple (caused by limit cycle oscillation, LCO) is inevitable.

The primary error source in analog LDOs is the offset voltage of the error amplifier (EA), while the output error in digital LDOs mainly stems from the quantizer resolution and quantization error. Increasing the quantizer resolution leads to an exponential increase in the area and power consumption of the quantizer and controller, whereas achieving high-precision analog amplifiers is more straightforward. Consequently, analog LDOs typically offer higher output voltage accuracy.

The analog power transistor usually operates in the saturation region where the output current is less sensitive to the V_{DS} voltage variation, providing good power supply rejection (PSR). In contrast, the power transistors in digital LDOs usually operate in the linear region, resulting in poorer PSR.

In terms of reliability, according to the description in Sect. 2.16, an analog power stage distributes the current and heat dissipation across the entire transistors. In the digital power stage, part of the power cells concentrates all the current and heat dissipation, potentially causing EM and self-heating issues. Therefore, an analog power stage offers better reliability compared to the digital ones.

X. Mao et al., *Fully-Integrated Low-Dropout Regulators*, Analog Circuits and Signal Processing, https://doi.org/10.1007/978-3-031-84916-9_4

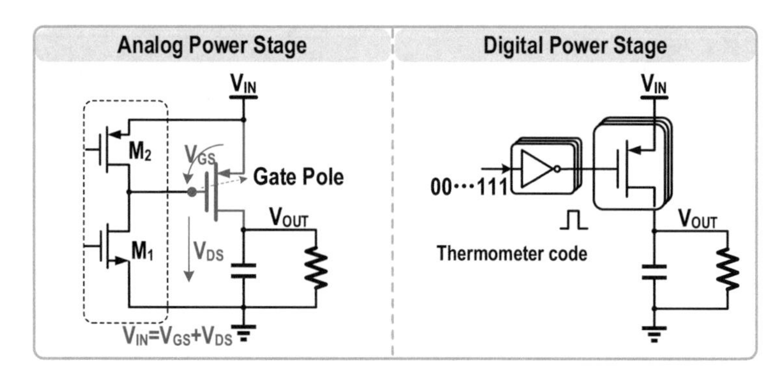

Fig. 4.1 An analog power stage and a digital power stage

Despite the disadvantages of digital LDOs compared to analog LDOs in terms of accuracy, ripple, power supply rejection, and reliability, digital LDOs exhibit advantages in easy configurable stability, low-voltage operation, easy distributed integration, small area, flexible large-signal response, and enjoyment of the process scaling.

For stability, analog LDOs require careful consideration of the gate pole in the power stage. We need to push the gate pole beyond the bandwidth for fast-transient applications. For digital LDOs, the inverter-based driver eliminates the gate pole. Hence, digital LDOs have an advantage in loop stability design, although they still need to account for the phase shift introduced by the operating clock frequency and propagation delay.

The gate voltage of the analog power transistors should not be too low, requiring a VDS margin for proper operation of the driver circuit. At low input voltage conditions ($V_{IN} < 0.6$ V), the analog LDO loop gain and performance will be degraded. In contrast, digital circuits can operate in the deep subthreshold region, making them more suitable for low-voltage operation.

The analog power stages are usually integrated centralized to prevent interference. Since the digital control signals have high immunity to interferences, the digital power stages can utilize the distributed integration to reduce the IR drop in high-current applications.

Digital power transistors operate in the fully turned-on state, while analog power transistors operate in the partially on state. With the same operation voltage and load capability, the digital power stage is more compact. Furthermore, as digital LDOs heavily rely on digital circuits, they can benefit more from process scaling in terms of area, power consumption, and performance.

Additionally, digital signals can be directly read from registers, enabling the system to monitor the real-time operating status of the digital LDO and the load conditions. In contrast, analog LDOs can only provide voltage and current information, requiring analog-to-digital conversion to obtain useful data in a digital system. Therefore, digital LDOs can find suitable applications in large-scale digital systems and microprocessors.

4.1 Shift Register-Based LDO

4.1.1 Typical Architecture

In 2010, Okuma proposed and presented the first digital LDO that uses the shift register control method [1]. Figure 4.2 depicts the circuit of this structure. This digital LDO consists of a comparator, a serial-in parallel-out bidirectional shift register (BI-SR), and a set of 256b PMOS power switches.

The BI-SR serves as the core control circuit, and its circuit implementation is illustrated in Fig. 4.3. It consists of a *D* flip-flop (DFF) and a selector. The comparator output controls the selector, which chooses whether the high bit (Q_{K+1}) or the low bit (Q_{K-1}) is used as the D-terminal (D_K) input to the local DFF. Thus, the comparator output determines the counting direction in the shift register, whether it shifts to the left or to the right.

Figure 4.4 illustrates the operation of the shift register-based LDO (SR-LDO). Initially, the shift register output is initialized to 1, and the power switches are all turned off. When the comparator detects that V_{OUT} is less than V_{REF}, the comparator output is high, the shift register decreases output decreases bit by bit, and the power switches are turned on sequentially. When the comparator detects that V_{OUT} is higher than V_{REF}, the comparator output is low, the shift register output increases bit by bit, and the power switches are turned off sequentially. In steady state, the output code varies between one or several neighboring code values to maintain the output voltage stable around V_{REF}. The operating waveform of the shift register LDO is shown in Fig. 4.5.

An error of a few *mV* in V_{OUT} and V_{REF} is sufficient to toggle the comparator output, so the shift register LDO can theoretically achieve a high-output voltage accuracy. However, SR-LDO suffers from poor transient performance, as it can only turn on or off one power switch per cycle. For an N-bit power stage (thermometer code), the search time takes more than N × cycle. Increasing the frequency reduces the

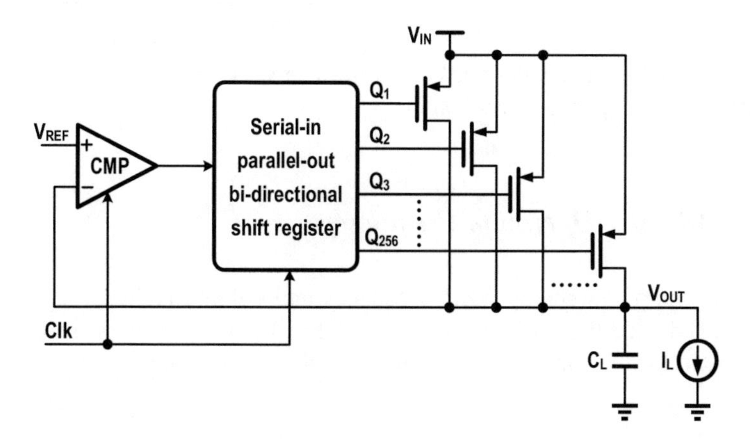

Fig. 4.2 Circuit implementation of the shift register LDO in [1]

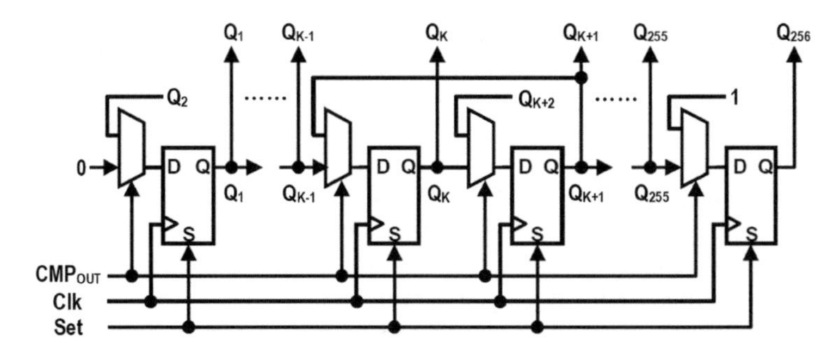

Fig. 4.3 Circuit implementation of the bidirectional shift register

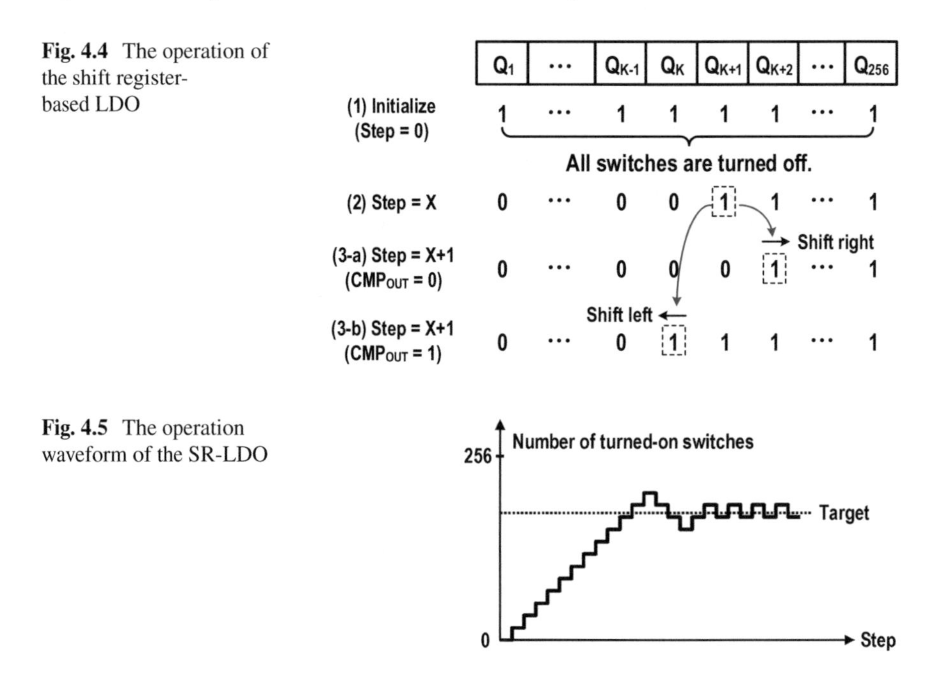

Fig. 4.4 The operation of the shift register-based LDO

Fig. 4.5 The operation waveform of the SR-LDO

search time and improves the transient performance; however, a higher frequency also implies larger power consumption.

4.1.2 Adaptive Operation Frequency

The relationship between operating frequency and stability is discussed in [2, 3]. Figure 4.6a depicts the small-signal model of the SR-LDO, where the shift register is equivalent to an integration loop. The loop transfer function is given as follows:

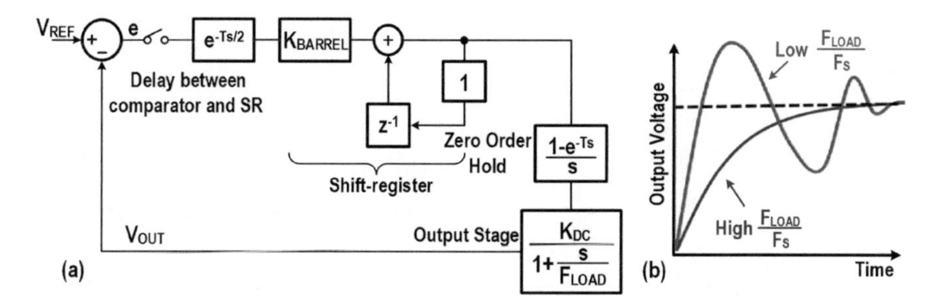

Fig. 4.6 (**a**) Small-signal model, (**b**) over-damped and under-damped response

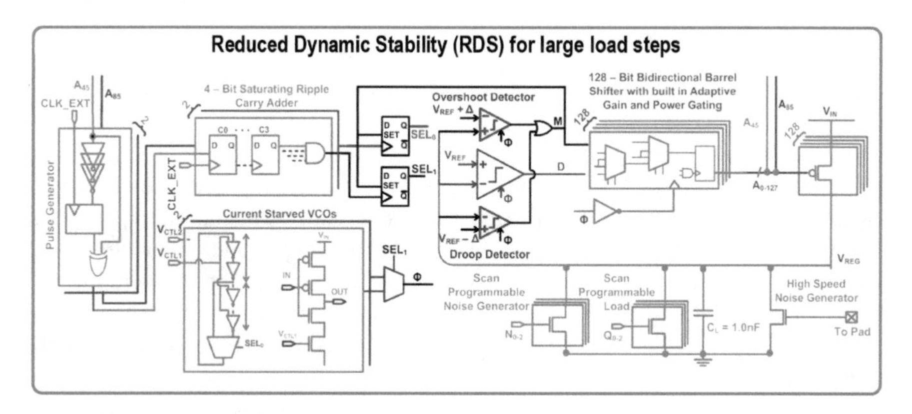

Fig. 4.7 The block diagram of the structure in [2]

$$H(z) = \frac{K_{\text{BARREL}} K_{\text{DC}} \times z^{0.5}}{(z-1)\left(z - e^{-F_{\text{LOAD}}/F_S}\right)} \tag{4.1}$$

The loop stability is related to the output poles as well as the operating frequency. A very wide range of loads implies a wide range of output poles. For a fixed frequency, the output poles at heavy-load conditions are far away from the integral poles, exhibiting an over-damped response. Light-load conditions shift the output poles to lower frequencies and closer to the integral poles, resulting in insufficient phase margins and exhibiting under-damped or even oscillation. Figure 4.6b demonstrates the over-damped and under-damped response.

For performance over a wider load range, the study [2] introduces three operation frequencies (F_{HIGH}, F_{NOM}, F_{LOW}), depending on the code value of the power switches. Additionally, Bin Nasir et al. [2] propose a reduced dynamic stability (RDS) approach to enhance the transient performance. The block diagram of the structure of [2] is given in Fig. 4.7.

Two comparators are utilized to detect the upper threshold ($V_{\text{REF}} + \Delta$) and the lower threshold (V_{REF}-Δ) of the output voltage, respectively. During load transient,

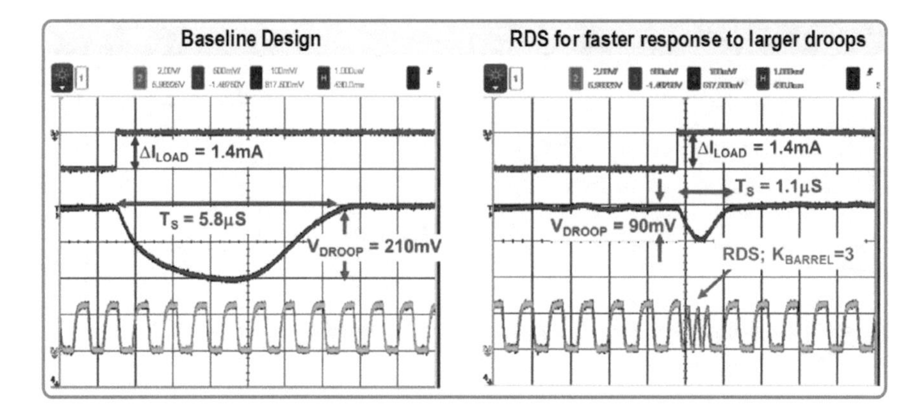

Fig. 4.8 The measured waveforms of load transient responses of baseline and with RDS [2]

when the output voltage is detected to be below the lower threshold (V_{REF}-Δ) or above the upper threshold (V_{REF} + Δ), the controller's operating frequency is instantaneously increased (400 MHz), reducing the code search time and improving the transient response speed. When V_{OUT} goes within the upper and lower thresholds (V_{REF}-Δ < V_{OUT} < V_{REF} + Δ), the frequency returns to normal. The measured waveforms are presented in Fig. 4.8, clearly demonstrating that with RDS, the transient dropout is significantly reduced, and the transient recovery time is significantly shortened.

4.1.3 Successive Approximation Recursive LDO

By increasing the operating frequency during transients, transient performance can be improved. However, for a typical SR LDO, it still requires multiple cycles to adjust to the proper code value, which does not satisfy the need for fast transient response. Based on the successive approximation recursive principle, Salem et al. [4] proposed a recursive LDO. Figure 4.9 depicts the block diagram of the recursive LDO. It utilizes 7 binary-weighted PMOS switches instead of the conventional 128 equal-size PMOS switches.

For a full-range 0-to-I_{MAX} load step, in the first cycle, the conventional linear approach can only turn on one power switch at a time, while the recursive LDO turns on half of the on-state capacity (≈64 power switches). In the second cycle, the conventional linear scheme continues to turn on one switch, while the successive approximation will turn on another quarter of the switches (≈32 power switches). The recursive LDO can determine the proper value in 7 cycles, while the conventional linear scheme requires 128 cycles. Figure 4.10 gives the transient waveforms of the recursive LDO structure and the SR LDO. The recursive LDO can effectively improve the transient performance compared to the typical SR structure.

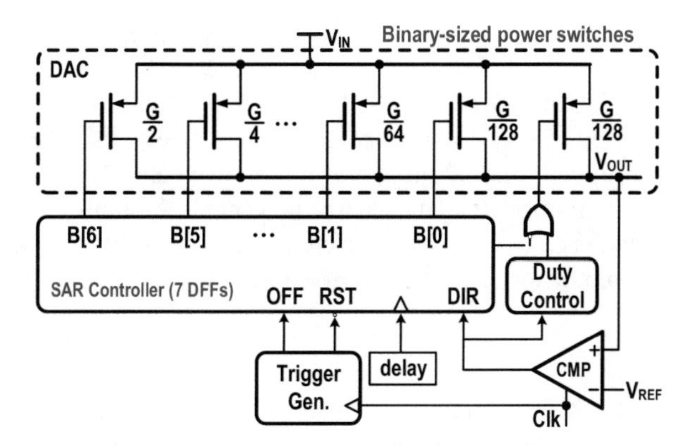

Fig. 4.9 The block diagram of the structure in [4]

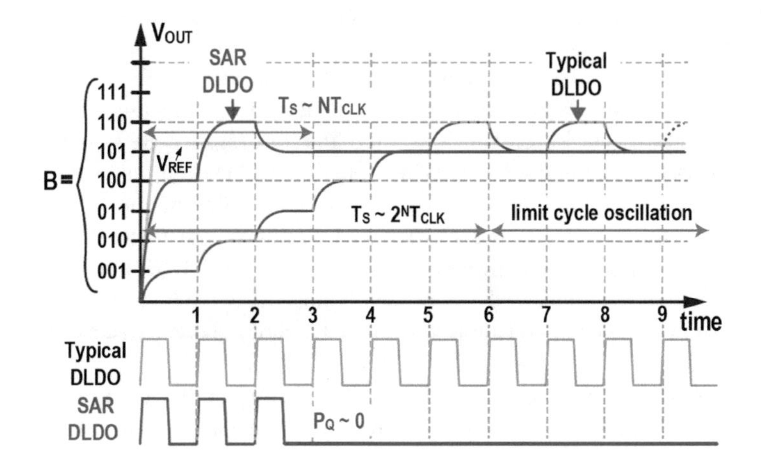

Fig. 4.10 The transient waveforms of the recursive LDO structure and the SR LDO

In addition, Salem et al. [4] introduced a duty cycle-controlled switch to improve the load range and reduce the output ripple. The switching control will be introduced in Chap. 5.

4.2 Coarse-Fine-Tuning DLDO

4.2.1 Two Shift Register Architectures

To achieve a wide load range and high-output voltage accuracy, the typical SR LDOs need many bits in the power switch and controller. However, this leads to two major problems: (1) slow transient response and (2) bidirectional shift registers that are area and power intensive.

A coarse-fine-tuning structure has been proposed to balance cost, accuracy, ripple, and transient performance [5–9]. A typical coarse-fine-tuning DLDO (CF-DLDO) structure block diagram is shown in Fig. 4.11. The power stage is divided into two parts, the coarse and fine parts. There are 63 groups of coarse power cells and 32 groups of fine power cells. The strength of each coarse power cell is 16 times that of each fine power cell. Assuming that the current of each coarse switch is I_{UNIT_C}, and the current of each fine switch is I_{UNIT_F}, the output current at a certain moment can be expressed as

$$I_{OUT} = K_C \times I_{UNIT\_C} + K_F \times I_{UNIT\_F} \tag{4.2}$$

where K_C is the number of coarse cells turned on, and K_F is the number of fine cells turned on. In steady state, the V_{OUT} voltage is between V_{REF_H} and V_{REF_L}, since only the fine power cells operate and a small ripple can be achieved as well as high-output voltage accuracy. Regarding load capability and output ripple, the DLDO in [5] can be equivalent to a 1056-unary-bit conventional SR-DLDO.

Figure 4.12 illustrates the transient response of the CF-DLDO in [5]. When V_{OUT} is higher than V_{REF_H} or lower than V_{REF_L}, the fine shift register (SR) stops counting, and the coarse SR starts counting. When the V_{OUT} voltage returns above V_{REF}, the coarse SR stops counting, and the fine SR carries out fine tuning. The operating frequency of the coarse SR is 500 MHz, while that of the fine SR is 50 MHz. Therefore, compared to the conventional SR LDO, the CF-DLDO in [5] can turn on more power cells simultaneously, significantly improving the transient performance without increasing power consumption. Table 4.1 shows a comparison between the CF-DLDO and the conventional SR-DLDO.

The work in [6] presented an all-digital CF-DLDO. It utilized a CMP-triggered oscillator to generate the clock signal (CLK_S) for the coarse loop. In steady state, the

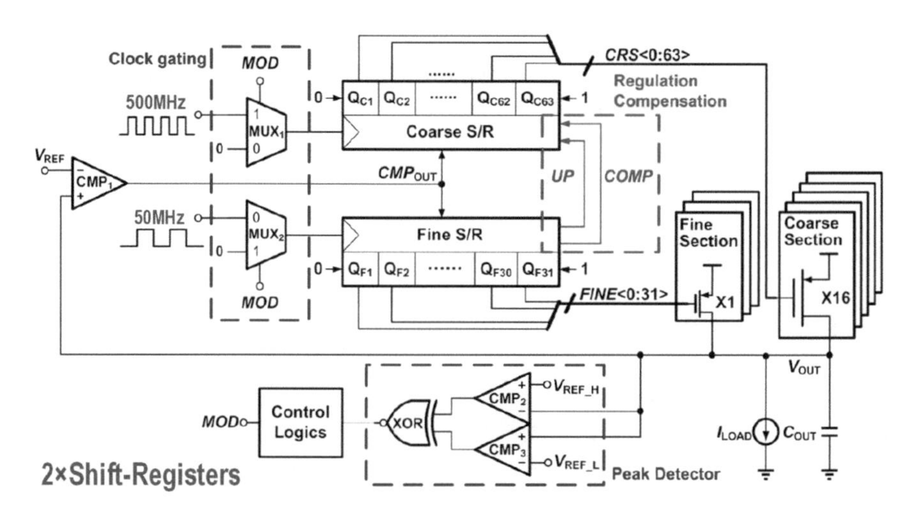

Fig. 4.11 Block diagram of the coarse-fine-tuning DLDO in [5]

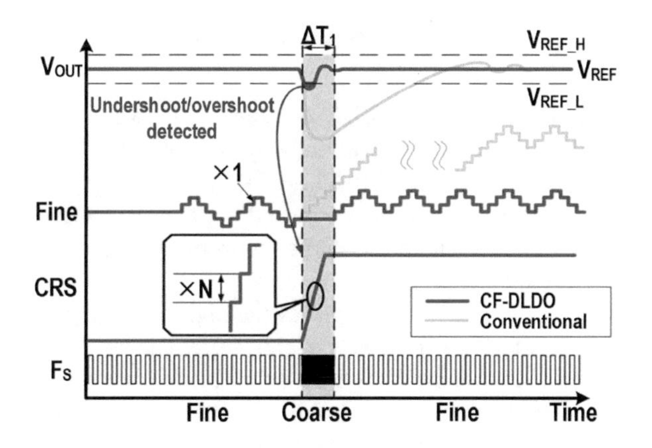

Fig. 4.12 Transient response of the CF-DLDO in [5]

Table 4.1 Comparison table between the CF-DLDO and the conventional SR-DLDO

Item	SR-DLDO	CF-DLDO	Remark
Number of bits of SR	1056	96	10× less
Regulation speed in one cycle	1	16	16× fast
Regulation speed in 20 ns	1	160	160× fast
Ripple	1	1	The same

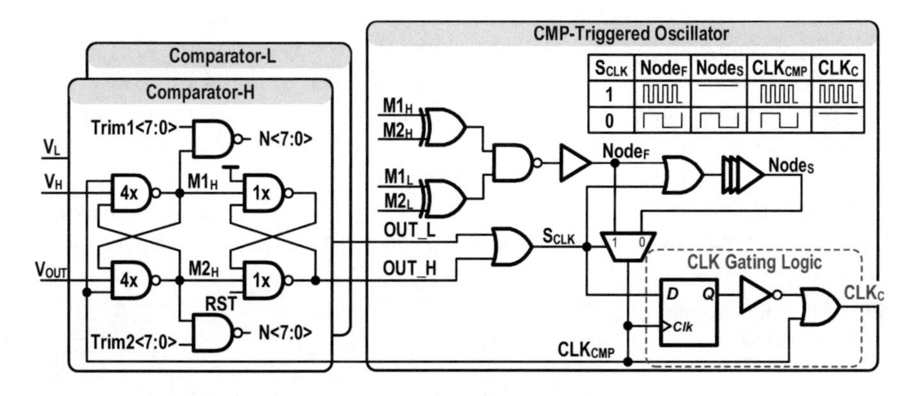

Fig. 4.13 All-digital comparator (CMP) and CMP-triggered oscillator

fine loop operates at 10 MHz, while the coarse frequency is about 300 MHz, with the control circuit clock, CLK_C, gated. Upon detection of a transient event, the clock CLK_S can be immediately raised to 1.99 GHz ($V_{IN} = 1$ V). This architecture balances transient performance and static power consumption. Figure 4.13 illustrates the circuits for the all-digital comparator and CMP-triggered oscillator. All circuits are designed using standard logic cells.

4.2.2 Shift Register with Flash ADC Architecture

A high-frequency SR LDO combined with a low-frequency SR LDO can form a CF-DLDO, which can effectively improve the transient performance compared with the conventional SR-DLDO. When a load transient occurs, the high-frequency SR LDO still needs to be adjusted to the proper value after dozens or even tens of cycles. A flash analog-to-digital converter (ADC) can detect the change of the output voltage within one cycle.

The work in [7, 10] replaces the high-frequency SR part with a flash ADC and proposes an SR + flash ADC architecture. The block diagram is depicted in Fig. 4.14. The fine loop comprises a 20-unary-bit bidirectional shift register and fine power switches. Each fine power switch can output 2 mA current. The coarse loop utilizes a 5-bit (thermometer code) flash ADC, and its outputs directly drive the coarse power switches. Each coarse switch can drive 40 mA current. Since the flash ADC can detect output voltage changes within one cycle, the transient response of the coarse loop is very fast. However, there is an accuracy issue.

In light-load conditions (<40 mA), the coarse power switches should be turned off, and the control code $C_{\mathrm{LPT}}[4{:}0]$ is 11111. The output voltage V_{OUT} needs to be much higher than V_{REF}. In heavy-load conditions (>200 mA), the coarse power switches should all be turned on, the control code value $C_{\mathrm{LPT}}[4{:}0]$ is 00000, and V_{OUT} needs to be much lower than V_{REF}. The V_{OUT} varies significantly with the load current and exhibits a large DC error, as shown in Fig. 4.15a. To mitigate this issue, the study [7] recommends to dynamically adjust the V_{REF} voltage according to $C_{\mathrm{LPT}}[4,0]$, which can effectively reduce the DC error, as illustrated in Fig. 4.15b. The coarse loop regulates the V_{OUT} to be near V_{REF} with a DC error within the least significant bit (LSB) error. The fine loop further improves the output voltage accuracy and reduces the output ripple with the shift register control.

The coarse loop utilizes a current-mirror flash ADC (CMF-ADC), as shown in Fig. 4.16. It contains a voltage-to-current G_{M} cell and a current-to-code converter.

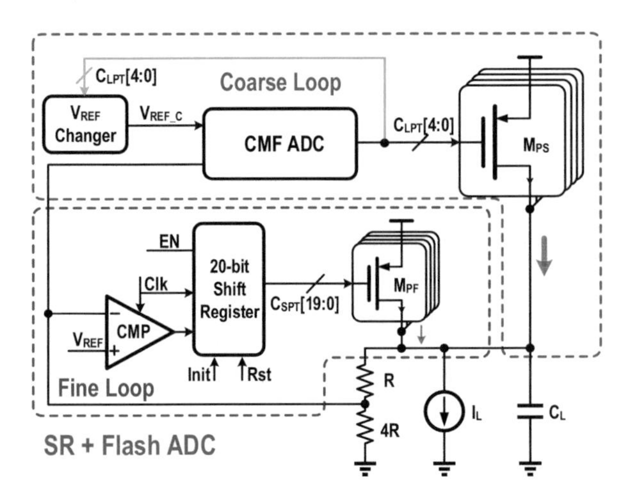

Fig. 4.14 Block diagram of the CF-DLDO architecture in [7]

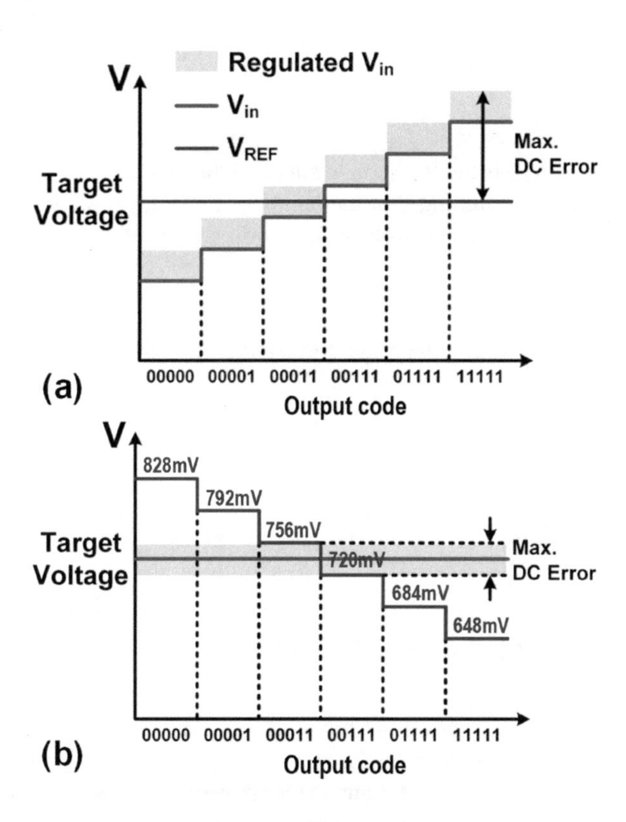

Fig. 4.15 DC error: (**a**) fixed V_{REF}, (**b**) V_{REF} changes with output codes

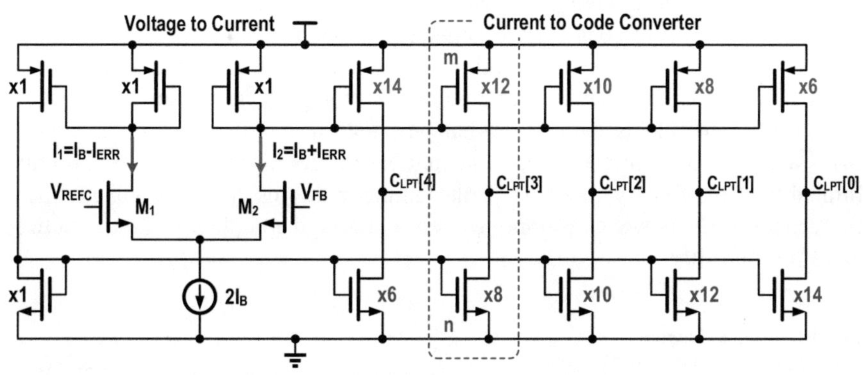

Fig. 4.16 Schematic of the current-mirror flash ADC

The input differential pair transistors M_1 and M_2 convert the voltage difference between V_{OUT} and V_{REF} into currents I_1 and I_2:

$$I_{ERR} = \left(V_{FB} - V_{REF}\right) \times g_m \tag{4.3}$$

$$I_1 = I_B - I_{ERR} \tag{4.4}$$

$$I_2 = I_B + I_{ERR} \tag{4.5}$$

where g_m is the transconductance of the input pair transistors, and the bias current is $2 \times I_B$. Assuming that the number of PMOS in a current comparator is m and the number of NMOS is *n*, at the flip-flop moment, we have

$$m \times I_2 = n \times I_1 \tag{4.6}$$

We can calculate the corresponding voltage threshold according to Eqs. (4.3), (4.4), (4.5), and (4.6):

$$V_{FB} = V_{REF} + \frac{I_B}{g_m} \times \frac{n-m}{m+n} \tag{4.7}$$

Setting $m + n = 20$, the values of m and n can be obtained respectively.

4.3 ADC-Based DLDO

Figure 4.17 shows the block diagram of a typical ADC-based DLDO [11–16]. It comprises a multi-bit quantizer, a proportional-integral-derivative (PID) controller, and power cells. The quantizer converts the difference between the output voltage V_{OUT} and the reference voltage V_{REF} into a digital code *e*(*n*), which is subsequently processed by the PID controller to generate the control code *P*(*n*) for the power cells. By employing a high-speed high-resolution quantizer and effective controller, the ADC-based DLDO can achieve high-output voltage accuracy and fast transient response. Commonly used quantizers can be categorized into voltage-domain quantizers and time-domain quantizers. The number of bits for the quantizer is usually limited to 6 bits (binary code). To further improve the resolution, it will exponentially increase the power consumption, area, and design complexity of the quantizer and PID controller.

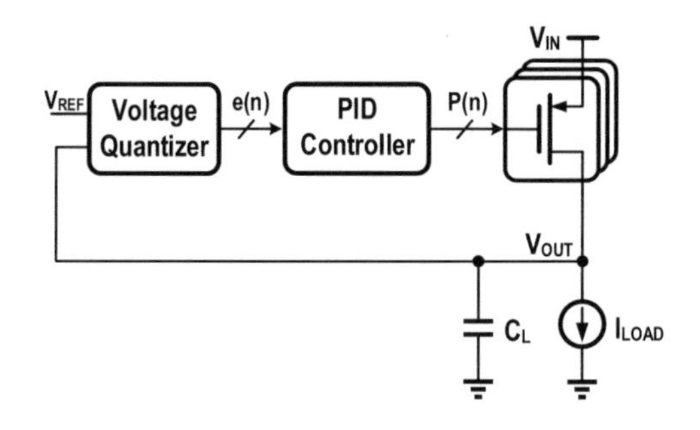

Fig. 4.17 Block diagram of the ADC-based DLDO

4.3.1 Voltage-Domain Quantizer

Flash ADC is the most commonly used voltage-domain quantizer [11, 12, 17, 18], as depicted in Fig. 4.18. A resistor-based voltage divider is employed to generate several reference voltages for the comparators. When the reference voltage of a comparator is higher than the output voltage V_{OUT}, the output of the comparator is logic "1." Conversely, when the reference voltage is lower than V_{OUT}, the comparator output is logic "0." Owing to the parallel structure, the flash ADC can detect changes in V_{OUT} within a single cycle. The output of the flash ADC is a thermometer code, which necessitates conversion to a binary code through a decoder for subsequent PID processing.

To mitigate the conversion latency of the comparator, the MOS transistor sizes of the comparator are relatively small, leading to the presence of offset voltages in the actual manufacturing. The offset voltage induces nonlinear errors in the quantization of the flash ADC and may even cause overlaps between adjacent quantization levels, as illustrated in Fig. 4.19. These nonlinear errors and overlaps can degrade the circuit's performance and potentially result in abnormal operation [19].

Thus, the calibration of the comparator is essential to enhance the robustness of the circuit. Figure 4.20 illustrates a discrete-time (DT) comparator with a wide input

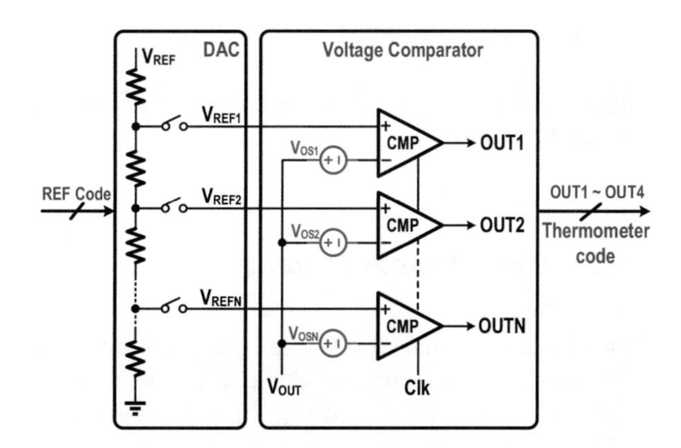

Fig. 4.18 Schematic of a typical flash ADC

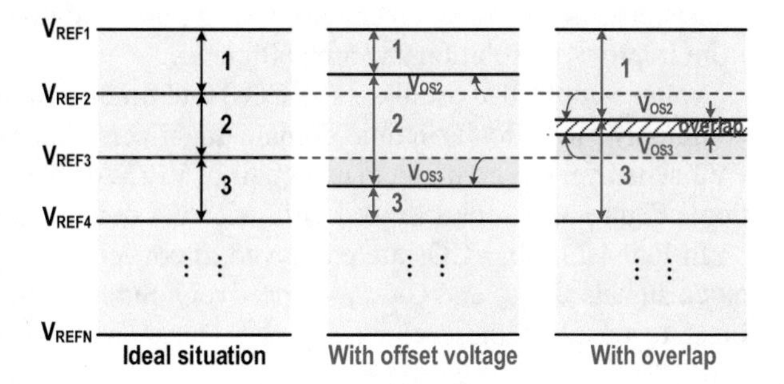

Fig. 4.19 The offset voltage of the comparators will cause a nonlinear error or overlap

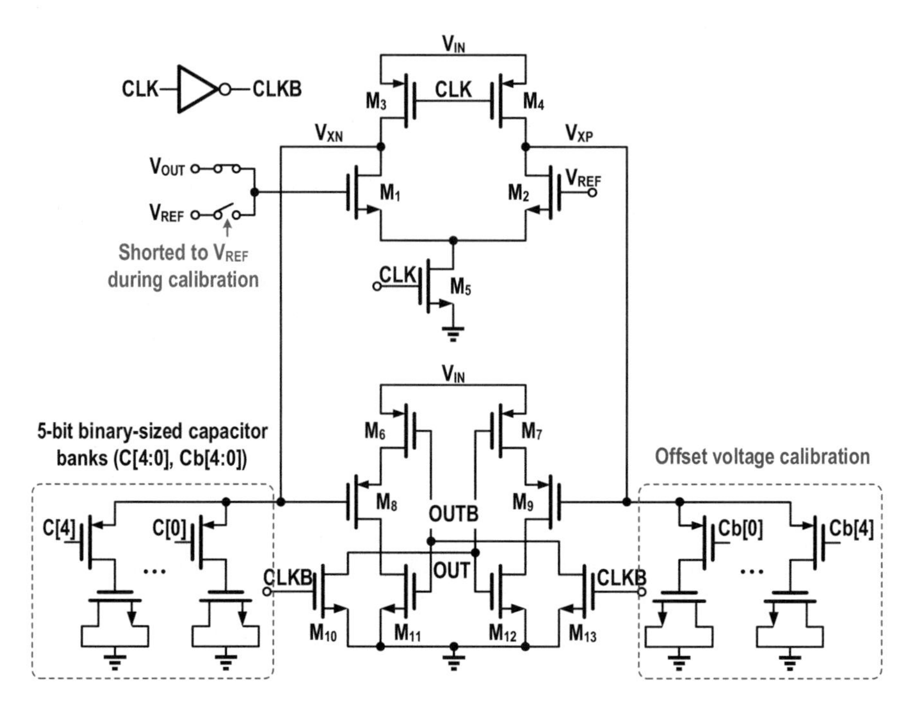

Fig. 4.20 The schematic of a discrete-time comparator

voltage range, where the threshold voltage can be adjusted by modifying the capacitance of V_{XN} and V_{XP}.

4.3.2 *Time-Domain Quantizer*

Time-domain quantizer mainly uses the gate delays of the transistors to convert voltage information into time-domain information such as clock and pulse, ultimately quantizing the signal into digital codes [13–16, 20]. Compared to flash ADCs, time-domain quantizers exhibit better linearity and require only a single reference voltage. Furthermore, their resolution is related to the gate delay of the transistors. As the process technology scale, the time-domain quantizer will benefit from improved resolution and area efficiency.

Voltage-controlled oscillators (VCOs) and time-to-digital converters (TDCs) are commonly used circuits in time-domain quantizers. Since both VCOs and TDCs are PVT sensitive, it is common to use a pair of VCOs or TDCs to cancel the PVT variations. Figure 4.21 shows several different time-domain quantizers.

In Fig. 4.21a, b, VCOs are employed to convert the V_{REF} and V_{OUT} voltages into clock signals CK_{REF} and CK_{OUT}, respectively. Subsequently, a time quantizer is utilized to quantize these clock signals. These two architectures do not need an

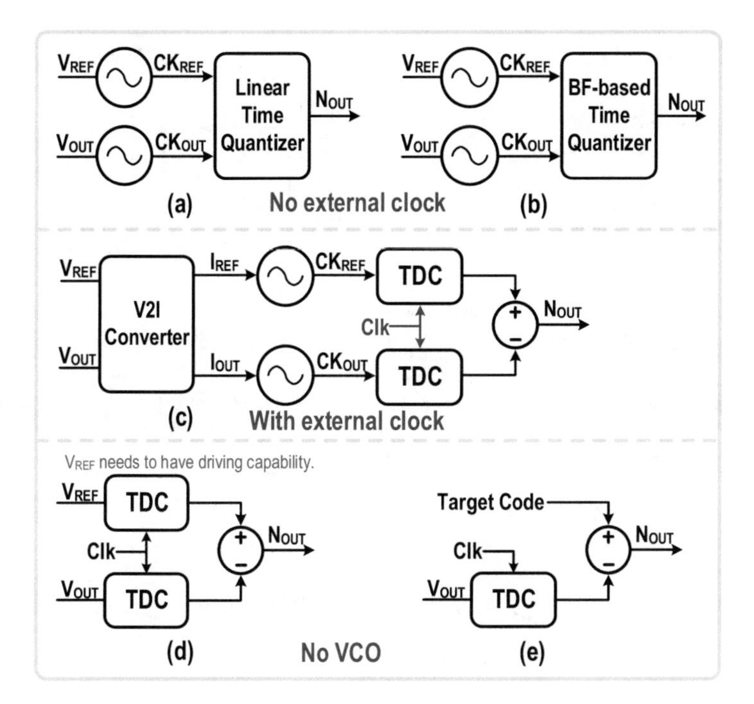

Fig. 4.21 Several time-domain quantizers used in (**a**) [13], (**b**) [14], (**c**) [15], and (**d**) and (**e**) [16]

external clock source. Figure 4.21a incorporates a linear quantizer, wherein the sampling clock CK_S for the time quantizer is generated by dividing the reference clock CK_{REF}. The output N_{OUT} exhibits a proportional relationship with the VCO frequency (CK_{OUT}) or the output voltage $V_{OUT,}$ generating a linear input-to-output transfer function.

On the other hand, Fig. 4.21b incorporates a beat-frequency time quantizer, wherein the sampling clock CK_S is derived from the difference between CK_{REF} and CK_{OUT}. Therefore, the frequency of CK_S changes adaptively with the difference between V_{OUT} and V_{REF}. In the steady state, when V_{OUT} and V_{REF} are close, the voltage difference is minimal, resulting in a slight frequency difference and a low sampling frequency. This helps to reduce the current consumption of the DLDO. However, when a load transient occurs, a change in V_{OUT} leads to a larger voltage difference and a correspondingly higher frequency difference. Consequently, the sampling frequency increases significantly, which shortens the transient response and recovery times, thereby enhancing transient performance. Figure 4.22 compares the two time quantizers, shown in (a) and (b).

The sampling clocks employed in Fig. 4.21a, b are significantly lower than the VCO output's clock frequency and cannot respond immediately to the changes in V_{OUT}. Consequently, they are suitable for application scenarios with low-speed transient steps. Figure 4.21c directly utilizes two delay-line TDCs to quantify CK_{REF} and CK_{OUT}, with the difference as the quantizer output. Compared to Fig. 4.21a, b,

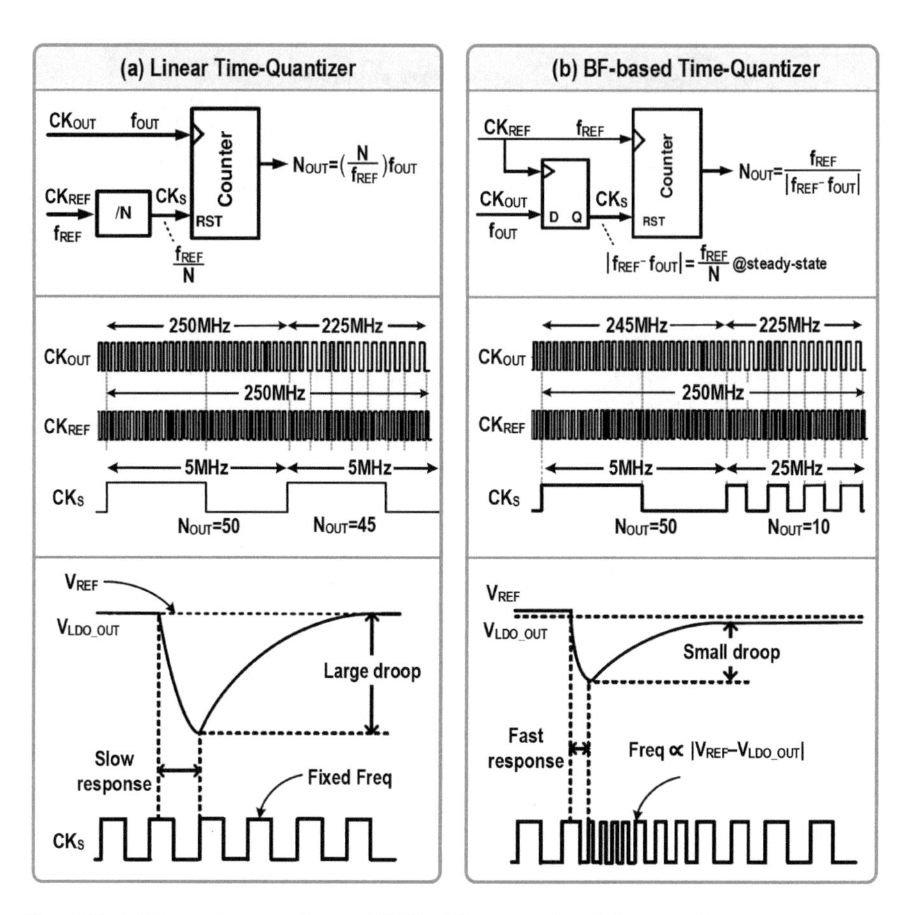

Fig. 4.22 (**a**) Linear time quantizer and (**b**) beat-frequency-based time quantizer

this circuit design is more intricate, but it offers faster detection speeds. Figure 4.21d directly employs two inverter-chain TDCs to quantize voltage information without the need for VCO conversion. This architecture is more compact and exhibits a faster response. However, since the quantized voltage is often utilized as the power supply for the TDC, the detected voltage V_{REF} also needs to have driving capability. Figure 4.21e incorporates a trilinear interpolation technique to dynamically adjust the target code value according to the process (P), voltage (V), and temperature (T) conditions, serving as the reference code value.

Figure 4.23 (left) shows the schematic of the VCO [14]. It consists of a voltage-to-current converter and a current-controlled ring oscillator. The current is proportional to the input voltage (V_{REF} or V_{OUT}). The desired oscillation frequency can be coarsely tuned by adjusting the amplification factor of the current. Furthermore, the frequency can be fine-tuned by adjusting the ring oscillator's capacitance.

Figure 4.24 (left) illustrates the schematic of the 5-bit TDC. The input clock pulse passes through the inverter delay chain every cycle. The DFFs are used to

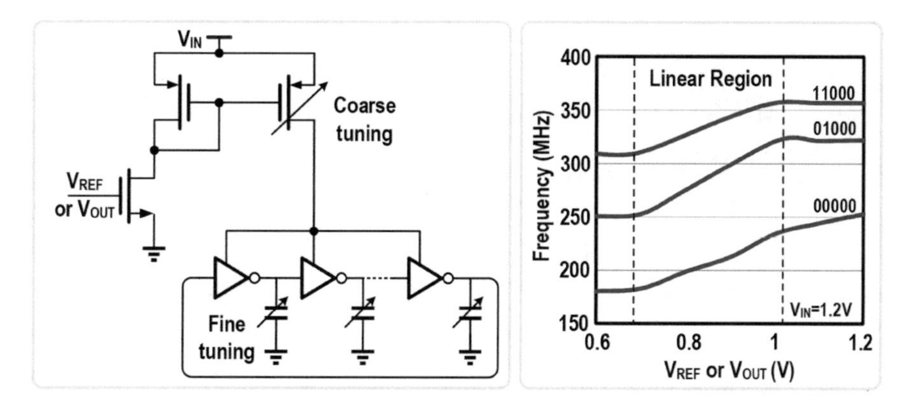

Fig. 4.23 The schematic of VCO and voltage-frequency curve

sample the inverter states at every rising edge of the input clock. If the clock edge and the data edge are too close, an LSB error could be introduced due to the metastability-induced bubble. By utilizing AND gates, the metastability-induced bubble can be suppressed. With advanced processes, the operating clock of the TDC can easily reach more than GHz, to achieve fast voltage conversion. Figure 4.24 (right) presents the simulated N_{OUT} code across V_{OUT} at different corners. The N_{OUT} code changes linearly in proportion to V_{OUT}.

4.3.3 PID Controller

Upon obtaining the digital error information, it needs to be input into the controller to derive the power stage's control code value $P[n]$. The proportional-integral-derivative (PID) control is the most widely adopted feedback control system, which consists of the proportional, integral, and derivative components:

$$P[n] = K_P \times e[n] + K_I \times \Sigma e[n] + K_D \times (e[n] - e[n-1]) \tag{4.8}$$

where K_P, K_I, and K_D represent the coefficients of the proportional, integral, and differential paths, respectively. Figure 4.25 shows the structure diagram of the classic position-type PID controller. The loop's stability and transient performance heavily depend on the three parameters K_P, K_I, and K_D.

The proportional (P) path output is proportional to the current error. Once the system has a deviation, the controller's proportional adjustment immediately takes effect to reduce the deviation. Increasing proportional action accelerates the adjustment speed and reduces the error. However, excessive proportional adjustment decreases system stability, as illustrated in Fig. 4.25(1)–(3). The standalone P controller exhibits a simple structure and fast response. However, it suffers from steady-state error.

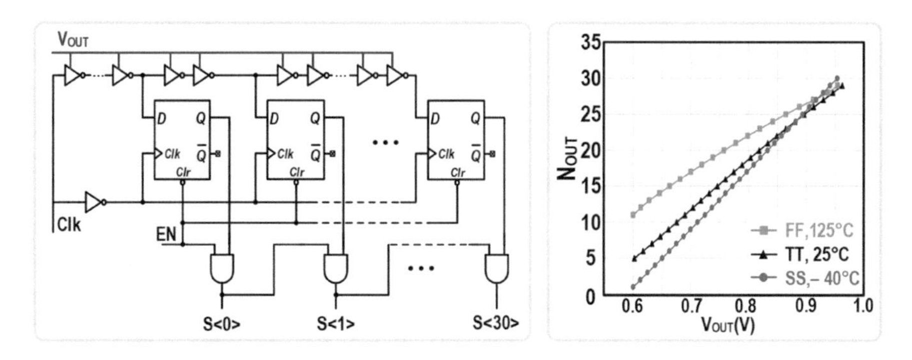

Fig. 4.24 The schematic of inverter-chain TDC and simulated results

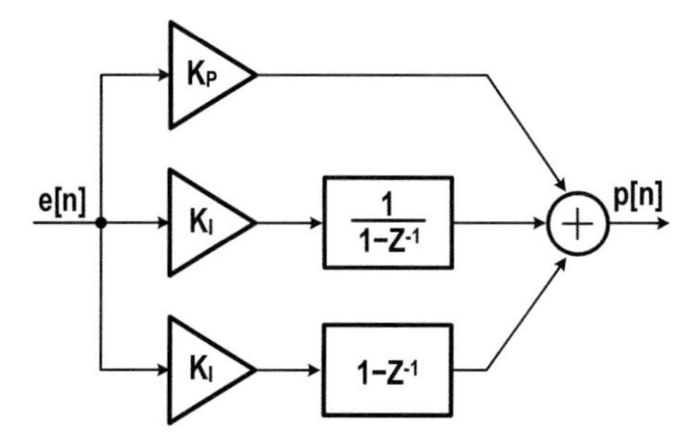

Fig. 4.25 The structure diagram of the position-type PID controller

The integral (I) path integrates the past information. When the system error occurs, the integral action of the controller is performed until the error disappears and the integral regulation ceases, at which time the integral regulation outputs a constant value. Therefore, the integral term is mainly used to eliminate the steady-state error of the system, as shown in Fig. 4.25(3)–(4). For example, shift register control is a typical integral control. The larger the integral gain, the faster it converges to the set value. However, it will deteriorate the loop stability, as depicted in Fig. 4.25(4)–(6). Since the integral term accumulates all the past errors, a larger K_I will make the overshoot larger.

The derivative (D) path predicts the future based on the rate of change. Differential regulation is predictive, which can predict the trend of deviation change and take corresponding actions. It can introduce effective early correction signals before the deviation signal becomes larger, produce advanced control effects, speed up the system response time, and improve the dynamic characteristics of the system, as shown in Fig. 4.25(7)–(9). The derivative term cannot be utilized independently because it is a regulation based on the system's rate of change. When the system given does not change, the controller's derivative action has no output and cannot produce any control effect. The differential term is often used in combination with the other two types of regulation to form a PD or PID controller.

Table 4.2 summarizes the effects of different PID parameters (K_p, K_i, K_d) on the system's step response [21]. First, we need to determine which system

Table 4.2 Effects of increasing a parameter independently

Parameter	Rise time	Overshoot	Settling time	Steady-State error	Stability
↑ K_P	Decrease ↓	Increase ↑	Small change ↗	Decrease ↓	Degrade ↓
↑ K_I	Decrease ↓	Increase ↑	Increase ↑	Eliminate ↓↓	Degrade ↓
↑ K_D	Minor change ↘	Decrease ↓	Decrease ↓	No effect →	Improve if K_D small ↑

characteristics need to be improved based on the transient response. According to Table 4.2, increasing KP can reduce the rise time and accelerate the recovery. Using K_D can reduce overshoot and settling time. Employing K_I can eliminate steady-state errors.

The above describes the considerations on adjusting K_P, K_I, and K_D to optimize system performance. Now, the question arises: How can we quickly and easily find a set of initial values? Next, we will introduce the manual tuning method and the Ziegler-Nichols method for PID parameter design.

Manual Tuning

First, we set the values of K_I and K_D to zero and increase K_P until the loop oscillates. Subsequently, we set the value of K_P to half that value for a "quarter amplitude decay"-type response. Then, we increase the value of K_I until the offset can be eliminated. It should be noted that too large a K_I may cause instability. Finally, we increase K_D until the system response is stable. The overall adjustment process can be referred to in Fig. 4.26(1)–(9).

Ziegler-Nichols Method

As in manual tuning, the K_I and K_D gains are first set to zero. The proportional gain K_P is increased until it reaches the ultimate gain K_U, at which the output of the loop oscillates constantly. The oscillation period is T_U. We can preliminarily calculate the corresponding parameter values from Table 4.3.

Figure 4.27 depicts the closed-loop small-signal analysis model for the ADC-based DLDO. The parameter K_S represents the ADC gain, G(z) is the transfer function of the PID controller, ZOH(s) is the zero-order hold, and P(s) is the transfer function of the output stage.

The open-loop transfer function can be derived from the above model:

$$H(s) = Ks \times G_C(z) \times \text{ZOH}(s) \times P(s) \tag{4.9}$$

The ADC gain is

$$Ks = \frac{1}{\text{ADC resolution}} \tag{4.10}$$

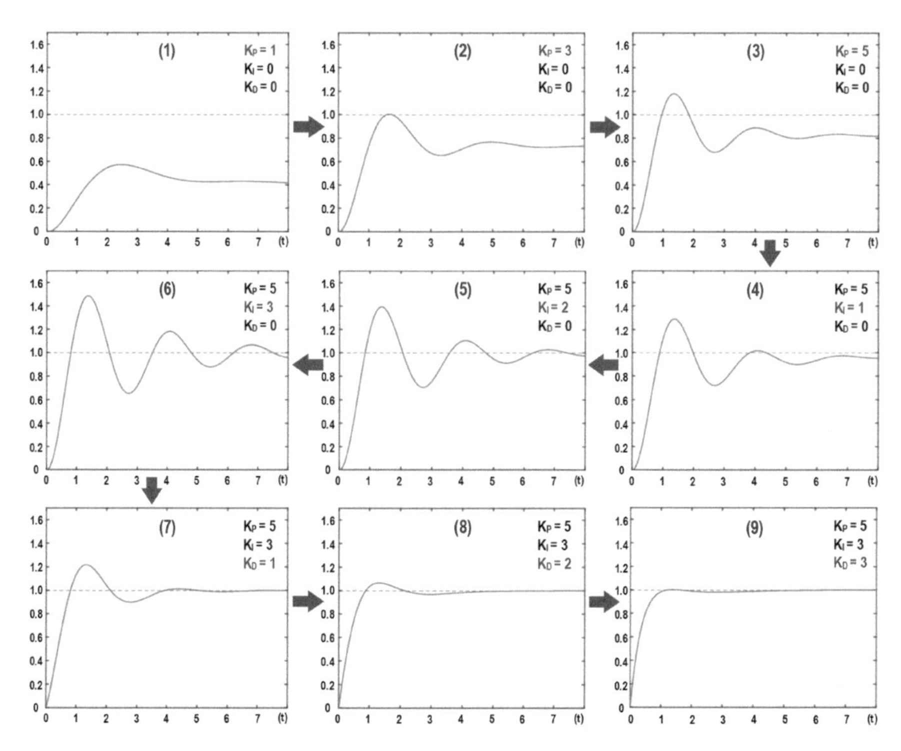

Fig. 4.26 Effects of varying PID parameters on the step response of a system, with (1)–(3) changing K_P, (4)–(6) changing K_I, and (7)–(8) changing K_D

Table 4.3 Ziegler-Nichols method

Control type	K_P	K_I	K_D
P	$0.5K_U$	–	–
PI	$0.45K_U$	$0.54K_U/T_U$	–
PID	$0.6K_U$	$1.2K_U/T_U$	$3K_UT_U/40$

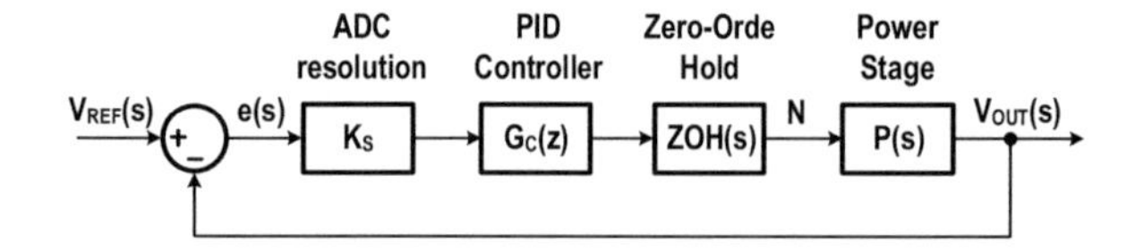

Fig. 4.27 Closed-loop small-signal analysis model of ADC-based DLDO

The transfer function of the PID controller is

$$G_C(z) = K_P + K_I \times \frac{z}{z-1} + K_D \times \frac{z-1}{z} \tag{4.11}$$

The transfer function of the ZOH is

$$\mathrm{ZOH}(s) = \frac{1 - e^{-Ts}}{sT} \tag{4.12}$$

For the digital power stage, I_{UNIT} is the unit current conducted through a single power unit:

$$P(s) = \frac{\partial V_{\mathrm{OUT}}}{\partial N} = \frac{I_{\mathrm{UNIT}} R_{\mathrm{O}}}{1 + sR_{\mathrm{O}} C_{\mathrm{L}}} \tag{4.13}$$

In addition to PID controllers, some other digital compensators can be employed in digital LDOs, such as a single-pole/single-zero (1P1Z) compensator [8] and a two-pole/two-zero (2P2Z) compensator [7, 11]. The architectures of these digital compensators are depicted in Fig. 4.28.

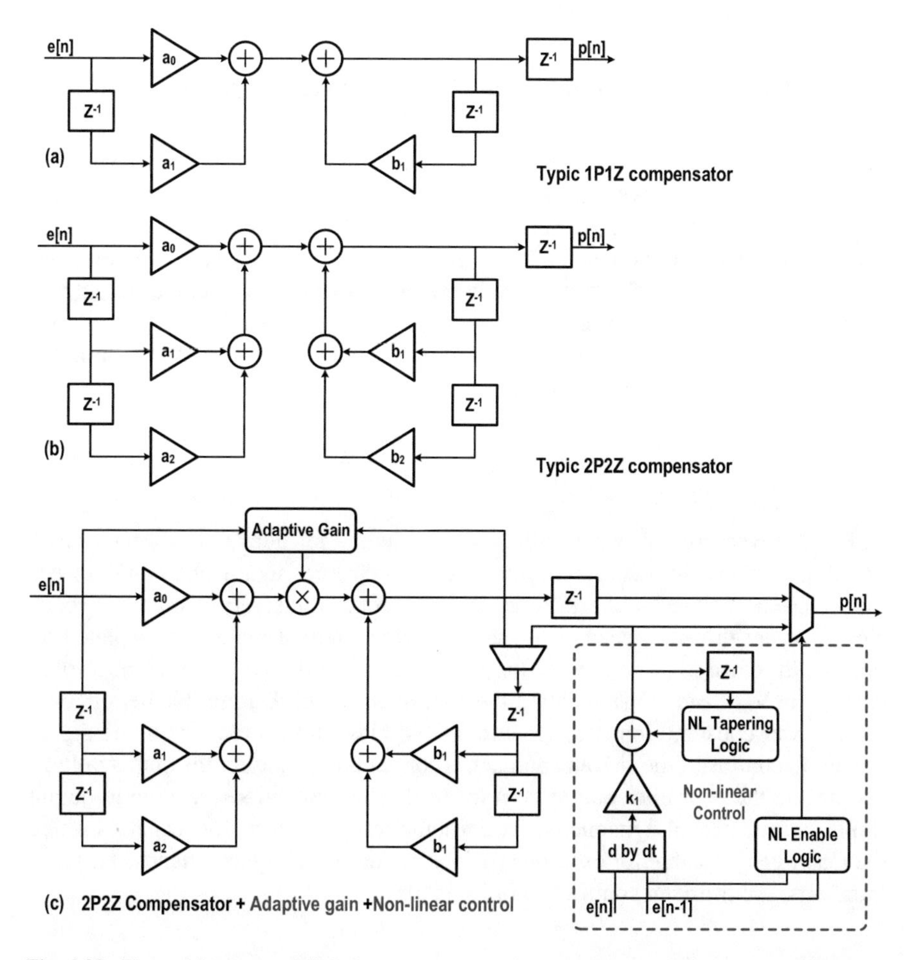

Fig. 4.28 The architectures of digital compensators, with (**a**) 1P1Z, (**b**) 2P2Z, and (3) 2P2Z with adaptive gain and nonlinear control

The 1P1Z compensator (Fig. 4.28a) introduces a pole and a zero. The integral pole is utilized to eliminate steady-state errors, and the zero compensates for the LDO's output pole, increasing the phase margin. The 1P1Z compensator can be approximated as a PI controller.

The 2P2Z compensator can provide two poles and two zeros to approximate a PID controller. In addition to the integral pole and the output capacitor compensation zero, as in the 1P1Z case, another zero is employed to increase the bandwidth, and a high-frequency pole is introduced for roll-off to suppress the effects of high-frequency noise. Compared to the 1P1Z compensator, the 2P2Z exhibits an improved transient response but is more complex to implement. Furthermore, digital compensators can be combined with specific algorithms or nonlinear control techniques to enhance the loop response and other performance metrics, as illustrated in Fig. 4.28c.

4.4 Event-Driven DLDO

4.4.1 Time-Driven Control and Event-Driven Control

Time-driven control is a synchronous control method that is commonly used in LDOs [11, 12]. It employs a fixed-frequency clock, and the circuit's operating status and control code value are refreshed once per clock cycle. When a load transient occurs, and the output voltage changes, the flash ADC or TDC can detect the voltage change within one clock cycle. Subsequently, the post-stage controller adjusts the control code value accordingly. As illustrated in Fig. 4.29a, for a very steep load transient, the delay of one clock cycle will result in a significant initial drop in the output voltage V_{OUT}, denoted as ΔV_{OUT_INI}:

$$\Delta V_{OUT\_INI} = \frac{\Delta T \times \Delta I_{LOAD}}{C_L} \tag{4.14}$$

Event-driven control is asynchronous, and its trigger does not rely on a clock [23–27]. Data updates occur only when the input changes significantly and exceeds the set threshold time. Unlike the time-driven approaches that rely on high-frequency clocks, event-driven mechanisms achieve very short latency, as depicted in Fig. 4.29b. Once V_{OUT} returns to the set value, event-driven observes no further changes in V_{OUT} and stops updating the controller output. During this period, only static power consumption (leakage current and bias current) remains in the digital circuit. In contrast, time-driven approaches continuously monitor the output voltage and update the controller output even in steady-state conditions, leading to output ripple and substantial dynamic power consumption. Therefore, for scenarios where the load remains stable for extended periods but occasionally experiences bursts of transients, event-driven control is more suitable.

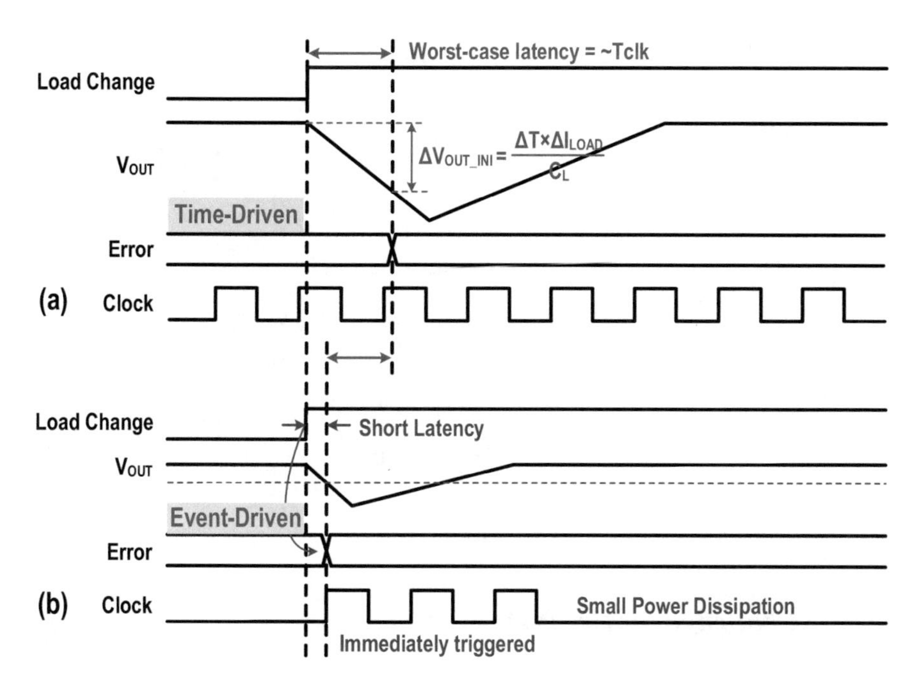

Fig. 4.29 Comparison between (**a**) time-driven control and (**b**) event-driven control

4.4.2 Continuous-Time Comparator

Event-driven systems necessitate continuous-time (CT) comparators. While traditional analog amplifier-based comparators can perform real-time comparisons, they operate at relatively high voltages, and their circuit design and calibration are complex. In contrast, inverter-based comparators, employing in [23–27], are simpler, can function at lower V_{DD} voltages, and support a wide input voltage range.

The circuit of the inverter-based CT comparator is depicted in Fig. 4.30, and it operates in two stages. The inverter serves as a high-gain operational amplifier. During the calibration stage, the input capacitor C_{IN} is connected to V_{REF}. The input and output of the inverter are short-circuited, biasing the input voltage to half of the power supply voltage V_{DD} ($V_{TRIP} = 0.5 \times V_{DD}$). At this point, the voltage bias, V_{OS}, between V_{REF} and V_{TRIP} is stored in C_{IN}. Due to potential charge leakage in C_{IN}, a specific calibration frequency is necessary. A larger capacitor can help reduce the calibration frequency. During calibration, the output value of the comparator is disregarded by the post-stage controller. Once calibration is complete, the comparison stage begins. In this stage, C_{IN} is connected to V_{OUT}, and the input and output of the inverter are disconnected. The inverter amplifies the voltage difference, $V_{OUT} - V_{REF}$, to produce a rail-to-rail output.

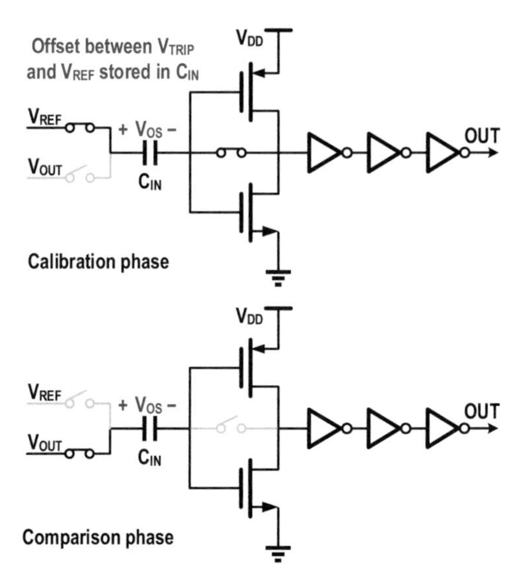

Fig. 4.30 Inverter-based CT comparator

4.4.3 Challenges of Event-Driven Control

There are two major challenges in implementing event-driven control: explicit-time coding and sticking problem.

Explicit-Time Coding

Take the integral term in a PID control as an example. The integral formula of traditional synchronous control is

$$u_I[K] = u_I[K-1] + K_I \times e[K] \tag{4.15}$$

Compared to the synchronous control, the asynchronous circuit in event-driven control is more complicated. In asynchronous control, whenever the output voltage V_{OUT} crosses a reference voltage level, the CT quantizer outputs a corresponding code value. Consequently, the time interval between adjacent code values is nonuniform, as depicted in Fig. 4.31.

We need to consider the nonuniform sampling time in the integral term. Therefore, the integral formula of asynchronous control needs to be changed to

$$u_I[K] = u_I[K-1] + K_I \times e[K] \times (t[k] - t[k-1]) \tag{4.16}$$

The asynchronous integral term requires two additional hardware: (1) a TDC that quantizes the sampling time length ($t[K]$-$t[K$-1]) and (2) an additional real multiplier that handles the nonuniform time intervals. Furthermore, to mitigate the impact of calibration time on sampling, additional logic circuits are required to process the quantizer output value. These supplementary designs will increase the complexity and cost of the circuit implementation.

Fig. 4.31 Conventional event-driven control

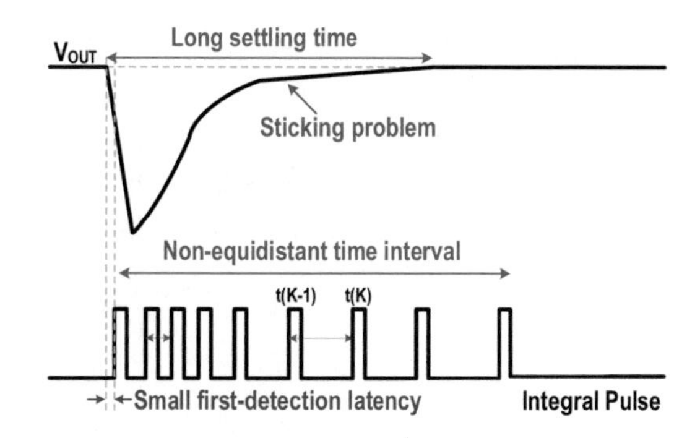

Sticking Problem

The sticking problem is another issue inherent to traditional asynchronous control. Since the trigger rate is related to the voltage change rate, when the output voltage is close to the set value V_{REF}, fewer samples are performed, thereby reducing the V_{OUT} adjustment speed, which further limits the generation of the trigger signal. Consequently, the controller cannot update the output, causing the output voltage V_{OUT} to change slowly or even stabilize at a non-set value, as illustrated in Fig. 4.31. To address this problem, the study [24] employs a watchdog counter and a timed trigger, but it increases the area and power consumption of the controller.

4.4.4 Event-Driven Self-Triggering Control

To solve the challenges in traditional event-driven control, Kim et al. [25, 27] proposed a technology that combines asynchronous triggering and synchronous operation.

Figure 4.32 shows the structure of the hybrid CT/DT quantizer. First, a steady-state region is set, and the CT comparator is used to detect whether the V_{OUT} voltage exceeds the upper and lower thresholds. Once either threshold is exceeded, DLDO activates the ring oscillator to generate a periodic clock signal, and DLDO will enter the synchronous working mode. Figure 4.33 demonstrates the waveforms of the self-triggering control.

During the output voltage recovery process, the introduction of the synchronous clock eliminates the need for time-interval quantization, simplifying the multiplier design. When V_{OUT} returns to the set zone region, the clock can be deactivated after waiting for several cycles, which helps the output voltage to quickly stabilize to the target value.

Compared to the traditional event-driven method, the detection time for the initial drop is significantly shorter, and static power consumption remains low when the clock is deactivated in a steady state. Even when compared to a quantizer that utilizes all CT comparators, Fig. 4.32 employs only two CT comparators. Additionally, since the DT comparator only has a static leakage current during steady state, the static power consumption of the hybrid self-triggering solution is

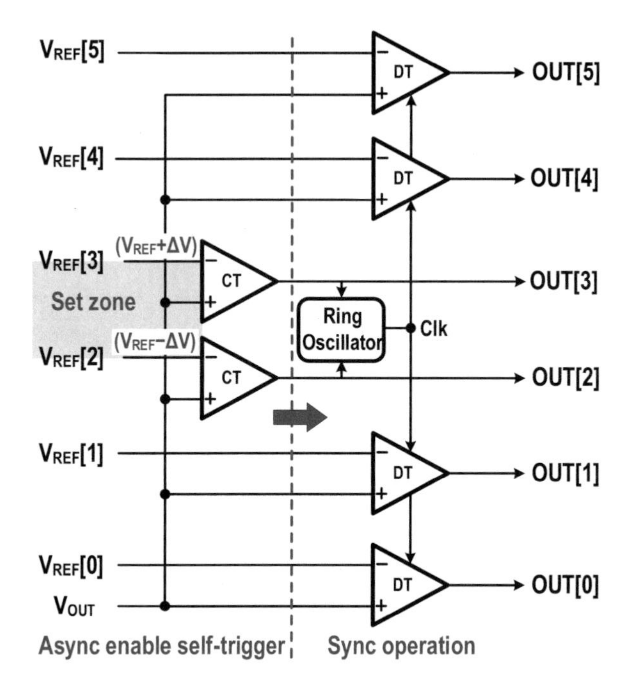

Fig. 4.32 The structure of CT and DT quantizer

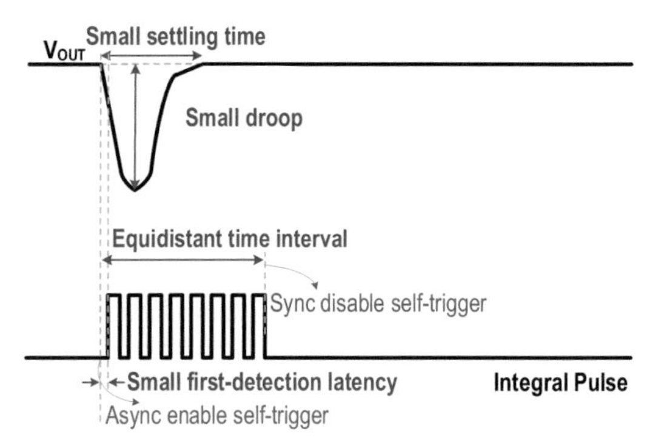

Fig. 4.33 The event-driven self-triggering control

further reduced. Self-triggering control preserves the advantages of traditional event-driven control while simultaneously addressing explicit-time coding and the sticking problem by using a synchronized clock during dynamic adjustments.

4.4.5 Parallelized PI Controller

Figure 4.34a introduces the block diagram of traditional PI control. P control is used for fast adjustment, while I control eliminates steady-state errors. The conventional approach sums the outputs of the P-part and I-part, subsequently controlling the

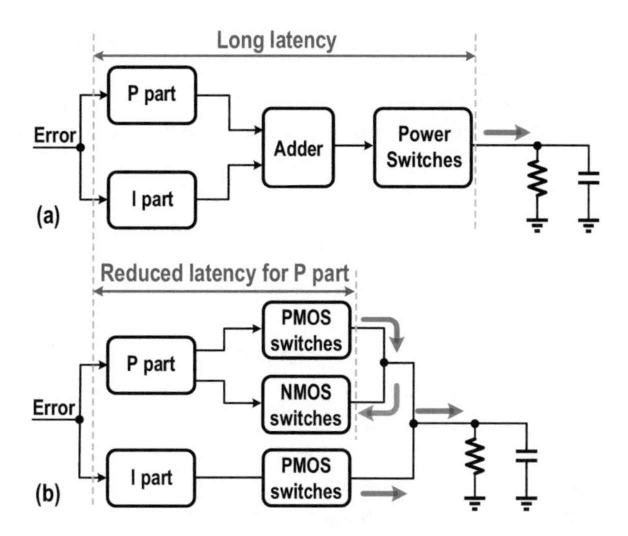

Fig. 4.34 (**a**) Traditional PI control and (**b**) parallelized PI control

power switch via a decoder. However, the adder and decoder operations are complex and introduce a certain amount of latency, which impacts loop response time.

The P control directly influences the transient performance of the DLDO. To minimize the latency from P control to the power stage, the author replaces the original adder and single power stage with two power stages. The P-part's power stage operates in binary form. During transients, the current provided by the P part and the current from the I part are added to obtain the total output current (as depicted in Fig. 4.34b). This current summation is realized through physical current addition rather than logic circuits, thereby shortening the control path and reducing the latency [24].

Furthermore, the P-part circuit employs a bit shifter instead of a real multiplier, further simplifying the circuit and reducing delay. When the output voltage has overshoots, an NMOS power stage generates the negative P-part value, effectively discharging the excess energy to ground through the NMOS, thereby reducing the overshoot on V_{OUT} and accelerating the overshoot recovery.

4.5 Computational DLDO

Most DLDOs work based on linear control technology. Some literatures have explored nonlinear control technique to improve transient performance [4, 6, 11]. In this section, we introduce two novel computational DLDOs. The first design in [28] utilizes a charge and discharge algorithm to compute the target value. The second method in [29] employs time-based exponential control and a slope detector to determine the control code quickly.

When a transient event occurs, a traditional DLDO adjusts the control code value based on the difference between the output voltage and the target voltage. In contrast, computational DLDOs control the circuit by considering the difference between the output current and the load current. These two approaches have distinct control strategies. An essential consideration in the design of computational DLDOs is to detect the load value directly or focus on the relative difference between the output current and the load current.

4.5.1 Charge and Discharge Algorithm

Linear control DLDOs dynamically adjust the control code value of the power switches according to the V_{OUT} changes. In [28], the authors calculate the amplitude of the load step through the charging and discharging processes and directly provide the corresponding code value. Figure 4.35 illustrates the block diagram of the computational DLDO presented in [28]. Several thresholds are set through multiple comparators. For small load transients, the shift register loop adjusts automatically. For larger load transients, when the undershoot and overshoot of the output voltage exceed the set threshold, the corresponding calculation process is triggered.

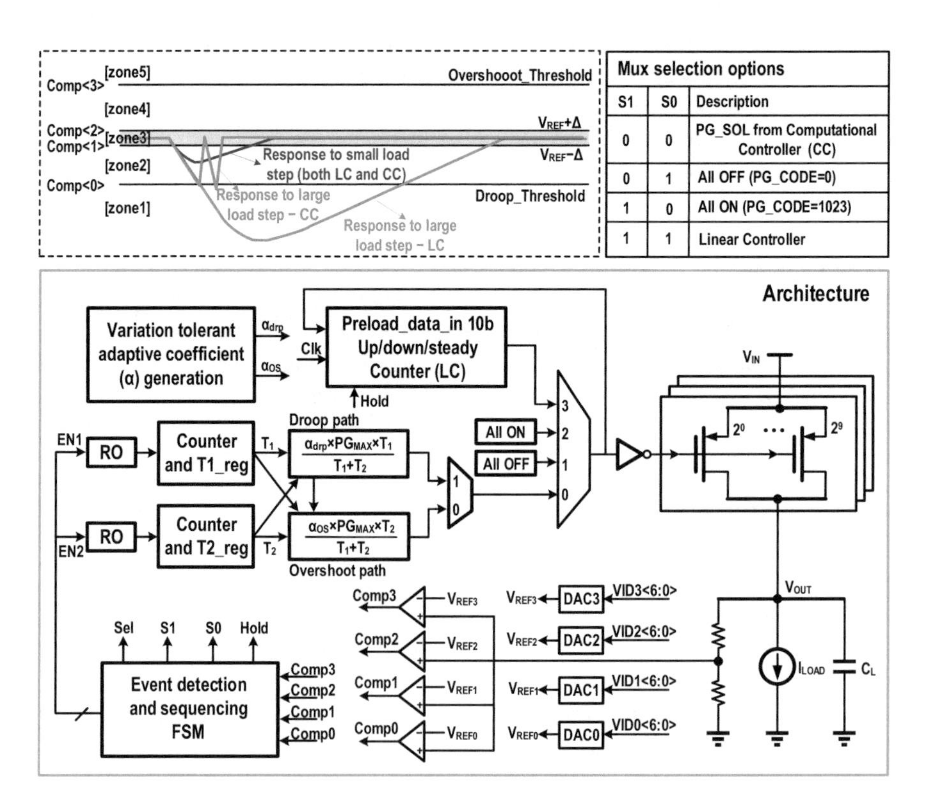

Fig. 4.35 The block diagram of the computation DLDO structure in [28]

As depicted in the upper right corner of Fig. 4.35, the system is divided into four working states according to the indication signals S1 and S0: 00 represents the calculated code value taking over control; 01 represents all power switches being turned off; 10 represents all power switches being turned on; and 11 represents the DLDO being in linear control.

Load Step-Up

Figure 4.36 shows the timing diagram of the operation principle during load step-up. The control process is divided into six stages. In the first stage, a load transient occurs, and the V_{OUT} voltage decreases. In the second stage, when V_{OUT} is lower than the comp<0> threshold, S1 = 1, S0 = 0, all power switches are turned on, the output current is greater than the load current, and the output voltage rises. A counter obtained the rise time T_1. According to the charge balance equation, we have

$$Q_1 = V_{DRP} \times C_{OUT} = \left(I_{MAX} - I_{LOAD}\right) \times T_1 \tag{4.17}$$

In the third stage, when V_{OUT} rises to the threshold set by COMP<1>, S1 = 0 and S0 = 1, turning off all power switches. As the load current discharges, the output

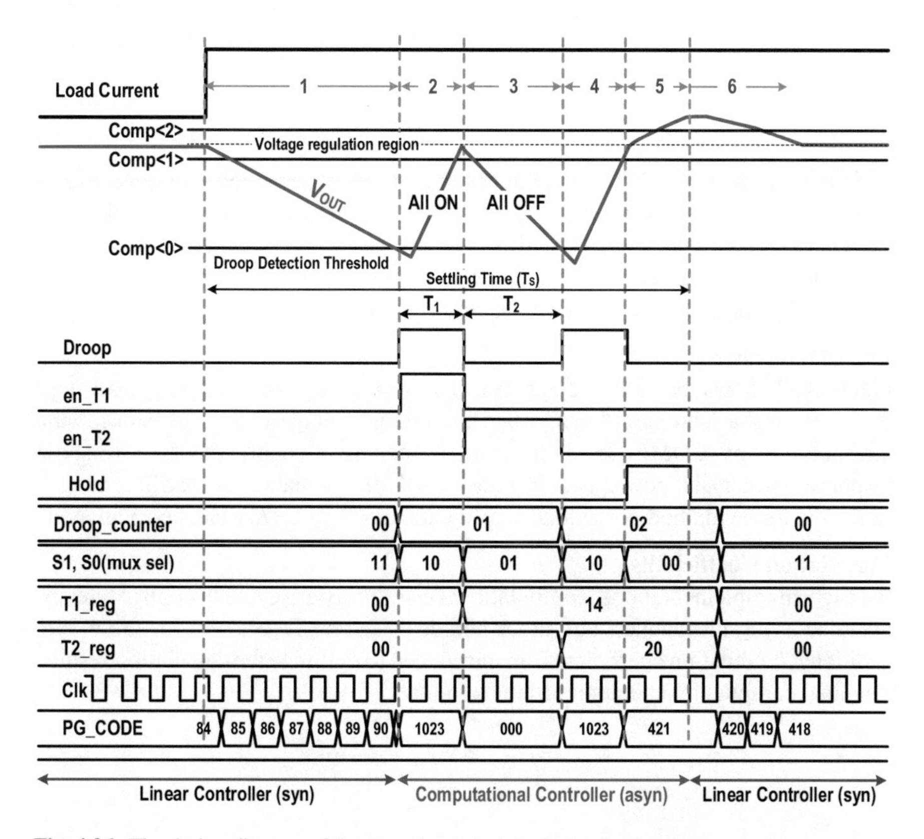

Fig. 4.36 The timing diagram of the operation principle during load step-up

voltage drops. When the voltage decreases to the threshold set by comp<0>, the fall time T_2 is recorded:

$$Q_1 = V_{\mathrm{DP}} \times C_{\mathrm{OUT}} = I_{\mathrm{LOAD}} \times T_2 \tag{4.18}$$

At the end of the third stage, we can calculate the number of power switches that should be turned on. I_{UNIT} is the current of a single power switch, N_{MAX} is the total number of all power switches, and K_{DRP} is the number of power switches that need to be turned on (match the load current I_{LOAD}). We have

$$I_{\mathrm{MAX}} = N_{\mathrm{MAX}} \times I_{\mathrm{UNIT}} \tag{4.19}$$

$$I_{\mathrm{LOAD}} = K_{\mathrm{DRP}} \times I_{\mathrm{UNIT}} \tag{4.20}$$

$$\left(N_{\mathrm{MAX}} - K\right) \times I_{\mathrm{UNIT}} \times T_1 = K \times I_{\mathrm{UNIT}} \times T_2 \tag{4.21}$$

According to Eqs. (4.17), (4.18), (4.19), (4.20), and (4.21), the target K value can be calculated:

$$K_{\mathrm{DRP}} = N_{\mathrm{MAX}} \times \frac{T_1}{T_1 + T_2} \tag{4.22}$$

$$K_{\mathrm{OS}} = N_{\mathrm{MAX}} \times \frac{T_2}{T_1 + T_2} \tag{4.23}$$

In the fourth stage, all power switches are turned on, resulting in a rapid rise of V_{OUT}. In the fifth stage, when V_{OUT} rises to the threshold set by COMP<1>, the control code K takes effect and is held for two cycles. In the sixth stage, the circuit enters the linear regulation mode (shift register mode) for fine adjustment. Finally, the output voltage stabilizes within the target region.

Load Step-Down

Figure 4.37 shows the timing diagram of the operation principle during load step-down. During a load step-down, when the output voltage overshoot exceeds the threshold set by COMP<3>, it triggers the computation process for overshoot response. Initially, all power switches are turned off, and then they are all turned on. Referring to the method in the load step-up, the target value K_{OS} can be calculated.

Modulation Coefficients

In practical implementations, computational errors may arise due to control latency, manufacturing mismatches, and other nonideal factors. To account for these, two additional modulation coefficients are introduced to optimize the calculation results. The final computational equations are as follows:

$$K_{\mathrm{DRP}} = \alpha_{\mathrm{DRP}} \times N_{\mathrm{MAX}} \times \frac{T_1}{T_1 + T_2} \tag{4.24}$$

$$K_{\mathrm{OS}} = \alpha_{\mathrm{OS}} \times N_{\mathrm{MAX}} \times \frac{T_2}{T_1 + T_2} \tag{4.25}$$

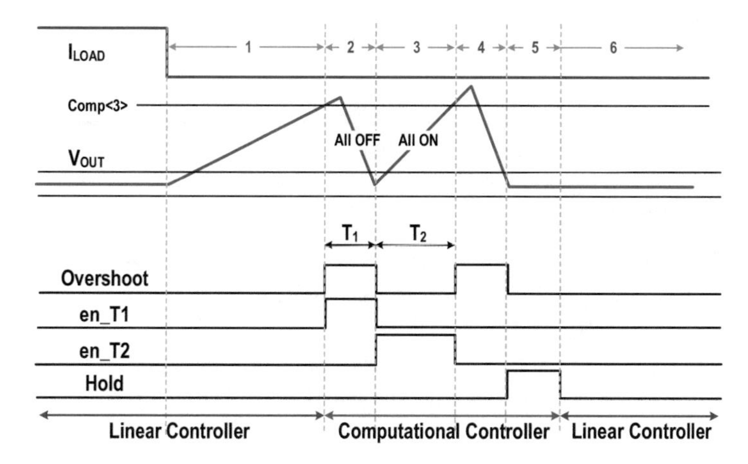

Fig. 4.37 The timing diagram of the operation principle during load step-down

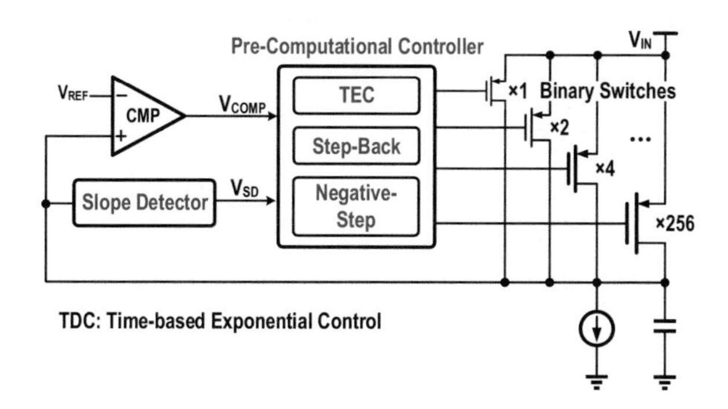

Fig. 4.38 The block diagram of the computational DLDO in [29]

4.5.2 Time-Based Exponential Control

In their work, Lee et al. [29] proposed a time-based exponential control (TEC) method. Figure 4.38 illustrates the overall architecture of the computational DLDO. The DLDO comprises a main comparator, a slope detector (SD), a pre-computational controller, and an 8-bit binary-weighted PMOS switch array. The main comparator compares the output voltage with the reference voltage, while the slope detector is used to compare the load current with the output current. The pre-computational controller involves three components: time-based exponential control, step-back control, and negative-step control.

In the conventional shift register control, power switches are adjusted one by one, as depicted by the blue waveform in Fig. 4.39. For an N-bit binary power stage, it may require up to 2^N steps to reach the target code value, resulting in a slow transient response. Contrastingly, the red waveform in Fig. 4.39 illustrates the time-based exponential control proposed in [29]. When the load current increases or decreases, the control code increments twice as much as in the previous cycle. During undershoot or overshoot conditions, the TEC exponentially increases in

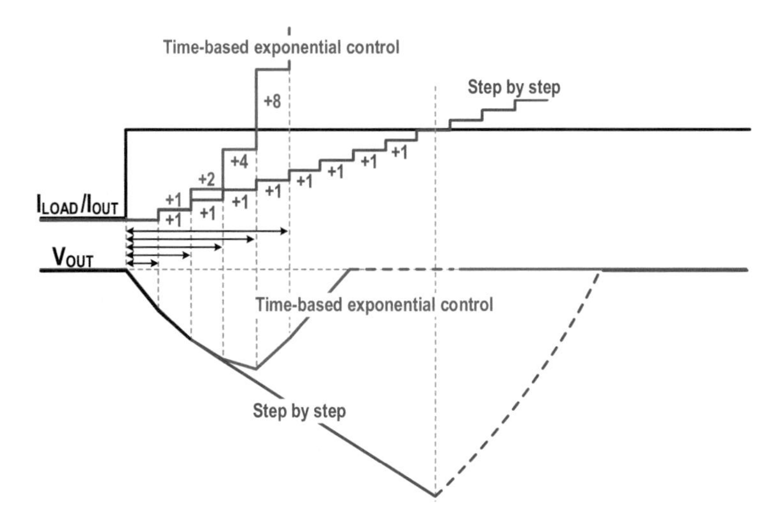

Fig. 4.39 Time-based exponential control (red line) and step-by-step regulation (blue line)

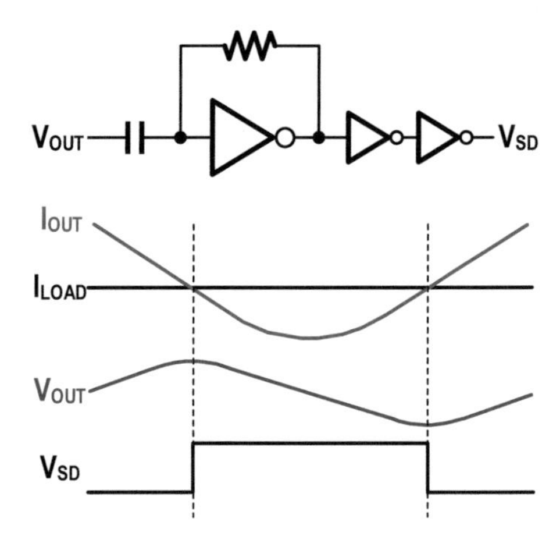

Fig. 4.40 The circuit and working principle of the slope detector

each cycle, allowing I_{OUT} to track I_{LOAD} within N clock cycles. Here, N represents the binary bits for the power switches.

When the output current exceeds the load current, the output voltage begins to rise. At this point, the output voltage remains lower than V_{REF}. If the control code continues its previous exponential increase, the output current could significantly surpass the load current, resulting in a large voltage overshoot. To address this problem, the system employs a slope detector to adjust the control method.

Figure 4.40 illustrates the circuit and working principle of the slope detector. SD comprises a V_{OUT} coupling capacitor, an inverter, a connecting resistor, and an output buffer. When V_{OUT} increases, it indicates that I_{OUT} exceeds I_{LOAD}, and the output V_{SD} is 0. Conversely, when V_{OUT} decreases, it signifies that I_{OUT} is less than I_{LOAD}, causing the output V_{SD} to change to 1. Therefore, SD serves as both a slope detector for V_{OUT} and a comparator for the output current relative to the load current.

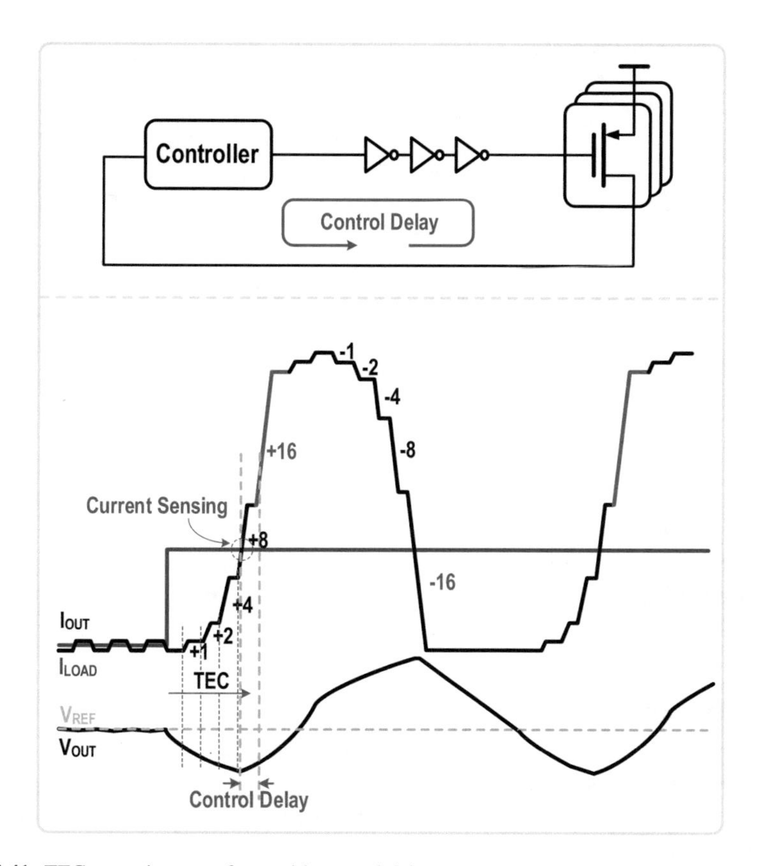

Fig. 4.41 TEC operation waveform with control delay

When the voltage undershoots, and the output current is less than the load current ($V_{COMP} = 0$, $V_{SD} = 1$), the system operates in TEC fast mode. Upon detecting that the output current exceeds the load current ($V_{COMP} = 0$, $V_{SD} = 1$), the output voltage enters the recovery stage, and the system should enter the slow SR regulation mode. However, since V_{OUT} remains below the target value, the output current continues to rise. Furthermore, due to control loop delays, the mode may not switch immediately after detecting $V_{SD} = 1$, resulting in the next cycle potentially remaining in TEC working mode. Consequently, the output current can significantly increase, leading to a pronounced LCO phenomenon, as depicted in Fig. 4.41.

To address this issue, Lee et al. [29] introduced a pre-computational controller based on the TEC method. This controller incorporates step-back and negative-step control, mitigating the impact of control delay to enhance system stability.

The working principle is illustrated in Fig. 4.42. When the system exits the TEC working mode, the control code value steps back to the previous cycle, reducing the current overshoot and minimizing the influence of control delay. However, only step-back control cannot completely eliminate existing overshoots because the

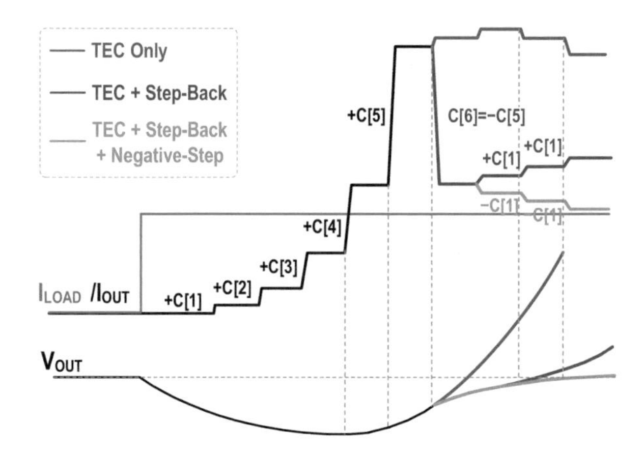

Fig. 4.42 TEC operation waveforms with step-back and negative-step control

output current already surpasses the load current. The negative-step control can gradually reduce the control code value by −1 per step until the output current aligns with the load current. By using step-back and negative-step control, DLDO mitigates the impact of control delay and enhances system stability.

4.6 Hybrid LDO

Due to the different control methods of analog LDO and digital LDO, there are differences in some characteristics between these two. For example, the output voltage quality (output noise/ripple and PSR) of analog LDOs is better than that of digital LDOs. However, digital LDOs can benefit from process scaling. Table 4.4 provides a qualitative comparison of the characteristics of general analog LDOs and digital LDOs, with blue font indicating better or more advantageous features. Both control methods have inherent advantages and disadvantages. In specific scenarios, a hybrid structure that combines analog LDOs and digital LDOs (as shown in Fig. 4.43) may achieve good overall performance [30–40].

In the following subsections, we will introduce several representative hybrid LDO structures developed in recent years.

4.6.1 Passive Analog-Assisted Digital LDO

Analog control can provide continuous-time immediate response. Passive analog-assisted (AA) technology typically involves using passive capacitor devices to introduce dynamic information of the output voltage into the power stage of digital LDO, enhancing the transient performance.

Table 4.4 The comparison of the characteristics of analog LDOs and digital LDOs

Topology	Analog LDO	Digital LDO
Regulation Fineness	Continuous (Voltage)	Discrete (Number)
Quantization Error	No	Yes
Ripple	No	Yes
PSR	Good	Poor
Self-heating and EM	Low	High
Input Voltage	Medium to High	Low
Gate Pole	Yes	No
Distributed Power Transistors	No	Yes
Benefits from Technology Scaling	Fewer	More
Current Density	Low	High

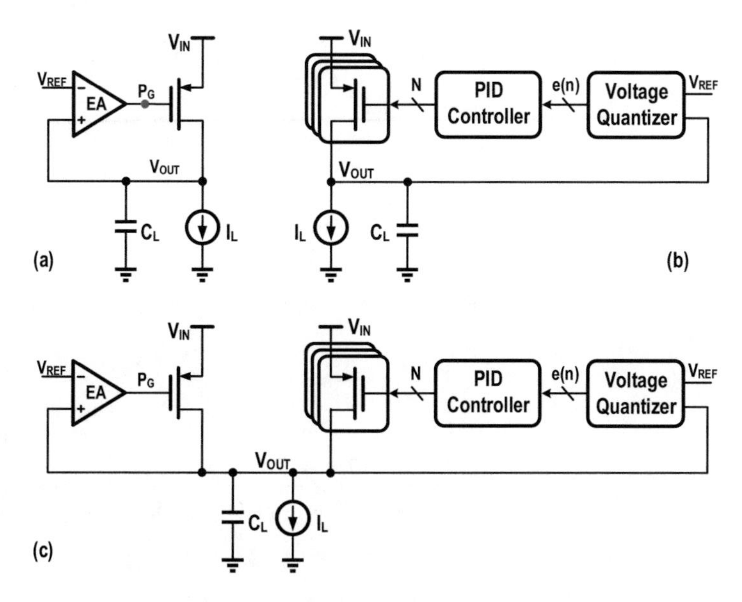

Fig. 4.43 (**a**) An analog LDO, (**b**) a digital LDO, and (**c**) a hybrid LDO

Figure 4.44 illustrates a classic passive analog-assisted technique in [22, 23]. The baseline is a classic shift register control loop. The NMOS of the gate driver connects to the ground (GND) through a large resistor, R_C. In addition, a coupling capacitor C_C links the V_{SSB} and V_{OUT} nodes. During steady state, the gate voltage of the "ON" power switches remains biased to GND ($V_{SSB} = 0$ V) via the resistor R_C.

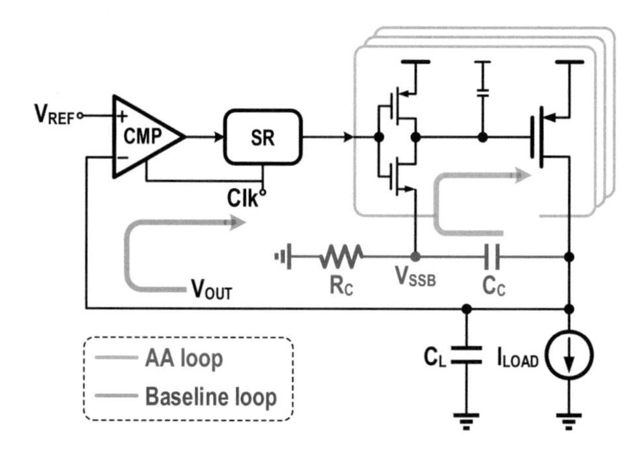

Fig. 4.44 The passive analog-assisted scheme in [22]

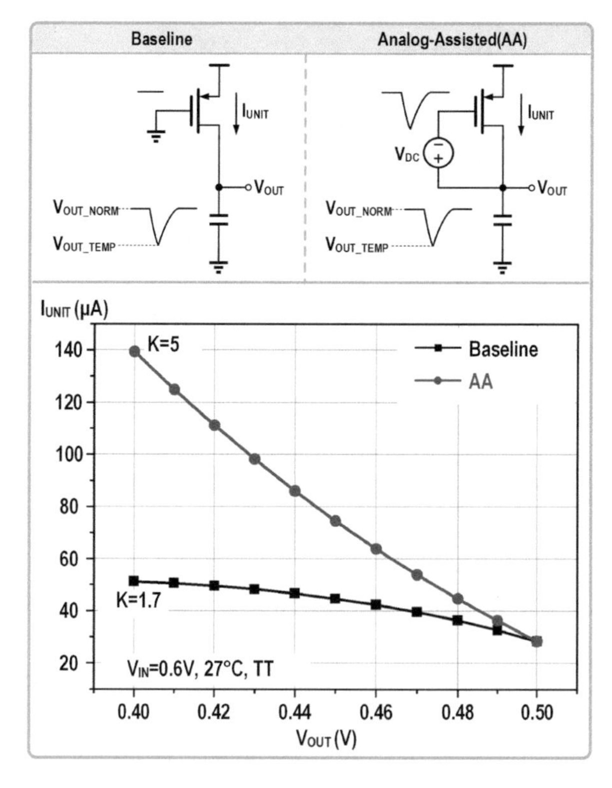

Fig. 4.45 The instantaneous current comparison

However, when a load transient occurs, the V_{OUT} droop couples to the V_{SSB} node through the capacitor C_C, resulting in a negative bias voltage for V_{SSB}. Consequently, the V_{GS} of the "ON" power switches increases instantly, significantly boosting the unit current I_{UNIT}.

Figure 4.45 compares the power switch behavior between the baseline and AA scheme. In the baseline design, a transient-caused V_{OUT} drop only affects the V_{DS} of

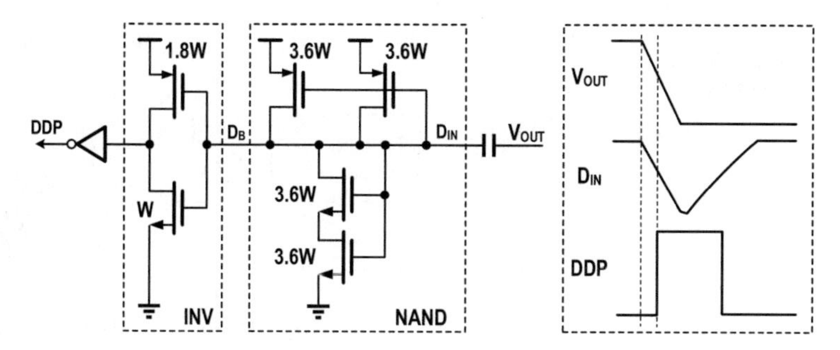

Fig. 4.46 Droop detection circuit in [5]

the power switches, impacting I_{UNIT}. In contrast, the V_{OUT} transient drop in the AA scheme increases both V_{GS} and V_{DS} of the power switches, leading to a significant increase in unit current and improved transient performance. We define K as the ratio of the maximum instantaneous current to the steady-state current during a transient. When $V_{IN} = 0.6$ V and $V_{OUT} = 0.5$ V, the baseline I_{UNIT} experiences only a 1.7-fold instantaneous increase, while the AA scheme achieves a 5-fold increase.

The previously mentioned AA loop utilizes capacitor coupling to convey dynamic information from V_{OUT}, forming an analog loop. Additionally, this coupling information can be amplified and shaped to generate a digital pulse signal for nonlinear digital control. Figure 4.46 illustrates a droop detection circuit from [5]. The detection circuit employs standard logic units. The coupling capacitor transmits voltage droop information to the input of the droop detector. Notably, the transistor width ratio of the NAND gate exceeds that of the INV gate. In a steady state, the detector's output (DDP) remains at 0. However, during transient states, the V_{OUT} droop causes the input voltage (D_{IN}) to decrease, leading to an increase in its output (D_B) and, ultimately, a change in the detector output DDP to 1.

It should be noted that NAND and INV consume some static current, and the input threshold voltage of NAND varies with process, voltage, and temperature, which may affect the quiescent current and transient response performance.

4.6.2 Active Analog-Assisted Digital LDO

Passive analog-assisted techniques only provide dynamic information during transient states, whereas active analog-assisted techniques utilize analog operational amplifiers (op amps) and incorporate analog LDOs or analog working states to enhance the performance of digital LDOs.

Figure 4.47 illustrates the structure diagram of an active analog-assisted DLDO proposed in [24]. The digital DLDO employs a coarse-fine switched-capacitor (SR) structure. The SR control is a digital integral control, which offers high gain and precision at low voltages but exhibits relatively slow transient response. Furthermore, due to the digital power transistor operating in the linear region and

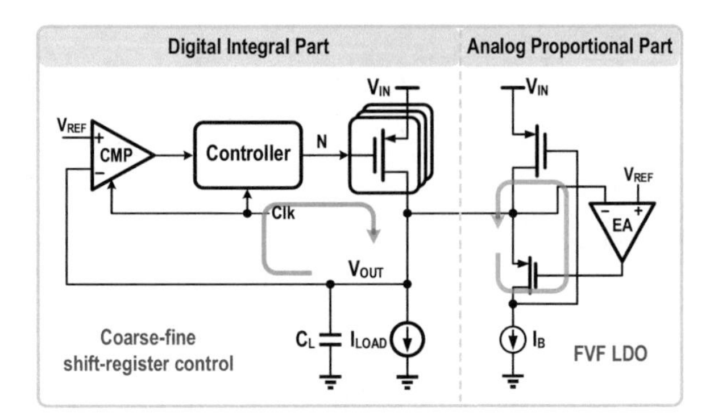

Fig. 4.47 An analog-proportional and digital-integral hybrid LDO in [24]

its discontinuous working states, the PSR is relatively poor. To address this, Kim et al. [24] proposed augmenting the digital LDO with an analog flipped-voltage follower (FVF) LDO. Since the FVF structure has a wide bandwidth and requires only $3 \times V_{DS}$ or $V_{GS} + V_{DS}$ voltage headroom, it maintains gain at low VIN voltages, serving as a fast-response analog proportional loop. Thus, the structure depicted in Fig. 4.47 constitutes a hybrid LDO with analog-proportional and digital-integral components.

After connecting the two in parallel, the analog LDO can share a portion of the current. The power transistors of the FVF LDO can be configured to operate in saturation mode, with a wider loop bandwidth, which can improve the overall PSR. In addition, in the steady state, the output current (I_A) of the analog LDO dynamically adjusts to compensate for the digital LDO's output current (I_D), reducing the overall output current ripple (I_{OUT}) and the limit cycle oscillation. In transient scenarios, digital circuits often introduce a delay due to clock cycles for detection. For load transients with steep rising edges, a significant voltage droop can occur during this delay. However, the fast analog feedback loop can continuously monitor the V_{OUT} and rapidly increase the analog LDO's output current, minimizing transient voltage droop.

Figure 4.48 also illustrates an example of using AA techniques to improve the PSR of a DLDO [25]. In this structure, N identical digital PMOS units and $N + 1$ identical analog LDO blocks are employed. Each unit can independently support a maximum load current I_L. One ALDO is always active to provide a fraction of the current smaller than that of an individual DLDO block, thereby minimizing output voltage ripple.

The ALDO has its main poles as output poles and uses an R-C feedforward path to improve PSR. More analog LDO tiles means better PSR. The modular hybrid LDO (MHLDO) can be configured either as a pure ALDO, meeting the PSR requirements with high I_Q, or as a pure DLDO, minimizing the power overhead for digital or memory domains. By controlling the combination of active ALDO/DLDO blocks, the PSR is satisfied while minimizing quiescent current I_Q.

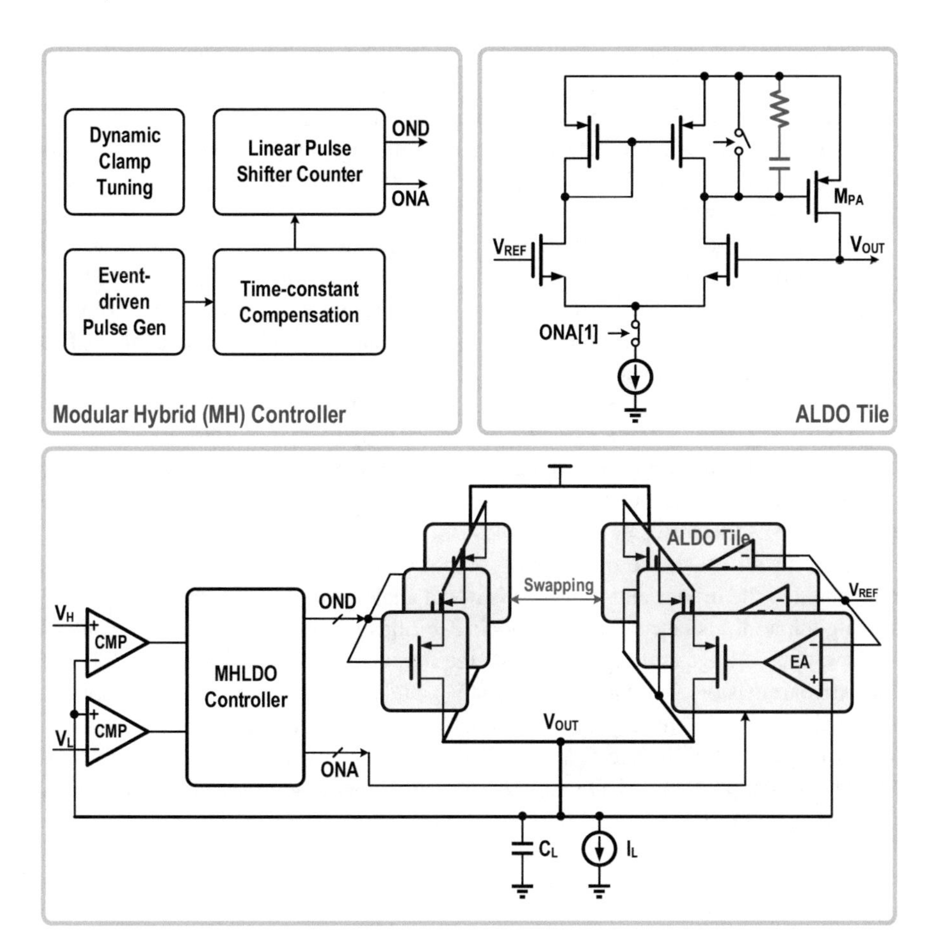

Fig. 4.48 A modular hybrid LDO with programmable PSRR in [25]

4.6.3 Digital-Assisted Analog LDO

Analog LDOs exhibit excellent steady-state characteristics, and some low-current analog LDOs have achieved commendable transient performance. For rapid load changes, the voltage drop at the output can be expressed as

$$\Delta V_{\mathrm{OUT}} = \frac{\Delta I_{\mathrm{L}}}{C_{\mathrm{L}}} \times \left(t_{\mathrm{BW}} + t_{\mathrm{SR}}\right) \tag{4.26}$$

where ΔV_{OUT} is the output voltage change, ΔI_{L} represents the amplitude of the load step, C_{L} is the output capacitance, t_{BW} is the response time associated with the loop bandwidth, and t_{SR} denotes the response time related to the slew rate.

Achieving fast transient response requires a wide loop bandwidth and a high slew rate (fast charging and discharging of internal nodes). To provide a large output current requires a larger power transistor. In high-bandwidth designs, pushing the gate pole of the power transistor to high frequencies becomes challenging. Additionally, the equivalent parasitic poles of the driver itself can impact loop stability. Therefore, designing an analog LDO with high current and fast transients presents great challenges. Typically, fully integrated analog LDOs have load capacities of below 150 mA.

Digital circuits are compact, exhibit low steady-state power consumption, and have high dynamic slew rates. Moreover, with process advancements, the advantages of digital technology become even more pronounced. Therefore, exploring digital techniques to enhance the transient performance of high-current analog LDO designs is a promising avenue [26, 27].

Figure 4.49 illustrates the structural block diagram of the digital-assisted ALDO from [26]. It comprises three parts: (1) an analog loop for steady-state operation, (2) a digital loop for responding to large transient steps, and (3) a finite-state machine controller to decide the LDO's control mode. Unlike conventional digital LDOs, the gate voltage of the active power switches is not 0 V, but V_A or V_B.

In the steady state, the digital loop output code is frozen, the EA output V_A is connected to the driver, and the LDO is in the analog loop hold state (S[1:0] = 10). During a load transient, the digital loop's 3-bit flash ADC quickly detects changes in the output voltage and rapidly adjusts the output code value. The gate voltage of the active transistors switches to V_B ($V_B < V_A$), and the LDO enters in digital control state (S[1:0] = 10). The digital loop's coarse adjustment prevents further drops in V_{OUT} voltage and stabilizes it near V_{REF}. When the output voltage stops dropping and LCO ripples appear, the LDO enters analog adjustment mode (S[1:0] = 01). Then, the ADC's clock is gated, and the output code value is frozen. The high-gain EA finely adjusts the value of V_A to equalize V_{OUT} with V_{REF}. Finally, the LDO returns to the steady state (S[1:0] = 10). The working timing diagram during a transient is shown in Fig. 4.50.

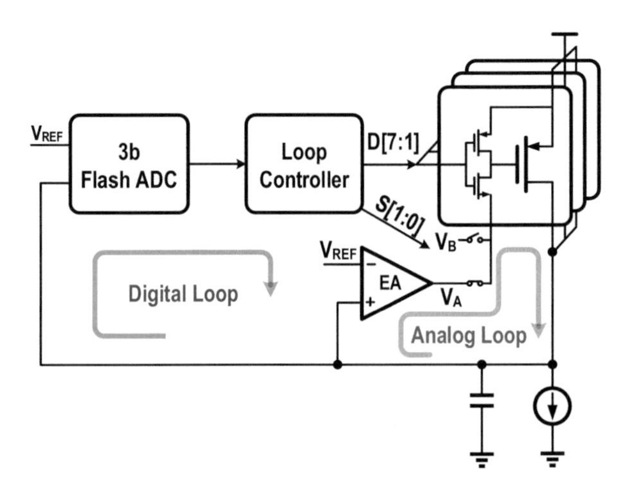

Fig. 4.49 The digital-assisted analog LDO in [26]

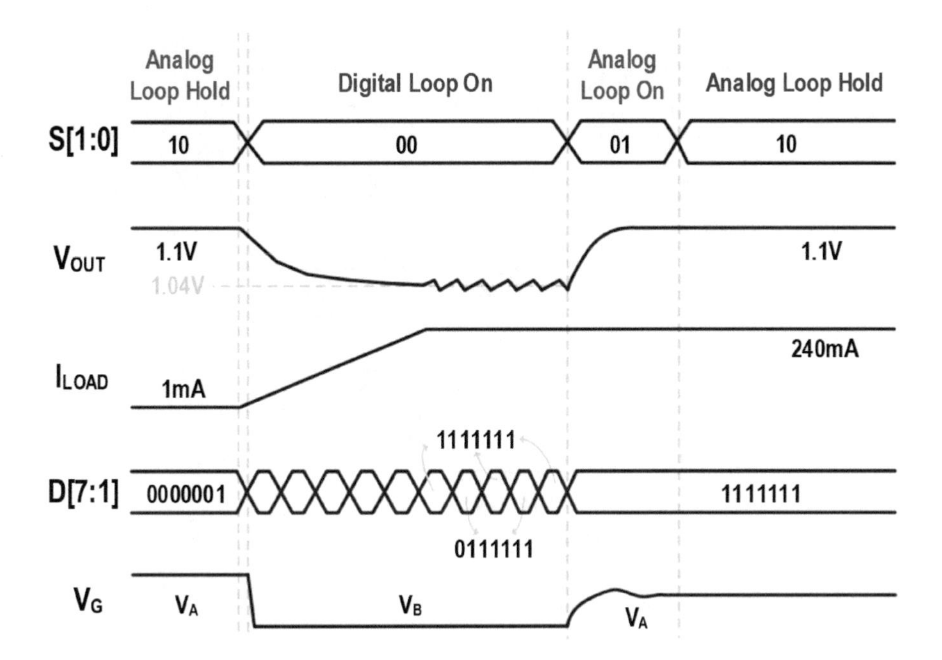

Fig. 4.50 The timing diagram of the digital-assisted analog LDO for a load step

In the structure in [26], the transient response speed of the LDO depends on the digital loop, which accordingly reduces the bandwidth requirements for the analog loop. This structure shares similarities with the coarse-fine digital LDO; however, the distinction lies in the fine adjustment employing analog control.

4.6.4 Analog-Digital Merged Control LDO

Digital power stages have no gate pole and support distributed integration, making them suitable for high-current fully integrated applications. Although digital LDOs exhibit some output ripple due to limit cycle oscillation, the digital loads are not sensitive to the small output ripple. However, digital LDOs also have drawbacks. Firstly, the output accuracy of traditional digital LDO is limited by the quantization resolution. Increasing the bit of the ADC leads to proportional increases in power consumption, area, and design complexity of the quantizer and PID controller. Secondly, ADCs often require calibration, further adding to the design complexity.

In contrast, analog amplifiers can easily achieve good accuracy through cross-matched transistors. And analog LDOs can obtain high DC accuracy using high-gain error amplifiers (EAs). In SoC systems, while the power input voltage can be as low as 0.6 V, the analog circuits can utilize a stable 1.8 V analog power supply to ensure optimal performance. Therefore, we can combine the advantages of analog

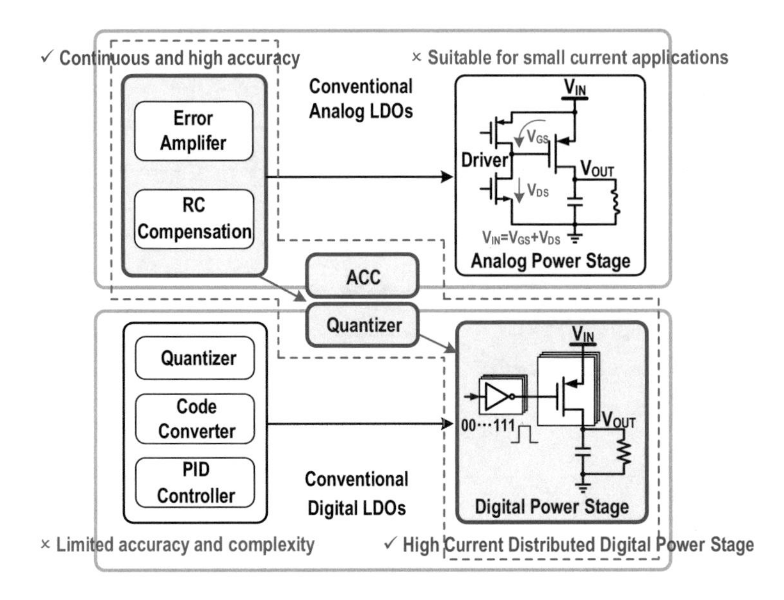

Fig. 4.51 The block diagram of analog LDO and digital LDO

control with a digital power stage and present an analog-digital merged control LDO (Fig. 4.51).

Figure 4.52 shows the structure of the analog-digital merged control LDO proposed in [28]. It consists of an EA with an RC compensation network, a voltage buffer, a 5-bit time-to-digital converter (TDC), and the digital power stage. The EA amplifies the error information between VOUT and VREF. Its output has a voltage buffer to enhance the driving capability and prevent kickback noise. The TDC quantizes the buffered EA output and directly drives the digital power stage.

Figure 4.53 depicts the operating principle of this circuit. In the steady state, the EA output gradually increases with the load, causing the TDC output code to increment and thus increasing the output current to stabilize the output voltage. During rapid load transients, the EA output rises quickly, leading to corresponding rapid changes in the TDC output and a swift increase in output current.

Although the TDC exhibits good linearity, it is sensitive to PVT and frequency variations. In the structure described in [28], the TDC is within the feedback loop, and the EA output can adjust to an appropriate operating point based on TDC variations. Similar to analog LDOs, variations in the driving MOS transistor do not affect the output voltage accuracy (as shown in Fig. 4.54).

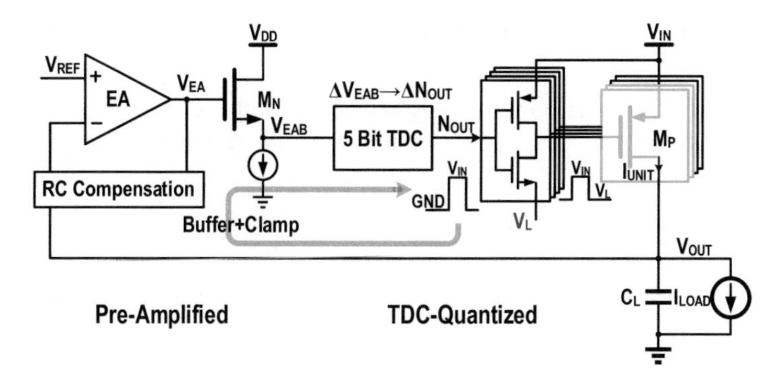

Fig. 4.52 The block diagrams of the analog-digital merged control LDO in [28]

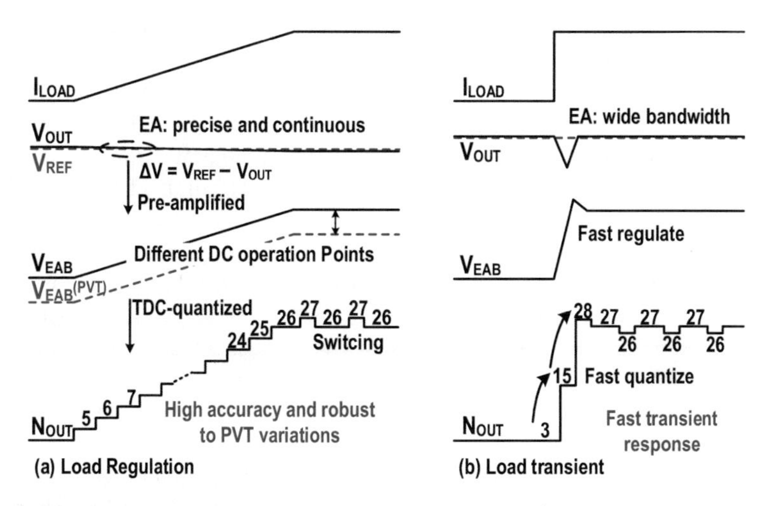

Fig. 4.53 The operation principle of the hybrid LDO in [28]

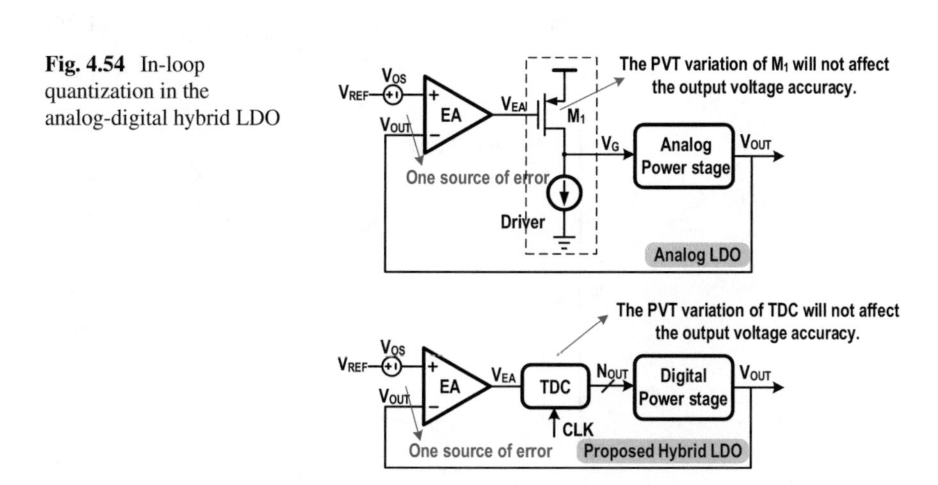

Fig. 4.54 In-loop quantization in the analog-digital hybrid LDO

4.7 Stability and Reliability Issues

As mentioned above, the power transistors of the digital LDO work in the linear region, and the unit current I_{UNIT} flowing through each switch can be expressed as

$$I_{\text{UNIT}} = \frac{1}{2}\mu_n C_{\text{OX}} \frac{W}{L}\left[2\left(V_{\text{GS}} - V_{\text{TH}}\right)V_{\text{DS}} - V_{\text{DS}}^2\right] \tag{4.27}$$

where μ_n is the mobility of the charge carriers; C_{OX} is the gate-oxide capacitance per unit area; and W and L are the width and length of the power transistor, respectively. $V_{\text{DS}}=V_{\text{IN}}-V_{\text{OUT}}$ is the voltage difference between the power transistor's drain and source terminals. The I_{UNIT} exhibits a linear relationship with V_{DS}. As the drain-source voltage difference V_{DS} increases, I_{UNIT} will increase significantly. For certain applications with wide input-output ranges, I_{UNIT} may vary by more than tenfold, as depicted in Fig. 4.55.

In addition, in digital systems, the demand for supply voltage level depends on the load performance demand, i.e., the digital load current is directly proportional to its supply voltage. When $V_{\text{IN}} = 1$ V and $V_{\text{OUT}} = 0.5$ V, the power transistor strength is maximized, but the load current demand is very low. The digital LDO can provide current dozens of times greater than the load demand.

All current and heat in an analog LDO are dispersed across all power transistors, whereas the current and heat in a digital LDO are concentrated on parts of the power transistors. Consequently, the reliability issue under large V_{DS} voltage must be considered for digital LDOs. Figure 4.56a compares the heat and current distribution of the LDO power switches for the same load current, with $V_{\text{OUT}} = 0.95$ V and $V_{\text{OUT}} = 0.5$ V. With a large voltage difference, the heat and output current of the digital LDO become more concentrated.

For the same load current, the first power switch of the digital LDO is always on. Assuming $V_{\text{IN}} = 1$ V and $V_{\text{OUT}} = 0.5$ V, the power loss on this power transistor is

$$P_{500\text{m}} = V_{\text{DS}}\left(500\text{m}\right) \times I_{\text{UNIT}}\left(500\text{m}\right) = 50 \times P_{50\text{m}} \tag{4.28}$$

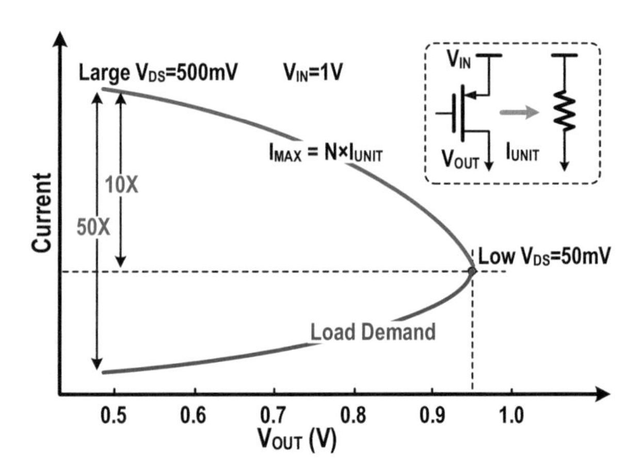

Fig. 4.55 Output capability and load demand

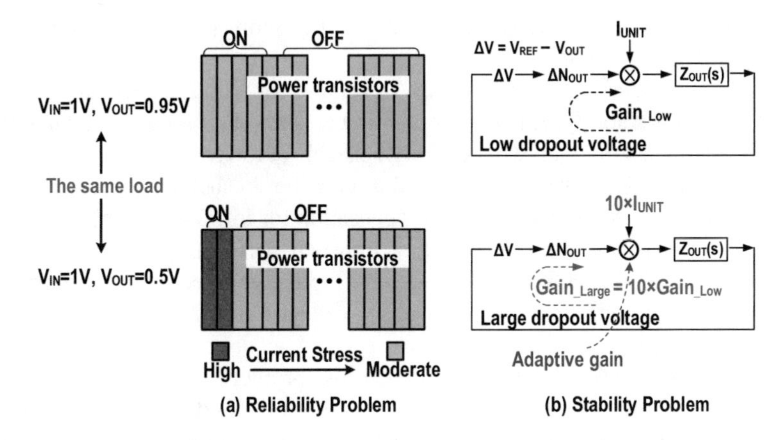

Fig. 4.56 Large V_{DS} voltage issues: (**a**) reliability and (**b**) stability

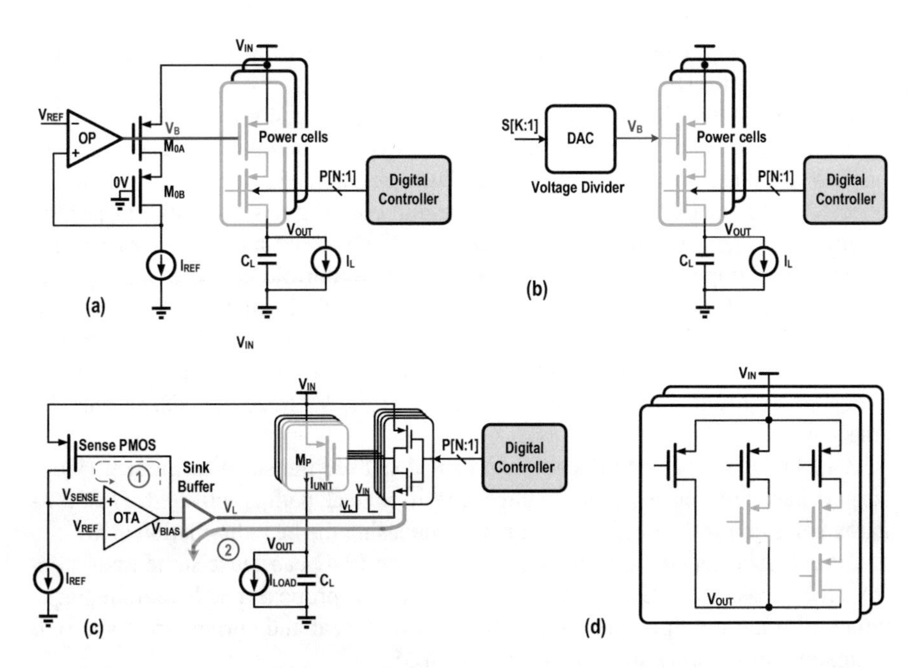

Fig. 4.57 Several techniques to limit I_{UNIT} changes: (**a**) analog stacked transistors in [15], (**b**) DAC-based stacked transistors in [41], (**c**) single-transistor ACC technique in [36], (**d**) segmented stacked transistor technique in [12]

In this condition, the power consumption on the first power switch is 50 times higher than that with V_{OUT} = 0.95 V, which may lead to self-heating and electromigration (EM) issues (Fig. 4.57).

In addition to reliability concerns, the increase in I_{UNIT} also augments the digital LDO loop gain, thereby broadening the loop bandwidth. This makes the impact of high-frequency poles and loop delays on stability more pronounced, potentially resulting in loop instability.

Therefore, for digital LDOs, a large V_{DS} voltage condition represents the worst-case scenario for stability. If the compensation ensures stability under a large voltage difference, then the loop will certainly be stable at lower V_{DS} voltage conditions as I_{UNIT} decreases. However, it is essential to note that in such a scenario, the compensation will inevitably lead to overcompensation for low V_{DS} voltage conditions, sacrificing loop bandwidth and affecting transient performance. To address this issue, several digital LDOs have adopted an adaptive gain technique, dynamically adjusting the gain coefficient based on the V_{DS} voltage to mitigate the impact of I_{UNIT} variations on stability.

Since the root cause of stability and reliability problems under large V_{DS} conditions is that I_{UNIT} changes with V_{DS}, fixing I_{UNIT} can solve both reliability and stability problems. The study [15] employs a stacked PMOS transistor structure as the power unit. The top PMOS is controlled by an analog loop to limit the maximum current flowing through it. The bottom PMOS acts as the actual switch, controlled by the digital loop. These power units effectively form an array of current digital-to-analog converters (DACs).

The work [41] follows a similar approach, but with a slight difference. In this case, the gate voltage of the top PMOS transistor is not generated by an analog loop. Instead, it is directly output by a voltage divider resistor, which can be dynamically selected and adjusted based on V_{DS} voltage and corner cases. Although as per the studies [15, 41], it can simultaneously resolve reliability and stability issues, achieving the same dropout and load capability requires a fourfold increase in the power transistor area. Mao et al. [36] propose a single-transistor auxiliary constant current (ACC) technique that enables a single transistor to function as both a constant current source and a switch. However, this technique requires the V_L supply to possess strong sink current capability, and the impact of the $\mathrm{V_L}$ trace impedance must be considered.

Zelikson et al. [12] introduced a segmented stacked transistor technique. Under large voltage differences, only the three-transistor stack path is activated. This limitation helps control the I_{UNIT} current while increasing the heat dissipation area.

In processor applications, the power unit of an LDO can reuse some transistors from the bypass switches. Muthukaruppan et al. [11] propose a code roaming technique. By time-multiplexing the power transistors, heat and current are dispersed, mitigating self-heating and EM issues (Fig. 4.58).

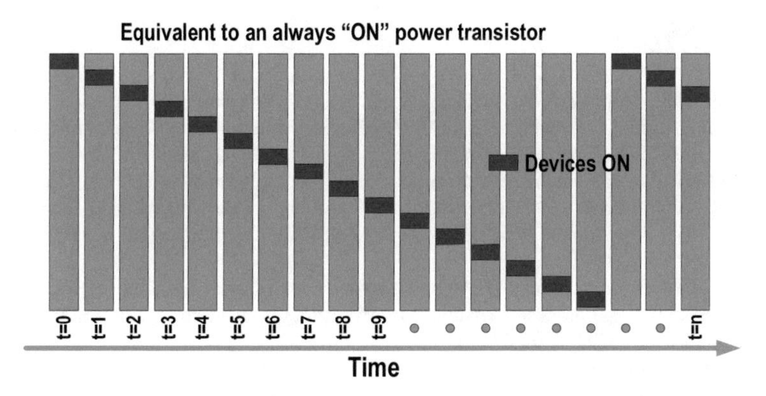

Fig. 4.58 Code roaming technique in [11]

4.8 Summary

DLDOs are known for their scalability, technology portability, and other benefits associated with digital-oriented design [42], especially playing an important role in powering digital circuits and fully integrated high-current fields. This chapter introduces several typical digital LDO architectures and discusses the reliability issues associated with digital LDOs.

The SR-based DLDO can obtain relatively high-output voltage accuracy, but its step-by-step adjustment method results in poor transient response.

The CF DLDO segments the power transistor into coarse and fine adjustment sections, managed by two independent control loops. The LDO's operational mode is dictated by voltage thresholds and FSM. This chapter discusses the SR + SR and SR+ flash ADC combinations.

The ADC-based DLDO represents the most classic digital control method, with performance contingent on the ADC and PID controller design. With redundant design and PID parameter adjustments, this DLDO can be tuned for various applications. However, it demands numerous power units, necessitating strong layout and routing capabilities.

Event-driven LDOs are designed for low static power applications. Combining asynchronous triggering and synchronous operation can simplify the controller design.

Computational DLDOs use nonlinear control to swiftly determine the control signal range through charge/discharge timing or time-index control and slope detection. Hybrid LDOs leverage both analog and digital control advantages to enhance LDO performance.

References

1. Y. Okuma, K. Ishida, Y. Ryu, X. Zhang, P.-H. Chen, K. Watanabe, M. Takamiya, T. Sakurai, 0.5-V input digital LDO with 98.7% current efficiency and 2.7-μA quiescent current in 65nm CMOS, in *IEEE Custom Integrated Circuits Conference*, (2010), pp. 98–101
2. S. Bin Nasir, S. Gangopadhyay, A. Raychowdhury, A 0.13μm fully digital low-dropout regulator with adaptive control and reduced dynamic Stability for ultra-wide dynamic range, in *IEEE International Solid-State Circuits Conference – (ISSCC) Digest of Technical Papers*, (2015), pp. 98–99
3. S. Bin Nasir, S. Gangopadhyay, A. Raychowdhury, All-digital low-dropout regulator with adaptive control and reduced dynamic stability for digital load circuits. IEEE Trans. Power Electron. **31**(12), 8293–8302 (2016)
4. Salem, L. G., Warchall, J., & Mercier, P. P. (2017). A 100nA-to-2mA successive-approximation digital LDO with PD compensation and Sub-LSB duty control achieving a 15.1-ns response time at 0.5V, in *IEEE International Solid-State Circuits Conference – (ISSCC) Digest of Technical Papers*, pp. 340–342.
5. M. Huang, Y. Lu, S.W. Sin, S.P. U, R.P. Martins, A fully integrated digital LDO with coarse-fine-tuning and burst-mode operation. IEEE Trans. Circuits Syst. II Express Briefs **63**(7), 683–687 (2016)
6. J. Oh, J.-E. Park, Y.-H. Hwang, D.-K. Jeong, A 480mA output-capacitor-free synthesizable digital LDO Using CMP-triggered oscillator and droop detector with 99.99% current efficiency, 1.3ns response time, and 9.8A/mm2 current density, in *IEEE International Solid-State Circuits Conference – (ISSCC) Digest of Technical Papers*, (2020), pp. 382–384
7. Y.-J. Lee, M.-Y. Jun, S. Singh, T.-H. Kong, D.-Y. Kim, K.-H. Kim, A 200-mA digital low drop-out regulator with coarse-fine dual loop in mobile application processor, in *IEEE International Solid-State Circuits Conference – (ISSCC) Digest of Technical Papers*, (2016), pp. 150–151
8. C. Cao, Y. Tang, X. Huang, Z. Zou, L. Zheng, A fully synthesizable capacitorless digital LDO for distributed power delivery network, in *IEEE International Symposium on Circuits and Systems (ISCAS)*, (2024)
9. J. Lee, J. Bang, Y. Lim, S. Yoo, Y. Lee, T. Seong, A fast-transient and high-accuracy, adaptive-sampling digital LDO using a single-VCO-based edge-racing time quantizer. IEEE Solid-State Circuits Letters **2**(12), 305–308 (2019)
10. Y.-J. Lee, W. Qu, S. Singh, D. Kim, K. Kim, A 200-mA digital low drop-out regulator with coarse-fine dual loop in mobile application processor. IEEE J. Solid-State Circuits **52**(1), 64–76 (2017)
11. R. Muthukaruppan, T. Mahajan, H.K. Krishnamurthy, S. Mangal, A. Dhanashekar, R. Ghayal, A digitally controlled linear regulator for per-core wide-range DVFS of atom cores in 14 nm tri-gate CMOS featuring non-linear control, adaptive gain and code roaming, in *Proceedings of 43rd IEEE European Solid State Circuits Conference*, (2017), pp. 275–278
12. M. Zelikson, K. Luria, L. Gil, Y. Brown, V. Goldenbeg, D. Kasif, A Digital Low-Dropout (LDO) linear regulator with adaptive transfer function featuring 125A/mm^2 power density and autonomous bypass mode, in *IEEE International Solid State Circuits Conference – Digest of Technical Papers*, (2023), pp. 230–231
13. M.Z. Straayer, M.H. Perrott, A 12-bit, 10-MHz bandwidth, continuous-time ADC with a 5-bit, 950-MS/s VCO-based quantizer. IEEE J. Solid-State Circuits **43**(4), 805–814 (2008)
14. S. Kundu, M. Liu, S.-J. Wen, R. Wong, C.H. Kim, A fully integrated digital LDO with built-in adaptive sampling and active voltage positioning using a beat-frequency quantizer. IEEE J. Solid-State Circuits **54**(1), 109–120 (2019)
15. T. Mahajan, R. Muthukaruppan, D.M. Shetty, S. Mangal, H.K. Krishnamurthy, Digitally controlled voltage regulator using oscillator-based ADC with fast-transient-response and wide dropout range in 14 nm CMOS, in *Proceedings of IEEE Custom Integrated Circuits Conference (CICC)*, (2017), pp. 1–4

16. S. Bang, W. Lim, C. Augustine, A. Malavasi, A fully synthesizable distributed and scalable all-digital LDO in 10 nm CMOS, in *IEEE International Solid State Circuits Conference – Digest of Technical Papers*, (2020), pp. 380–381
17. Z. Yuan, S. Fan, C. Yuan, L. Geng, A 100 MHz, 0.8-to-1.1 V, 170 mA digital LDO with 8-cycles mean settling time and 9-bit regulating resolution in 180-nm CMOS. IEEE Trans. Circuits Syst. II Express Briefs **67**(9), 1664–1668 (2020)
18. S. Weaver, B. Hershberg, U.-K. Moon, Digitally synthesized stochastic flash ADC using only standard digital cells, in *Symposium on VLSI Circuits-Digest Technical Papers*, (2011), pp. 266–267
19. T. Lyu, Z. Wang, J. Guo, A 2.5-A 3-ns-response-time calibration-free hybrid LDO using scalable self-clocked stochastic flash-ADC for in-loop quantization, in *IEEE Transactions on Circuits and Systems I: Regular Papers*, (2024)., early access
20. X. Wang, X. Liu, W.-H. Ki, A self-clocked and variation-tolerant unified voltage-and-frequency regulator for in-order executed digital loads. IEEE Trans. Circuits Syst. II Express Briefs **70**(11), 4627–4640 (2023)
21. K.H. Ang, G. Chong, Y. Li, PID control system analysis, design, and technology. IEEE Trans. Control Syst. Technol. **13**(5), 559–576 (2005)
22. J. Zhong, PID controller tuning: a short tutorial, 2006. [Online]. Available: https://web.archive.org/web/20150421081758/http://saba.kntu.ac.ir/eecd/pcl/download/PIDtutorial.pdf
23. D. Kim, M. Seok, Fully integrated low-drop-out regulator based on event-driven PI control, in *IEEE International Solid State Circuits Conference – Digest of Technical Papers*, (2016), pp. 148–149
24. D. Kim, J. Kim, H. Ham, M. Seok, A 0.5V-V_{IN} 1.44mA-class event-driven digital LDO with a fully integrated 100pF output capacitor, in *IEEE International Solid State Circuits Conference – Digest of Technical Papers*, (2017), pp. 346–347
25. D. Kim, S. Kim, H. Ham, J. Kim, M. Seok, 0.5V-V_{IN}, 165-mA/mm^2 fully-integrated digital LDO based on event-driven self-triggering control, in *IEEE Symposium on VLSI Circuits*, (2018), pp. 346–347
26. S. Kim, D. Kim, H. Ham, J. Kim, M. Seok, A 67.1-ps FOM, 0.5-V-hybrid digital LDO with asynchronous feedforward control via slop detection and synchronous PI with state-based hysteresis clock switching. IEEE Solid-State Circuit Letters **1**(5), 130–133 (2018)
27. S. Kim, D. Kim, Y. Pu, C. Shi, S.B. Chang, M. Seok, 0.5–1-V, 90–400-mA, modular, distributed, 3 × 3 digital LDOs based on event-driven control and domino sampling and regulation. IEEE J. Solid-State Circuits **56**(9), 2781–2794 (2021)
28. K. Ahmed, H. Krishnamurthy, C. Augustine, X. Liu, S. Weng, K. Ravichandran, J. Tschanz, V. De, A variation-adaptive integrated computational digital LDO in 22-nm CMOS with fast transient response. IEEE J. Solid-State Circuits **55**(4), 977–987 (2020)
29. D. Lee, T. Nomiyama, D. Jung, D. Kim, J. Lee, S. Kwak, A 10A computational digital LDO achieving 263A/mm2 current density with distributed power-gating switches and time-based fast-transient controller for mobile SoC application in 3nm GAAFET, in *IEEE International Solid State Circuits Conference – Digest of Technical Papers*, (2024), pp. 264–265
30. M. Huang, Y. Lu, U. Seng-Pan, R.P. Martins, An output-capacitor-free analog-assisted digital low-dropout regulator with tri-loop control, in *IEEE International Solid State Circuits Conference – Digest of Technical Papers*, (2017), pp. 342–343
31. X. Ma, Y. Lu, R.P. Martins, Q. Li, A 0.4V 430nA quiescent current NMOS digital LDO with NAND-based analog-assisted loop in 28nm CMOS, in *IEEE International Solid State Circuits Conference – Digest of Technical Papers*, (2018), pp. 306–308
32. M. Huang, Y. Lu, R.P. Martins, An analog-proportional digital-integral multiloop digital LDO with PSR improvement and LCO reduction. IEEE J. Solid-State Circuits **55**(6), 1637–1650 (2020)
33. X. Liu et al., 14.7 A modular hybrid LDO with fast load-transient response and programmable PSRR in 14nm CMOS featuring dynamic clamp tuning and time-constant compensation, in *IEEE International Solid State Circuits Conference – Digest of Technical Papers*, (2019), pp. 234–236

34. D. Zhou, J. Jiang, Q. Liu, E.G. Soenen, M. Kinyua, J. Silva-Martinez, A 245-mA digitally assisted dual-loop low-dropout regulator. IEEE J. Solid-State Circuits **55**(8), 2140–2150 (2020)
35. K.-S. Yoon, H.-S. Kim, W. Qu, Y.-S. Yuk, G.-H. Cho, Fully integrated digitally assisted low-dropout regulator for a NAND flash memory system. IEEE Trans. Power Electron **33**(1), 388–406 (2018)
36. X. Mao, Y. Lu, R.P. Martins, A 1.2-a calibration-free hybrid LDO with in-loop quantization and auxiliary constant current control achieving high accuracy and fast DVS. IEEE Trans. Circuits Syst. II Express Briefs **69**(11), 4443–4452 (2022)
37. M. Huang, Y. Lu, R.P. Martins, Review of analog-assisted-digital and digital-assisted-analog low dropout regulators. IEEE Trans. Circuits Syst. II Express Briefs **68**(1), 24–29 (2021)
38. J.-H. Jung, S.-K. Hong, O.-K. Kwon, A fast transient response hybrid LDO with highly accurate DC voltage using countable bidirectional binary search and soft swap switching. IEEE Trans. Circ. Syst. II: Express Briefs **67**(12), 3272–3276 (2020)
39. Y. Zhang, H. Song, R. Zhou, W. Rhee, I. Shim, Z. Wang, A capacitor-less ripple-less hybrid LDO with exponential ratio array and 4000x load current range. IEEE Trans. Circ. Syst. II: Express Briefs **66**(1), 36–40 (2019)
40. Y. Lim, J. Lee, Y. Lee, S. Yoo, J. Choi, A 320μV-output ripple and 90ns-settling time at 0.5V supply digital-analog-hybrid LDO using multi-level gate-voltage generator and fast decision PD detector, in *IEEE 44th European Solid State Circuits Conference (ESSCIRC)*, (2018), pp. 210–213
41. P.A. Meinerzhagen et al., An energy-efficient graphics processor in 14-nm Tri-Gate CMOS featuring integrated voltage regulators for fine-grain DVFS, Retentive Sleep, and V_{MIN} optimization. IEEE J. Solid-State Circuits **54**(1), 144–157 (2019)
42. Z. Wang, S.J. Kim, K. Bowman, M. Seok, Review, survey, and benchmark of recent digital LDO voltage regulators, in *IEEE Custom Integrated Circuits Conference (CICC)*, (2022), pp. 1–8

Chapter 5
Switching LDO

To satisfy the high current and ultrafast load transient requirements in server-grade microprocessors, fully turn on and off the power transistor, which provides an extra choice for linear regulators [1–6]. The hysteretic switching LDO topology offers the fastest response and transient recovery, compared to the analog and digital counterparts, to be discussed next.

5.1 Basic Switching LDO Architecture

5.1.1 Electrical Characteristic Comparison

Figure 5.1 depicts a basic hysteretic switching LDO and its operation principle. A high-speed comparator amplifies the error between V_{REF} and V_{OUT} to binary levels, driving the power transistor M_0. The sub-nanosecond propagation delay of the switching loop determines the fast transient response. When M_0 is turned on, the current I_{SW} flowing through the power transistor must be greater than the maximum load current I_{MAX} to ensure that the output capacitor is charged up. When M_0 is turned off, the load current discharges the output voltage. When a load change causes the output voltage to droop (an error of a few millivolts), all power transistors can be instantaneously turned on to arrest the output voltage droop. Thus, this kind of "linear" regulator can also be called as first-order switching converter that only has one output capacitor as the energy storage component, where second-order switching converter consists of one inductor and one capacitor.

Table 5.1 presents a comparison of the electrical characteristics of switching, analog, and digital LDOs. Analog LDOs continuously regulate the gate voltage of the power transistors, while digital LDOs adjust the on/off number of power transistors, which belongs to discrete control. Switching LDOs modulate the duty cycle of the power transistors. Within a specific single cycle period, the duty cycle remains

X. Mao et al., *Fully-Integrated Low-Dropout Regulators*, Analog Circuits and Signal Processing, https://doi.org/10.1007/978-3-031-84916-9_5

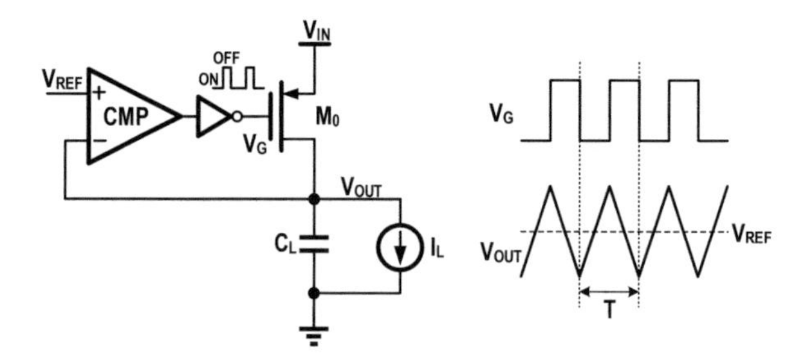

Fig. 5.1 A basic hysteretic switching LDO and its operation principle

Table 5.1 The comparison of switching LDO, analog LDOs, and digital LDOs

Topology	Analog LDO	Digital LDO	Switching LDO
Regulation Fineness	Analog Continuous (Voltage)	Digital Discrete (Number)	Analog Discrete (Duty cycle)
Quantization Error	No	Yes	No
PSR	Good	Poor	Medium
Self-heating and EM	Low	High	Low
Input voltage	Medium to High	Low	Low
Gate pole	Yes	No	No
Distributed power transistors	Hard	Easy	Easy
Benefits from Technology Scaling	Fewer	More	More
Area Efficiency	Low	High	High
Drive Current Consumption	Low	Medium	High
Ripple	No	Medium	Large

fixed, rendering it discrete. However, from a steady-state perspective, the duty cycle value can be continuously adjusted, which significantly differs from digital LDOs.

Analog LDOs can readily obtain high output accuracy, whereas the accuracy of digital LDOs is generally limited by the resolution of the quantizer. Switching LDOs can employ analog operational amplifiers and continuous duty cycle regulation to achieve high-output voltage accuracy.

The power transistors of both switching LDOs and digital LDOs operate in a similar state during conduction, exhibiting numerous shared characteristics.

Notably, both types have no gate poles, support low-input voltages, facilitate distributed integration, and benefit from process scaling. Consequently, some researchers and engineers colloquially refer switching LDOs as digital LDOs [7, 8].

The heat and current of a digital LDO are concentrated in parts of the power transistors. In contrast, all the power transistors of the switching LDO are active, resulting in the distribution of current and heat across all the power transistors. From the perspective of self-heating and EM considerations, the switching LDO exhibits greater similarity to an analog LDO, which makes it superior to the digital LDO.

In addition, since all the power transistors of the switching LDO operate in a fast bang-bang state, the ripple of the switching LDO is typically much larger than that of the digital LDO. Furthermore, the current consumed by the driver circuit is significantly higher in the switching LDO. Obviously, these two characteristics are the most significant disadvantages of switching LDO.

According to the comparative analysis above, it is evident that switching LDOs are more suitable for high-current application scenarios, especially for supplying power to large-scale digital circuits. Of course, the characteristics of switching control can also be combined with other architectures to obtain better performance.

5.1.2 Equivalent Model and Ripple Analysis

For fixed input and output voltages, the power stage of a switching LDO can be equated to a constant current source cascaded with a 0-Ω switch in steady state, as shown in Fig. 5.2. The duty cycle signal controls the charging and discharging of the output capacitor:

$$I_{\mathrm{SW}} \times T_{\mathrm{ON}} = I_{\mathrm{L}} \times T \tag{5.1}$$

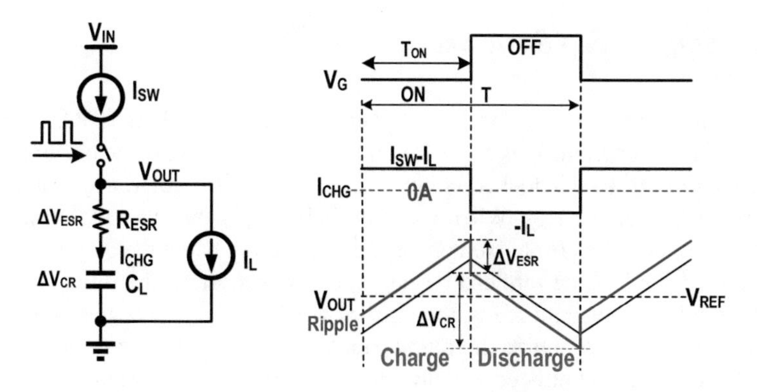

Fig. 5.2 The equivalent model of the switching power stage

where T is the switching period, and T_{ON} represents the transistor on-time in one period. The duty cycle D is

$$D = \frac{T_{\mathrm{ON}}}{T} = \frac{I_{\mathrm{L}}}{I_{\mathrm{SW}}} \quad (5.2)$$

According to (5.2), we can see that D is linearly proportional to the load current I_L in the steady state. The output ripple comprises two parts: the capacitor charging or discharging component V_{CR} and the equivalent series resistance (ESR) component V_{ESR}. The duty cycle D can be used to normalize the relationship between ripple and load current. For a single-phase control, the output ripple is given as follows:

$$\Delta V = \Delta V_{\mathrm{CR}} + \Delta V_{\mathrm{ESR}} = (1 - D) D \times I_{\mathrm{SW}} / (C_{\mathrm{L}} \times F) + I_{\mathrm{SW}} \times R_{\mathrm{ESR}} \quad (5.3)$$

where $F = 1/T$ is the switching frequency. When $D = 0.5$, the output ripple reaches the maximum ΔV_{MAX}:

$$\Delta V_{\mathrm{MAX}} = 0.25 \times I_{\mathrm{SW}} / (C_{\mathrm{L}} \times F) + I_{\mathrm{SW}} \times R_{\mathrm{ESR}} \quad (5.4)$$

The maximum value of the ripple is only related to the transistor strength I_{SW}, operating frequency F, and output capacitor C_L.

Assuming $I_{SW} = 1$ A, $F = 1$ GHz, $D = 50\%$, $R_{ESR} = 5$ mΩ, and $C_L = 25$ nF, the output ripple is about 15 mV. The most direct way to reduce ripple is to increase the output capacitance and operating frequency. Almost all switching LDO designs utilize advanced processes that reduce loop delay and increase operating frequency, and some employ deep-trench capacitors (DTCs).

5.2 Hysteretic Switching Control

5.2.1 Single-Loop Structure

The work in [1] presented a classic hysteretic controlled switching LDO. Figure 5.3 shows its block diagram. This regulator employs a constant on-time hysteretic control scheme. A fast self-clocked comparator compares the reference voltage V_{REF} with the output voltage V_{OUT}. When V_{OUT} falls below V_{REF}, the controller turns on the PMOS power switches for a fixed one-cycle interval to charge the output capacitor. The value of the load current can be estimated by measuring the interval between switching pulses. For light loads, the interval would be large. For heavy loads, the interval is small. To optimize the output ripple across a wide input-output voltage range, Kudva et al. [1] utilized a digital strength calibration technique.

The power switch is divided into 24 equally sized segments, and the digital calibration loop dynamically adjusts the strength of the power transistors based on the

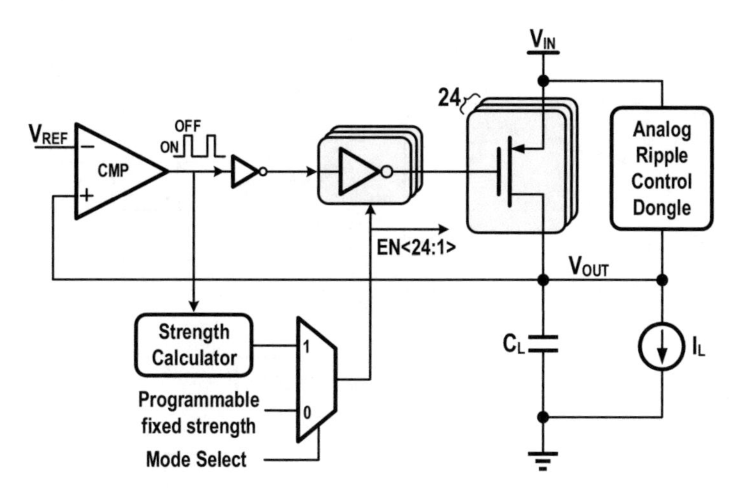

Fig. 5.3 The block diagram of the hysteretic controlled switching LDO in [1]

interval sizes. It is important to note that this ripple control loop responds slowly and may lead to droop during large load transients. For faster transient response, the calibration loop can be closed, and the preset switching strength code can be used, although this will result in additional ripple during light loads. Additionally, an analog ripple control module responds to high-frequency ripple and generates an opposing current to reduce output ripple.

To keep ripple small, in addition to the methods described above, it is more important that the regulator must operate at very high frequencies. Fabricated in 16 nm FinFET technology, the operating frequency can reach up to 1.8–4 GHz [1].

Figure 5.4 shows the fast-self-clocked comparator, its core circuits, and the operation timing diagram. The operating frequency can be adjusted through a programmable delay chain. And the intermediate signal of the delay chain is used to latch the comparator's output CMP_OUT. The input offset of the comparator impacts the accuracy of the output voltage of the LDO.

Figure 5.4b depicts the schematic of the comparator. The comparator has two input branches, A and B. And the transistors in A and B are skewed in size, with X > Y, so each branch has an offset, but of opposite sign. By adjusting the digital code TUNE, A and B have unequal currents, thereby eliminating the offset.

Figure 5.5 depicts another single-loop hysteretic switching LDO [2]. Both Kudva et al. and Saxena et al. [1, 2] originate from the same research team. Kudva et al. [1] employ a self-clocked comparator, while Saxena et al. [2] utilize a continuous-time comparator comprising a self-biased first stage and a complementary self-biased second stage [9], as shown in Fig. 5.6. Moreover, to reduce ripple, the power transistors are divided into fast and slow transistors. The comparator output duty cycle signal directly controls the fast transistor, while the slow transistor's gate control signal, V_{SLOW}, is RC filtered. The fast transistor only provides 20% of the load current.

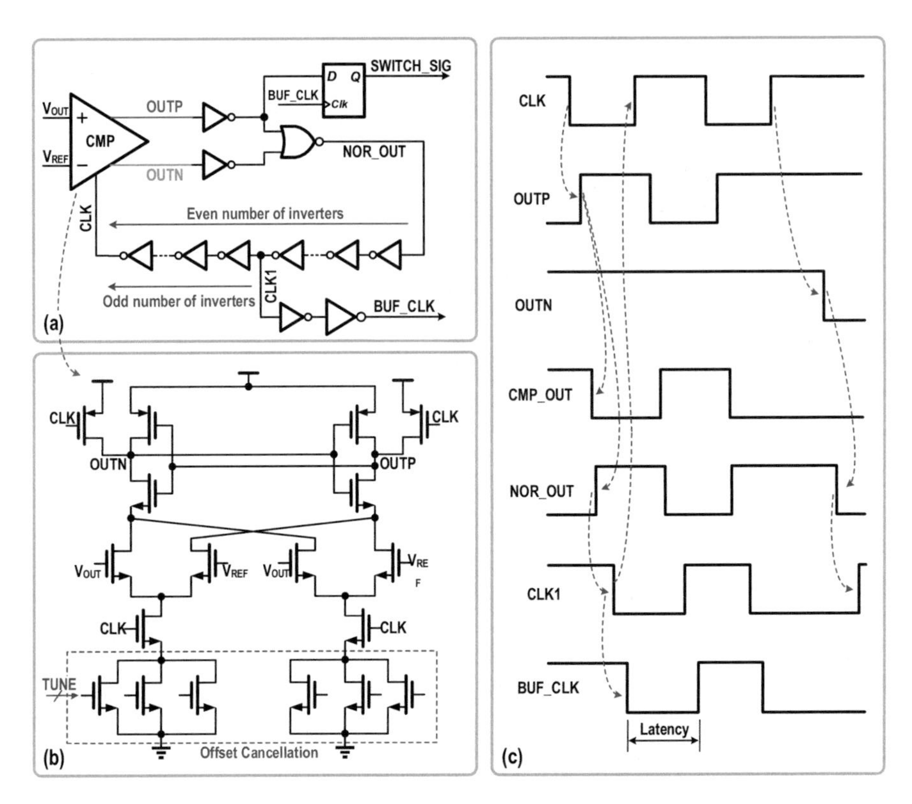

Fig. 5.4 (a) The fast-self-clocked comparator, (b) core circuits, and (c) timing diagram

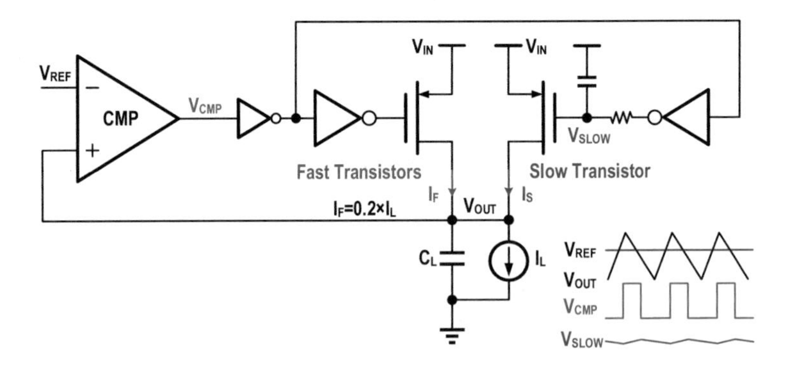

Fig. 5.5 The hysteretic controlled switching LDO in [2]

5.2.2 Dual-Loop Structure

The first work on switching LDOs was published by the team from IBM in 2012 [3]. Later on, the IBM team also published two distributed switching LDO architectures in the following years [4, 5], which were applied to the Power™ processors.

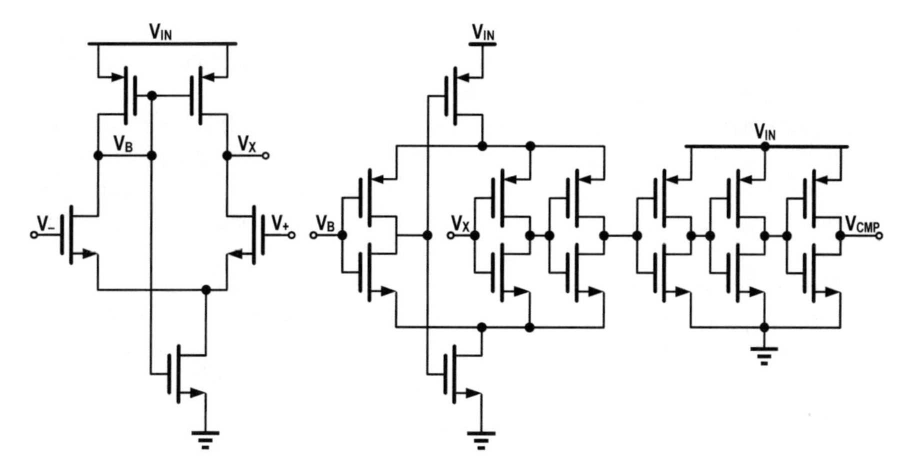

Fig. 5.6 Schematic of the continuous-time comparator

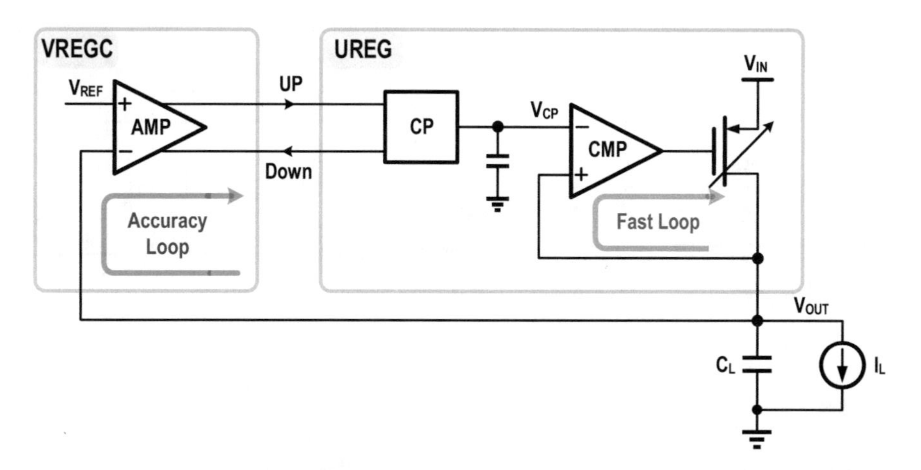

Fig. 5.7 The dual-loop switching LDO architecture in [3]

Bulzacchelli et al. and Deniz et al. [3, 4] added a pre-amplification before the comparator, forming a dual-loop structure. The comparator's reference voltage is no longer V_{REF}, but rather the amplified and charge pump (CP)-filtered signal VCP. Figure 5.7 gives the block diagram of the dual-loop structure. It consists of a high-accuracy voltage regulator controller (VREGC) and a fast-response microregulator (UREG). The VREGC amplifies the error information between V_{OUT} and V_{REF} and provides it to the charge pump in the form of up/down currents.

Assume that V_{CP} is too low initially and V_{OUT} is lower than V_{REF}. At this time, the UP current becomes larger, while the down current decreases, which causes V_{CP} to charge upwards, thereby raising V_{OUT}. As V_{OUT} approaches V_{REF}, VREGC adjusts the up/down currents to bring the loop into steady equilibrium. The outer amplifier loop can improve load regulation of the switching LDO. Moreover, it offers another advantage. Since V_{CP} can be adjusted to whatever value is needed for V_{OUT} to equal V_{REF}, the comparator offset can be automatically canceled. This allows the comparator to use smaller devices to achieve faster speeds.

The schematic of the UREG in [3] is depicted in Fig. 5.8. Instead of using the traditional differential input pair, it adopts a PMOS common-gate stage (M_1), where the source is connected to V_{OUT}, and the gate voltage is V_{CP}. The output error information is then amplified by two additional NMOS common-source stages, being converted into the duty cycle D information. Note that the duty cycle information is decided by the up and down current ratio. Also, the power supply for the charge pump and the amplifier is V_{OUT}, not V_{IN}, which can improve the power supply rejection ratio (PSR) of these circuits and prevent interference from V_{IN} noise. Finally, the amplified signal is transferred to the V_{IN} domain via a level shifter to ensure that the power transistors can be fully turned off.

Figure 5.9 depicts the schematic of VREGC in [3]. It comprises a high-gain transconductance amplifier with low offset and a common-mode feedback loop (CMFB). The error information between V_{OUT} and V_{REF} is amplified by the input pair M_1/M_2 and converted to up and down currents by M_5/M_6.

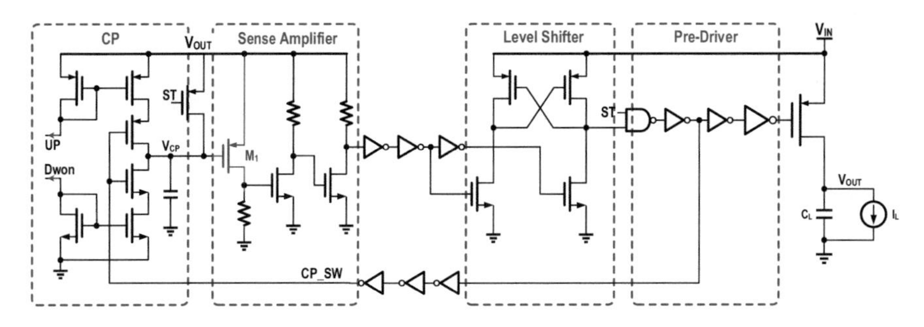

Fig. 5.8 Schematic of the UREG

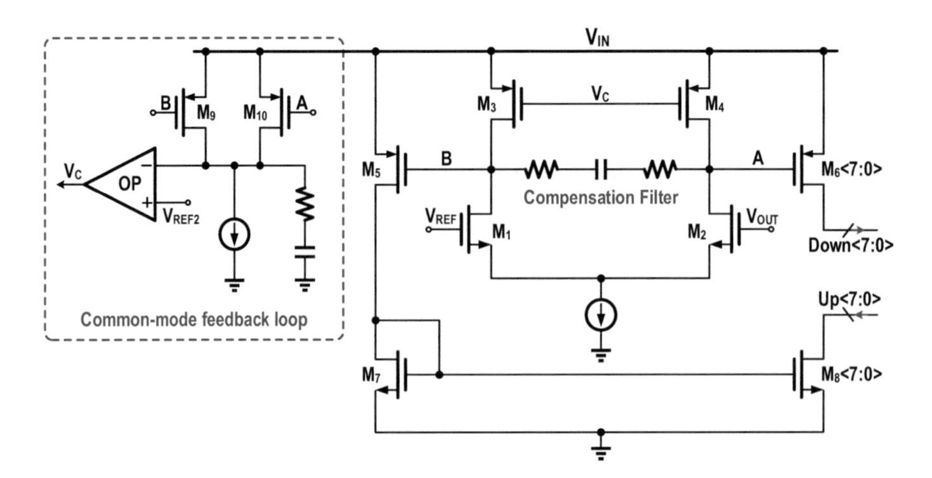

Fig. 5.9 Schematic of the VREGC in [3]

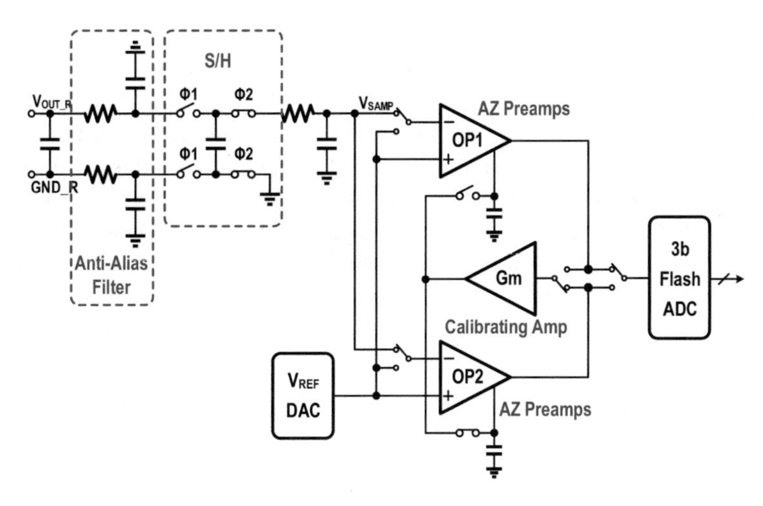

Fig. 5.10 Schematic of the VREGC in [4]

The common-mode feedback loop ensures that the sum of the up and down currents remains constant (50 μA). M_9/M_{10} and M_5/M_6 are matched so that the total currents flowing through M_9/M_{10} equal those flowing through M_5/M_6. An operational amplifier (OP) detects the current sum of M_9/M_{10} and regulates the voltage of V_C to stabilize the current sum of M_5/M_6. The CMFB is further stabilized by an RC compensation network connected between points A and B.

When powering large-scale digital circuits, especially in distributed applications, VREGC and UREG are located far apart. In a noisy processor environment, the analog up/down current signal may be subject to interference. The VREGC in [4] uses a 3-bit flash ADC to quantize the output of the amplifier and deliver the digital code value to UREGs. Subsequently, the digital signal is converted into up/down current in the local UREG. Figure 5.10 shows the block diagram of the VREGC in [4].

Due to the large load area, the load ground and the local ground are different, resulting in an IR drop between them. To reduce the error, the authors employ a sample/hold (S/H) circuit to offer differential sampling and convert it into a local single-ended voltage signal V_{SAMP}. The error between V_{SAMP} and V_{REF} is pre-amplified by the operational amplifier. The RC filter in front of the S/H filters out high-frequency ripples on V_{OUT_R} to prevent erroneous sampling by the S/H.

To improve the output voltage accuracy, the pre-amplifier employs auto-zeroed (AZ) technique to eliminate offset voltage. When one amplifier is in use, the other amplifier undergoes calibration, and they alternate their functions.

5.2.3 Ripple Reduction Technique

The voltage differential between the input and output of a switching LDO significantly impacts the inductor switching current (I_{SW}) flowing through the power switches. When the input-output voltage difference is substantial, the increase in I_{SW}

leads to larger output ripple. To maintain the ripple within an acceptable range, some switching LDOs employ two techniques [4–6], as illustrated in Fig. 5.11.

The first technique involves PMOS strength calibration: dynamically adjusting the strength of the power transistors based on the maximum load demand. These power transistors are divided into several equal units, and a finite state machine (FSM) is employed to set the control code value according to input-output voltage, process, temperature, and maximum load demand. In [5], a look-up table method was used. Figure 5.12 illustrates the I_{SW} comparison waveforms, and strength calibration allows control over the maximum value of I_{SW}. However, it is essential to consider the impact on the output voltage during code value switching.

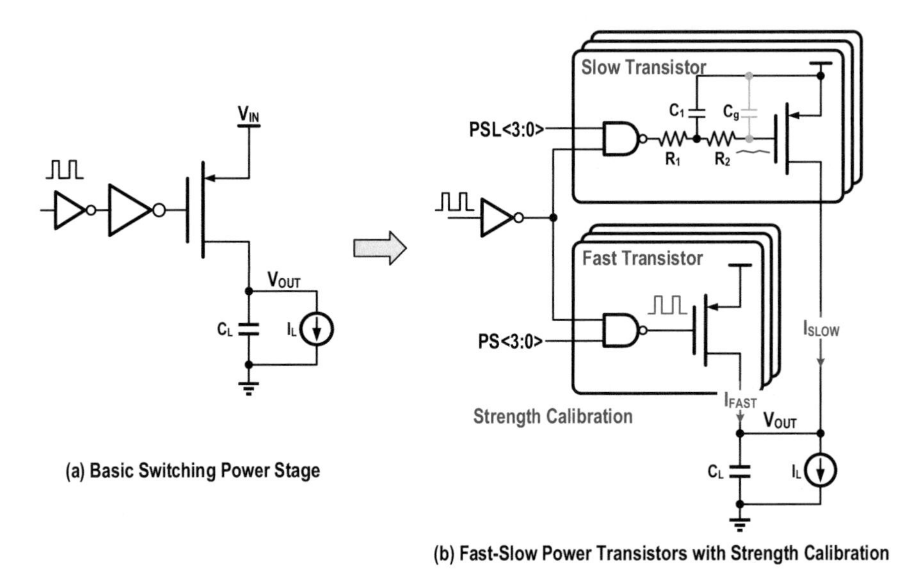

Fig. 5.11 (**a**) Basic switching power stage and (**b**) fast-slow power transistors with strength calibration [4]

Fig. 5.12 I_{SW} comparison waveforms

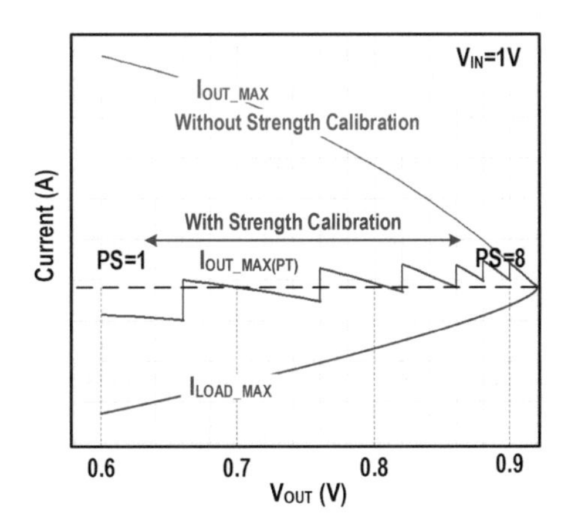

The second method involves using a fast-slow transistor. The fast transistor is directly controlled by a duty cycle signal and is employed to respond to dynamic current changes. The gate voltage of the slow transistor is a filtered signal, which is not fully modulated, resulting in smaller ripple contribution. Slow transistors can share some of the load current, and even for some moderately load transient steps, the slow transistors can also provide considerable dynamic current. In [2], the slow transistors shared almost 80% of the load current. The fast-slow transistor improves the trade-off between output ripple and current handling.

5.3 PWM Switching Control

5.3.1 RAMP-Based PWM Control

Whether using a clocked comparator or a continuous-time comparator, the charge/discharge cycle of a hysteretic switching LDO is not fixed. Then the question is how to design a synchronous switching LDO with a fixed charge/discharge period.

By replacing the input reference voltage V_{REF} of the comparator with a triangular wave, we can obtain a single RAMP-based pulse-width modulation (PWM) control structure, as shown in Fig. 5.13. The central value of the triangle wave is equal to V_{REF}, and a continuous-time comparator is used. The RAMP signal is compared with V_{OUT}, and the output of the comparator is a duty cycle signal with a fixed period. The period of charging and discharging is equal to the period of the RAMP signal.

RAMP-based PWM control is often employed in DC-DC converter designs. We set the amplitude of the RAMP to be much larger than the ripple amplitude of V_{OUT} and ignore the effect of output ripple; we can derive the duty cycle expression:

$$D = \left(\frac{\text{RAMP}}{2} + V_{\text{REF}} - V_{\text{OUT}} \right) / \text{RAMP} = \frac{1}{2} + \frac{V_{\text{REF}} - V_{\text{OUT}}}{\text{RAMP}} = \frac{I_{\text{L}}}{I_{\text{SW}}} \tag{5.5}$$

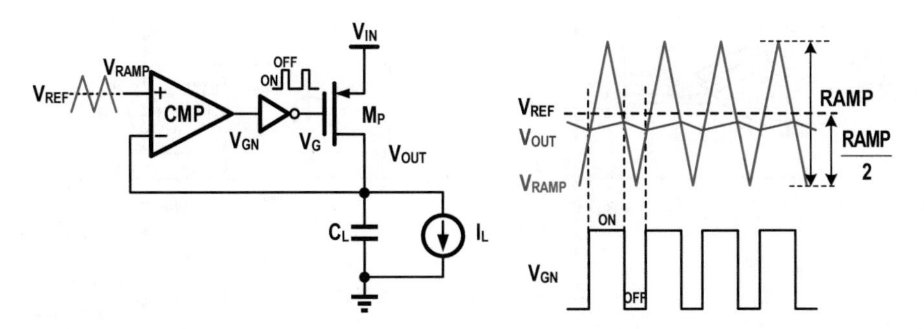

Fig. 5.13 RAMP-based PWM control

Referring to (5.4), the output ripple in single-phase PWM control is

$$\Delta V_{1\ \mathrm{PHASE}} = (1-D)D \times I_{\mathrm{SW}} / (C_{\mathrm{L}} \times F) + I_{\mathrm{SW}} \times R_{\mathrm{ESR}} \tag{5.6}$$

$$\Delta V_{1\ \mathrm{PHASE}} = (1-D)D \times V_{\mathrm{RM}} + I_{\mathrm{SW}} \times R_{\mathrm{ESR}} \tag{5.7}$$

$$V_{\mathrm{RM}} = I_{\mathrm{SW}} / (C_{\mathrm{L}} \times F) \tag{5.8}$$

Assuming that power transistor strength I_{SW}, output capacitance C_L, and switching frequency F are fixed, and neglecting PVT variations, the voltage regulator module V_{RM} maintains a constant value. We can normalize and derive the relationship between voltage ripple and duty cycle (which correlates with load current). In a single-phase switching control scheme, the ripple reaches its maximum when the duty cycle is 50%, corresponding to a load equal to half of the maximum load capacity.

5.3.2 Multiphase PWM Control

After fixing the operating frequency using RAMP-based PWM control, we can further use multiphase control to reduce the output ripple. Figure 5.14 shows the four-phase PWM control structure and its corresponding equivalent charging/discharging model.

The four-phase structure divides the power switches into four parts, and the charging current of each part becomes $0.25I_{SW}$. Since RAMP signals are interleaved, the charging is also interleaved. The charge/discharge current I_{CHG} of the output capacitor can be expressed as

$$I_{\mathrm{CHG}} = I_{\mathrm{SW\ PHASE0}} + I_{\mathrm{SW\ PHASE1}} + I_{\mathrm{SW\ PHASE2}} + I_{\mathrm{SW\ PHASE3}} - I_{\mathrm{L}} \tag{5.9}$$

Figure 5.15 illustrates the charging and discharging waveforms of four-phase PWM control. Consider a scenario with D = 30% and I_{LOAD} = $0.3I_{SW}$. Since the

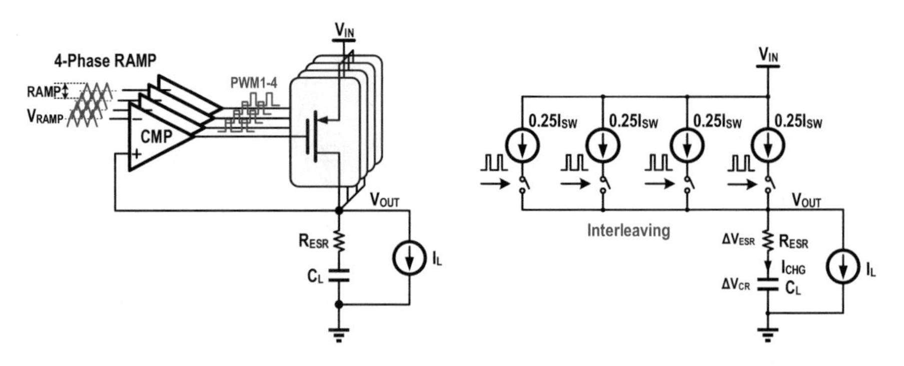

Fig. 5.14 The four-phase PWM control structure and its charging/discharging model

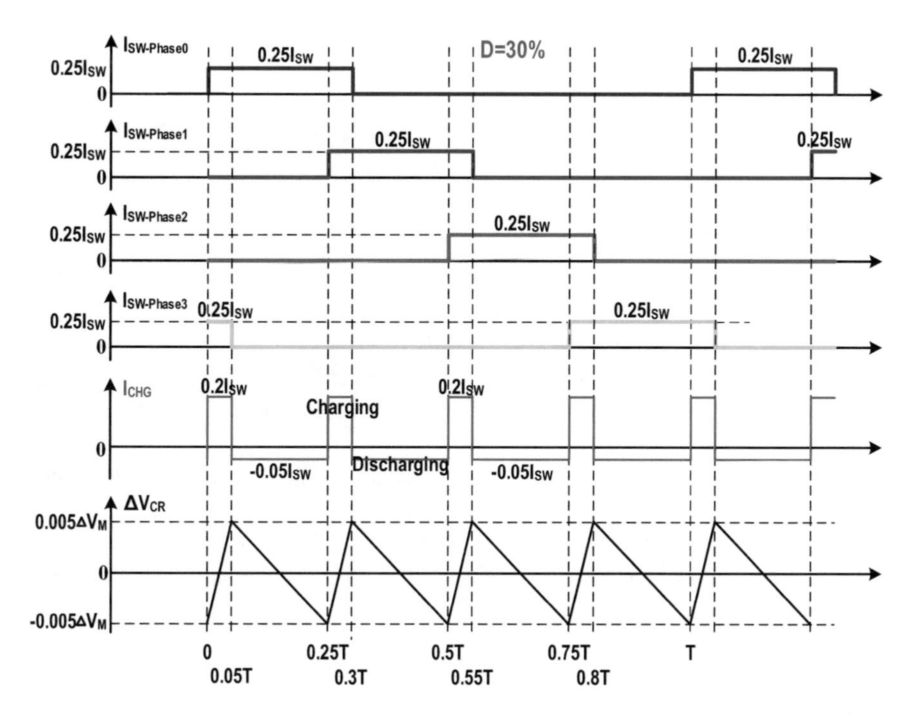

Fig. 5.15 Waveforms of charging current ICHG and ripple at $D = 30\%$

charging capacity of each phase is only $0.25I_{SW}$, which is less than I_{LOAD}, two groups of power switches must simultaneously charge the output capacitor for a specific duration.

The overlapping charging time D is 25%, with a charging current I_{LOAD} of $0.5I_{SW}$. Conversely, the discharge current I_{LOAD} is $0.25I_{SW}$. According to the charge balance principle, we can calculate the ripple amplitude when $25\% < D < 50\%$:

$$\Delta V_{4\ \mathrm{PHASE}} = (0.5 - D)(D - 0.25) \times V_{\mathrm{RM}} + 0.25 I_{\mathrm{SW}} \times R_{\mathrm{ESR}} \tag{5.10}$$

Extending this analysis, we can draw the charging and discharging waveforms for various duty cycles. By deriving the ripple calculation formula for each scenario, we obtain the following comprehensive expression:

$$\Delta V_{4-\mathrm{Phase}} = \begin{cases} (0.25 - D)D \times V_{\mathrm{RM}} + 0.25 I_{\mathrm{SW}} \times R_{\mathrm{ESR}} & 0 < D < 25\% \\ (0.5 - D)(D - 0.25) \times V_{\mathrm{RM}} + 0.25 I_{\mathrm{SW}} \times R_{\mathrm{ESR}} & 25\% < D < 50\% \\ (0.75 - D)(D - 0.5) \times V_{\mathrm{RM}} + 0.25 I_{\mathrm{SW}} \times R_{\mathrm{ESR}} & 50\% < D < 75\% \\ (1 - D)(D - 0.75) \times V_{\mathrm{RM}} + 0.25 I_{\mathrm{SW}} \times R_{\mathrm{ESR}} & 75\% < D < 1 \end{cases} \tag{5.11}$$

Based on the above formula, Fig. 5.16 compares the normalized ripples of one-phase and four-phase control. By implementing four-phase technology, the

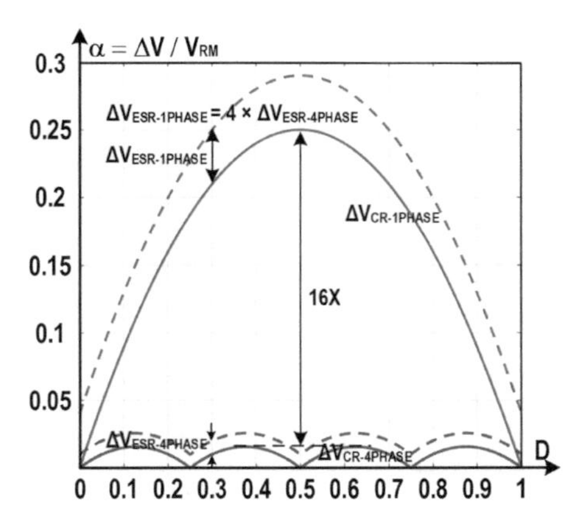

Fig. 5.16 Ripple comparison between the single phase and four-phase PWM control schemes

maximum component of the ripple is reduced by 16 times, and the ESR-related ripple component is reduced by 4 times.

Notably, when the duty cycle D is 0%, 25%, 50%, 75%, or 100%, the charging current at any time equals the discharging current, theoretically allowing the ripple to reach 0 mV.

5.3.3 *Current Balancing Analysis*

Multiphase control can significantly reduce output ripple, but we need to pay attention to the current sharing challenges of multiphase control. For four-phase RAMP-based PWM control, the load current can be expressed as

$$I_{\text{LOAD}} = \left(\frac{I_{\text{SW1}}}{4} \times D_1\right) + \left(\frac{I_{\text{SW2}}}{4} \times D_2\right) + \left(\frac{I_{\text{SW3}}}{4} \times D_4\right) + \left(\frac{I_{\text{SW4}}}{4} \times D_4\right) \tag{5.12}$$

Ignore the potential variations in the characteristics of the four-power transistor groups and assume that the current sharing is mainly caused by the differences in duty cycle among the phases:

$$\Delta I = \frac{I_{\text{SW}}}{4} \times \Delta D \tag{5.13}$$

According to expression (5.13), a small duty cycle difference will only lead to a small current sharing error. In RAMP-based PWM control, the expression of D is

$$D = \frac{1}{2} + \frac{V_{\text{BIAS}} - V_{\text{OUT}}}{RAMP} \tag{5.14}$$

where RAMP is the peak-to-peak amplitude of the ramp signal, and V_{BIAS} represents the midpoint voltage of the ramp signal. The primary sources of error in this system

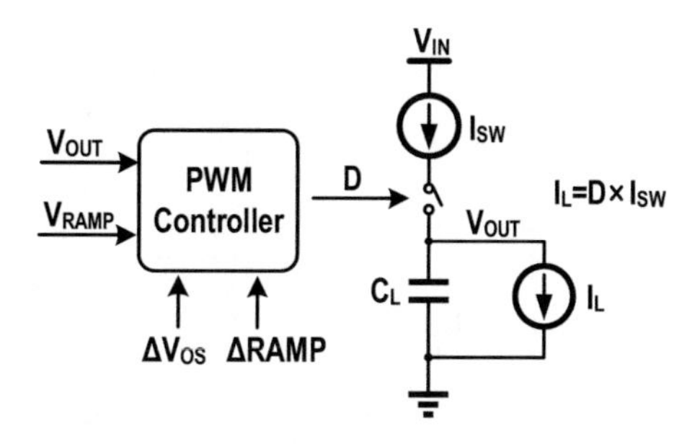

Fig. 5.17 Sources of the duty cycle error

are typically the input offset voltage (ΔV_{OS}) of the comparator and the amplitude error (ΔRAMP) of the ramp signal, as illustrated in Fig. 5.17:

$$D = \frac{1}{2} + \frac{V_{\mathrm{BIAS}} - V_{\mathrm{OUT}} + \Delta V_{\mathrm{OS}}}{\mathrm{RAMP} + \text{”}\,\mathrm{RAMP}} \tag{5.15}$$

$$\Delta D = \left(\frac{1}{2} + \frac{V_{\mathrm{BIAS}} - V_{\mathrm{OUT}} + \Delta V_{\mathrm{OS}}}{\mathrm{RAMP} + \text{”}\,\mathrm{RAMP}} \right) - \left(\frac{1}{2} + \frac{V_{\mathrm{BIAS}} - V_{\mathrm{OUT}}}{\mathrm{RAMP}} \right) \tag{5.16}$$

when

$$V_{\mathrm{BIAS}} - V_{\mathrm{OUT}} = -\frac{\mathrm{RAMP}}{2} \tag{5.17}$$

ΔD can reach the maximum value. Assuming that ΔRAMP = 7 mV, $\Delta V_{\mathrm{OS}} = 5$ mV, and RAMP = 100 mV, the duty cycle error $\Delta D = 7.94\%$.

According to the analysis above, multiphase ramp-based PWM control in switching LDO can achieve high current sharing accuracy without the need for additional auxiliary circuits. To further enhance current sharing accuracy, calibration of the comparator and ramp signal can be implemented.

5.3.4 High-Speed Comparator and RAMP Generation Circuit

The design of the comparator is critical for the switching LDO. Figure 5.18 illustrates a high-speed two-stage continuous comparator circuit.

To ensure consistent performance across a wide input-output voltage range, the comparator's input stage utilizes a 1.8 V AV_{DD} power supply, with IO devices employed for the transistors shown in blue. To enhance response speed, core devices are utilized for the bottom cross-coupled NMOS transistors, thereby reducing parasitic capacitance. The second stage amplifies the differential output of the first stage (V_{BN1} and V_{BN2}) into binary levels. In 28 nm process, the response latency of this comparator is less than 300 ps.

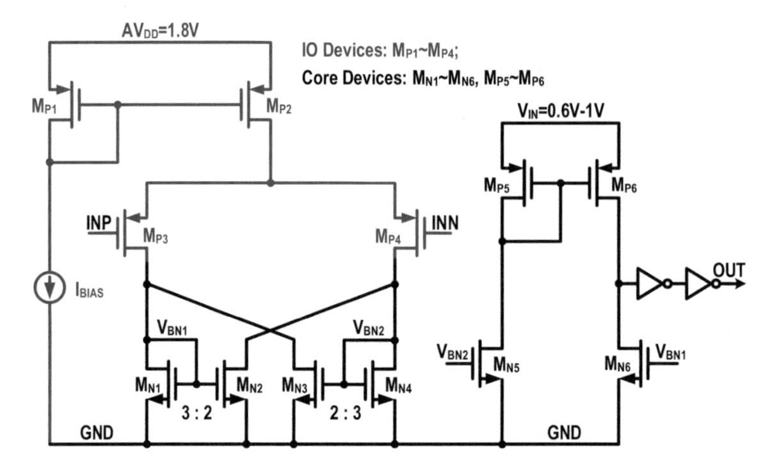

Fig. 5.18 Schematic of the two-stage comparator

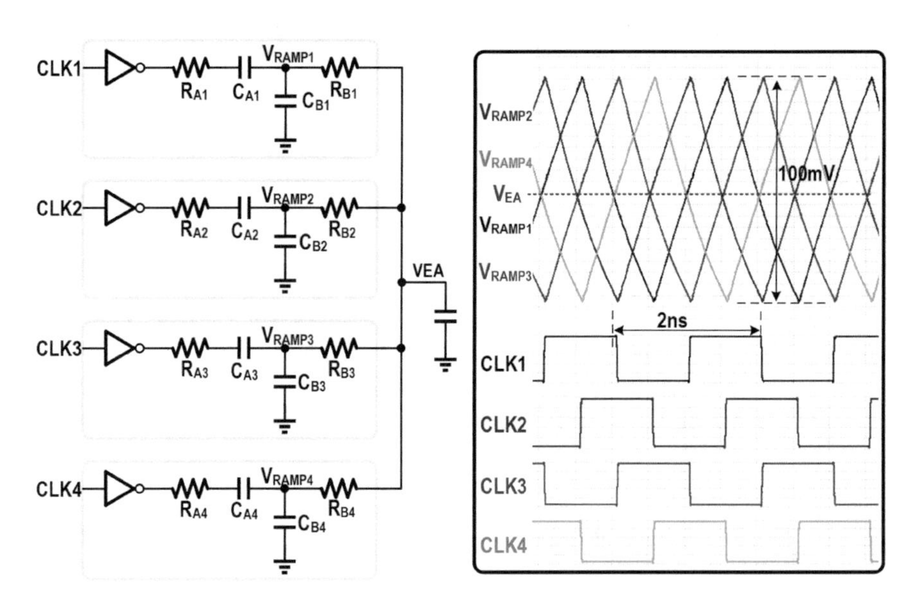

Fig. 5.19 Schematic of four-phase RAMP signals and simulated waveforms

We use a four-phase clock and RC coupling to generate RAMP signals. Figure 5.19 illustrates the conceptual diagram of the four-phase RAMP signals and their corresponding simulated waveforms. A 500 MHz clock signal is subjected to RC filtering, subsequently coupled with a DC voltage, ultimately generating the triangular waveform. The central voltages of V_{RAMP1} through V_{RAMP4} are determined by VEA, while their respective amplitudes can be modulated by adjusting RA1–RA4. The matching of resistors and capacitors will affect the amplitude of the RAMP signals.

5.3.5 Dual-Loop Four-Phase Structure

In the RAMP-based PWM controller, the comparator utilizes the triangular signal VRAMP as its input reference:

$$D = \frac{1}{2} + \frac{V_{\mathrm{BIAS}} - V_{\mathrm{OUT}}}{\mathrm{RAMP}} \tag{5.18}$$

According to the charge-balance principle,

$$D = \frac{T_{\mathrm{ON}}}{T} = \frac{I_{\mathrm{LOAD}}}{I_{\mathrm{SW}}} \tag{5.19}$$

Substituting Eq. (5.19) into Eq. (5.18), we can obtain

$$V_{\mathrm{OUT}} = V_{\mathrm{REF}} + \mathrm{RAMP} \times \left(\frac{1}{2} - \frac{I_{\mathrm{L}}}{I_{\mathrm{SW}}} \right) \tag{5.20}$$

The output voltage V_{OUT} varies with the load current I_{LOAD}, and the load regulation depends on RAMP. Figure 5.20 illustrates the duty cycle waveforms under light-load and heavy-load conditions.

To improve the output voltage accuracy, a high-gain loop is incorporated before the PWM controllers. Figure 5.21 depicts the dual-loop architecture. The primary loop (Loop 1) functions as a rapid PWM circuit, utilizing both V_{EA} and V_{OUT} information. Complementing this, the secondary loop (Loop 2) ensures highly accurate regulation. The EA integrates the error $\Delta V = V_{OUT} - V_{REF}$ to produce V_{EA}. Subsequently, a 500 MHz triangular RAMP signal is superimposed onto V_{EA}, yielding V_{RAMP}. The PWM signal is then generated through the comparison of V_{RAMP} and V_{OUT} by a high-speed comparator.

For a stability analysis, we break the V_{OUT} feedback signal path and divide the architecture into three parts: Loop 1, Loop 2, and output stage. For the output stage, we have

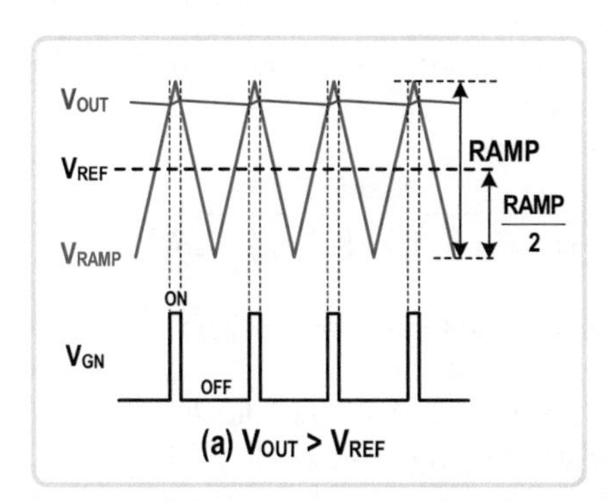

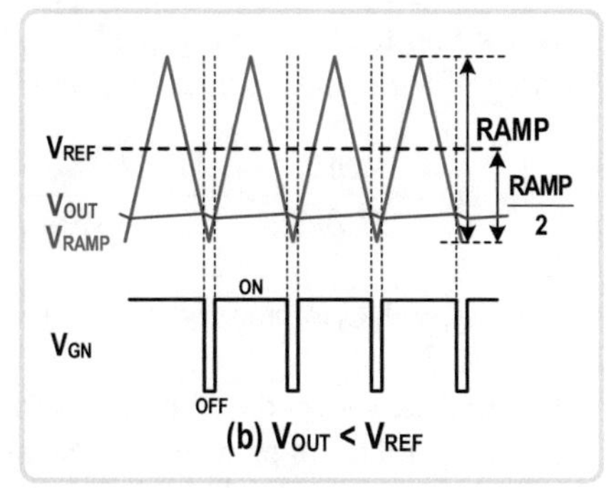

Fig. 5.20 The duty cycle waveforms under light-load and heavy-load conditions

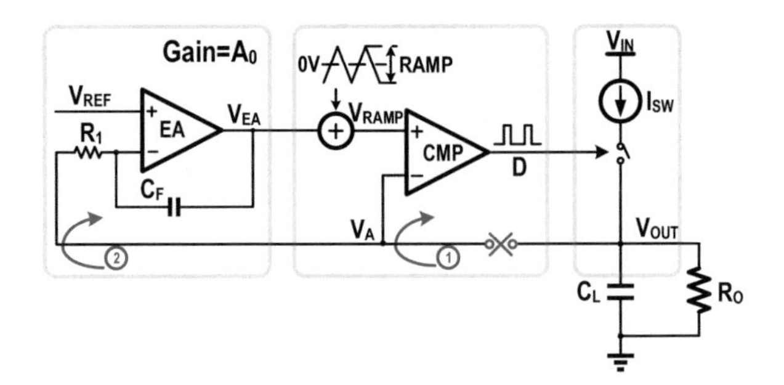

Fig. 5.21 A switching with dual-loop PWM-based control

$$V_{\text{OUT}} = D \times I_{\text{SW}} \times \frac{R_{\text{O}}}{1 + sC_{\text{L}}R_{\text{O}}} \tag{5.21}$$

$$\frac{\partial V_{\text{OUT}}}{\partial D} = \frac{I_{\text{SW}}R_{\text{O}}}{1 + sC_{\text{L}}R_{\text{O}}} \tag{5.22}$$

The transfer function of Loop 1 is

$$D = \left(V_{\text{EA}} + \frac{\text{RAMP}}{2} - V_{\text{A}}\right) / \text{RAMP} \tag{5.23}$$

$$\frac{\partial D}{\partial V_{\text{A}}} = \left(\frac{\partial V_{\text{EA}}}{\partial V_{\text{A}}} - 1\right) / \text{RAMP} \tag{5.24}$$

For the secondary loop (Loop 2), we can obtain

$$V_{\text{EA}} = \frac{-V_{\text{A}} \times A_0}{1 + (1 + A_0)sC_{\text{F}}R_1} \tag{5.25}$$

$$\frac{\partial V_{\text{EA}}}{\partial V_{\text{A}}} = \frac{-A_0}{1 + (1 + A_0)sC_{\text{F}}R_1} \tag{5.26}$$

where A_0 is the gain of the error amplifier.

According to expressions (5.22), (5.24), and (5.26), we can derive the transfer function of the complete dual-loop architecture:

$$H(s) = \frac{\partial V_{\text{OUT}}}{\partial V_{\text{A}}} = \frac{-(1 + A_0)I_{\text{SW}}R_{\text{O}}}{\text{RAMP}} \times \frac{1 + sC_{\text{F}}R_1}{(1 + sC_{\text{L}}R_{\text{L}})(1 + (1 + A_0)sC_{\text{F}}R_1)} \tag{5.27}$$

where $P_{\text{OUT}} = 1/C_{\text{L}}R_{\text{O}}$ and $P_{\text{EA}} = 1/(1 + A_0)C_{\text{F}}R_1$, and the left-half-plane (LHP) zero is $Z_1 = 1/C_{\text{F}}R_1$.

Figure 5.22 shows the Bode plot of the entire loop at different load conditions. The loop exhibits approximately consistent unit-gain bandwidth (UGB) across varying load conditions. This phenomenon can be attributed to the presence of the output pole within the bandwidth, coupled with the proportional adjustment of loop

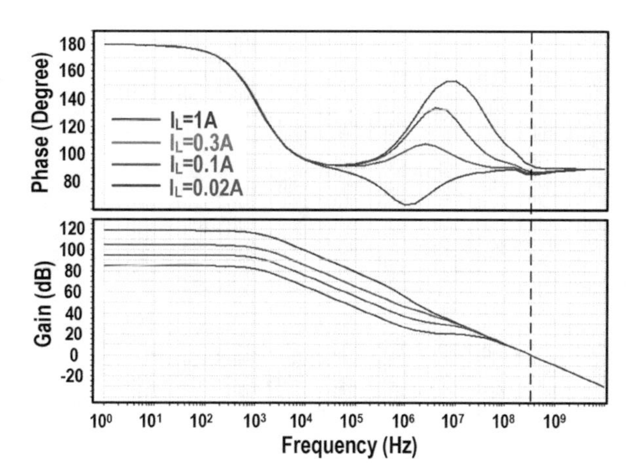

Fig. 5.22 Simulated Bode plot of dual-loop architecture

gain as the pole shifts. Leveraging Eq. (5.27), we can derive a more precise expression for the UGB:

$$\mathrm{UGF} = \omega_0 \approx \frac{I_{\mathrm{SW}}}{\mathrm{RAMP} \times C_{\mathrm{L}}} \tag{5.28}$$

According to Eq. (5.28), simultaneously increasing or decreasing the load current capability I_{SW} and output capacitance C_{L} in the same proportion does not alter the UGB. A constant UGB ensures the stability of the loop.

This architecture presents a trade-off between stability and transient performance. A low RAMP value results in excessive bandwidth, potentially leading to instability due to propagation delays and high-frequency parasitic poles. Conversely, increasing the RAMP value reduces the UGB, which adversely affects transient performance. Augmenting the output capacitance offers a viable solution to enhance both stability and transient performance simultaneously.

Simultaneous increase of the output capacitance and load capacity maintains the loop bandwidth, thereby ensuring system stability. This characteristic allows for the expansion of the load capacity in this architecture. Figure 5.23 shows the block diagram of the dual-loop switching LDO in [6]. The power transistor design incorporates current-limiting technology, as discussed in Sect. 4.7, and employs the fast-slow structure detailed in Sect. 5.2.3. Furthermore, the power switches are implemented as standard power cells, each capable of delivering a maximum current of 220 mA.

These power cells are positioned on the edge of a microprocessor in series or parallel, forming a power cell network. This modular approach facilitates easy reconfiguration of the architecture to accommodate varying load requirements without necessitating redesign of the main circuits and layouts.

The active voltage positioning (AVP) technique is often used in microprocessor power management [10]. This approach aims to relieve the load transient requirement on power supplies. The fundamental principle is to set V_{OUT} at a light-load condition to a voltage slightly higher than its nominal value and to set V_{OUT} at a

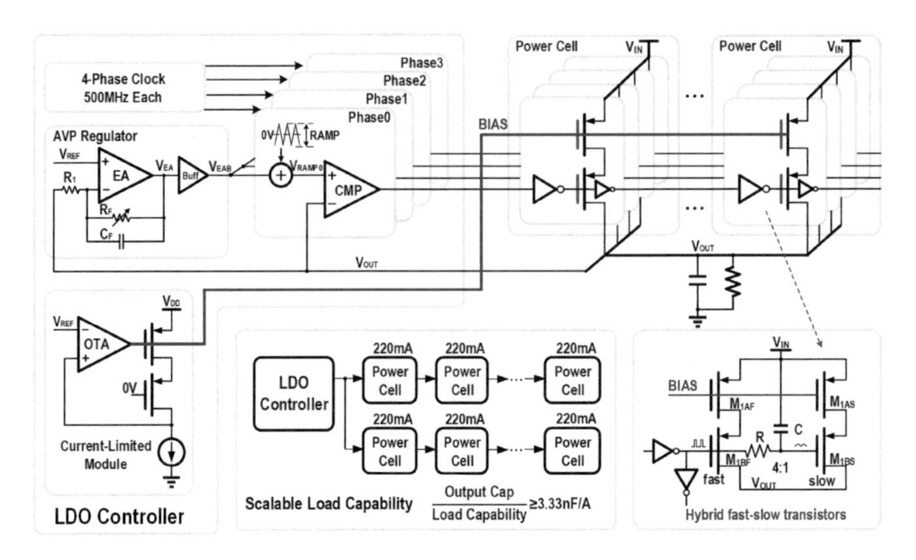

Fig. 5.23 Block diagram of the dual-loop LDO with scalable load capability in [6]

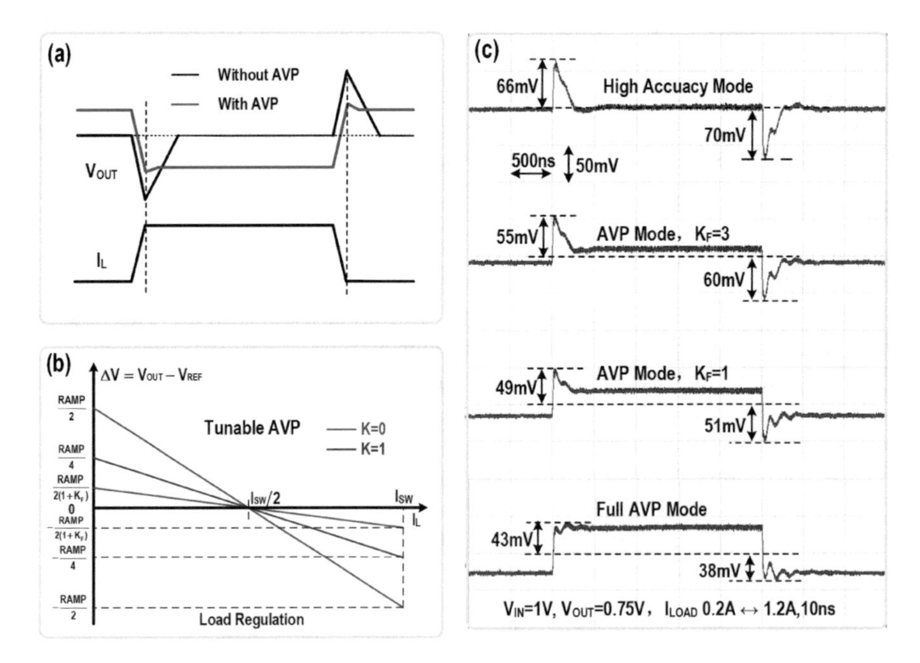

Fig. 5.24 (**a**) Active voltage positioning, (**b**) load regulation in AVP mode, and (**c**) transient response with tunable AVP function

heavy-load condition to a slightly lower voltage. Although this method sacrifices DC output accuracy, it offers significant advantages. AVP not only reduces the transient drops, but also reduces processor power consumption during heavy-load operations, as shown in Fig. 5.24a.

To implement a tunable AVP function, we incorporated a variable resistor, RF, into the EA circuit. By changing the resistance value of RF, different AVP effects can be achieved, and the LDO exhibits varying load regulations (Fig. 5.24b). Figure 5.24c shows transient measured waveforms at different AVP levels. In high-accuracy mode, the voltage for light and heavy loads is nearly the same, with the maximum transient drop. However, in full AVP mode, the voltage for light loads is noticeably higher than for heavy loads, but the corresponding drop is significantly reduced.

5.3.6 Single-Loop Structure with ACC Control

The dual-loop architecture presented in [6] achieves rapid load transient response due to its fast PWM loop. However, its dynamic voltage scaling (DVS) and load transient recovery speed are constrained by the slower loop. A single-loop structure can potentially mitigate this limitation [11]. Figure 5.25 illustrates both dual-loop and single-loop switching.

In the dual-loop structure, transient response time is primarily determined by the operating frequency and response delay of the PWM comparator, while DVS speed and transient recovery time are contingent upon the adjustment speed of the voltage V_{EA}. Conversely, the single-loop structure employs a sole V_{OUT} feedback path, wherein DVS speed, transient response time, and transient recovery speed all depend on the V_{EA}'s adjustment speed. Consequently, when a single-loop structure satisfies rapid transient response requirements, it inherently achieves faster DVS and shorter recovery times.

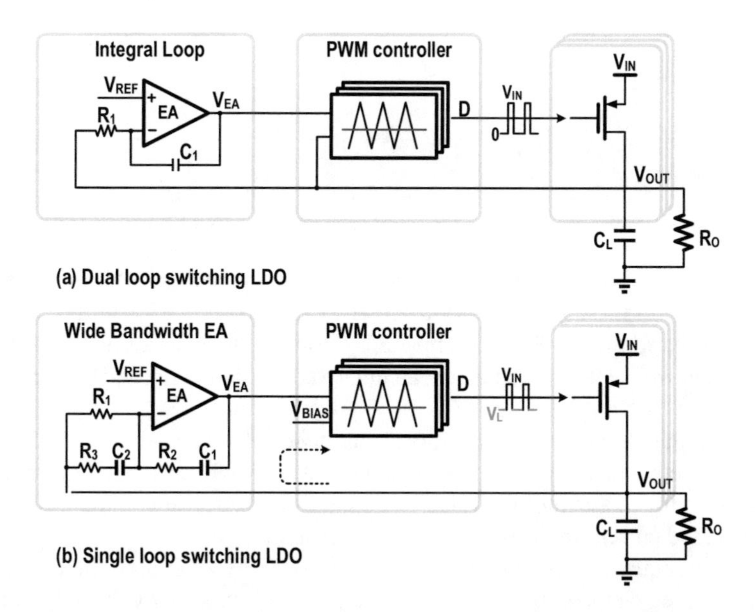

Fig. 5.25 (**a**) Dual-loop switching LDO and (**b**) single-loop switching LDO

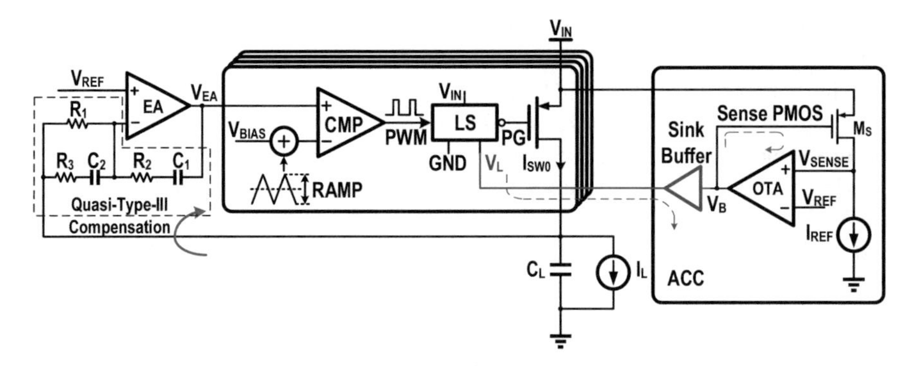

Fig. 5.26 Block diagram of the single-loop switching LDO with four-phase PWM control

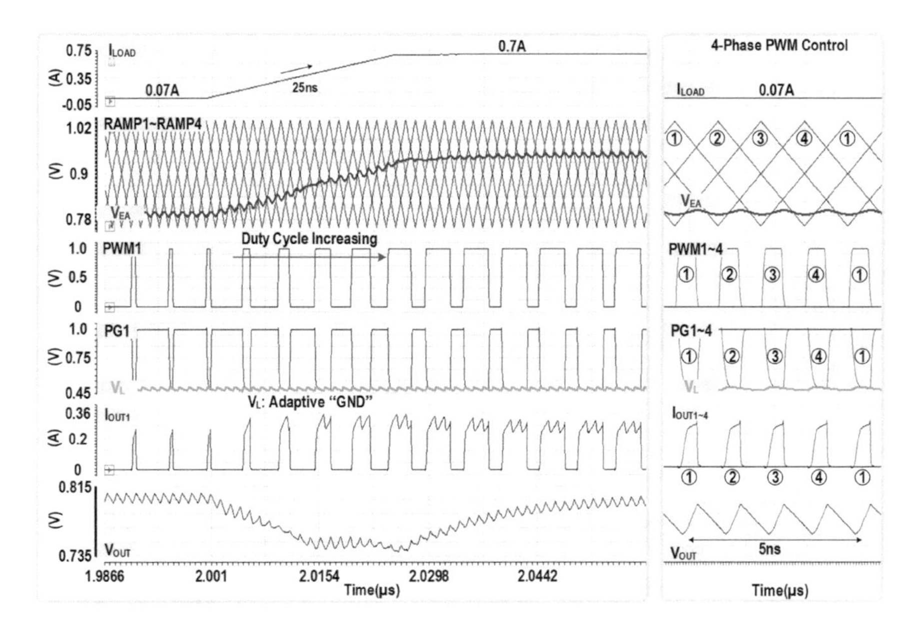

Fig. 5.27 Simulated waveforms of load transient and four-phase PWM control signals

Figure 5.26 illustrates the block diagram of the switching LDO incorporating single-loop four-phase PWM control. The EA utilizes a pseudo-type III compensation network, while the PWM comparator's triangle wave is centered at a fixed V_{BIAS}. The power switches implement an improved auxiliary constant current (ACC) technique, as discussed in Sect. 4.7.

The simulated load transient response and associated four-phase PWM control signals are depicted in Fig. 5.27. Notably, the power switch's gate voltage (indicated by the green trace) assumes the value V_{L} during its on-state, rather than GND. Upon a sudden load current increase, a momentary drop in V_{OUT} triggers a rapid elevation of V_{EA}. Consequently, the PWM controller responds by extending the duty cycle, thereby increasing the output current and facilitating the output voltage back to the target value.

5.4 Switching-Assisted

Since all the power transistors of the switching LDO operate at high frequency, the driving current is relatively large. To address this issue, Tsai et al. [10] replaced the high-speed comparator with a current-mode flash ADC and designed an event-driven digital LDO, as depicted in Fig. 5.28.

In the structure proposed by [10], the power transistors employ 4 bits (thermometer code), with 3 bits of power switches operating in digital mode and only 1 bit of power switches operating in switching mode. Consequently, the corresponding driving current and ripple are considerably reduced. Furthermore, since the duty cycle of the 1-bit power switch can be continuously adjusted, it enables fine-grained adjustment of the output current and supports a wider output current range. This is precisely the advantage of the switching operating mode.

Figure 5.29 illustrates the circuit of the current-mode flash (CMF) ADC and the power stage. The output voltage V_{OUT} generates a current I_{FB} through a local-feedback current generator. I_{FB} is mirrored to currents IPD [3:0], where I_{PD} has different mirroring ratios. The output voltage V_{CP} of the charge pump generates a current I_{PU}. I_{PU} is compared with IPD [3:0], and the output signal directly controls four groups of power switches.

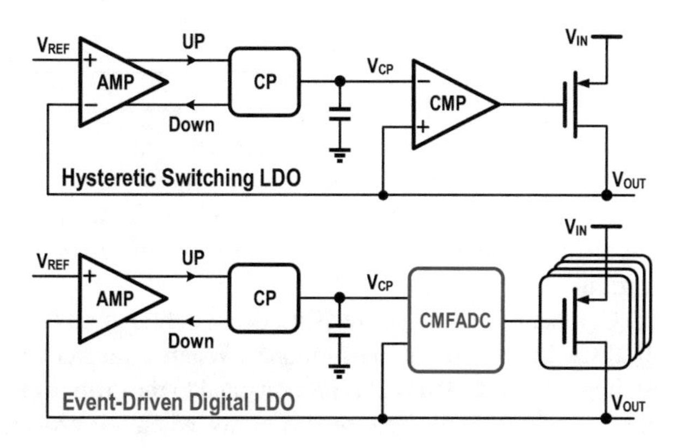

Fig. 5.28 Hysteretic switching LDO in [3] and event-driven digital LDO in [12]

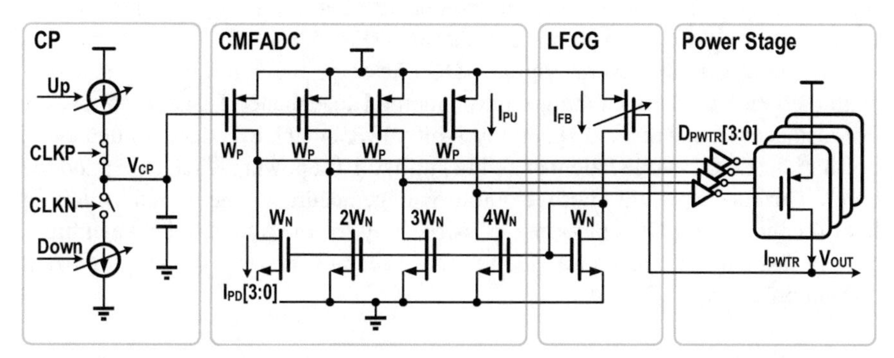

Fig. 5.29 Schematic of the CMFADC and the power stage in [12]

I_{LOAD}	State	PWTR[3]	PWTR[2]	PWTR[1]	PWTR[0]
Light	State A	Switching	OFF	OFF	OFF
↓	State B	ON D=0	Switching	OFF	OFF
	State C	ON	ON D=0	Switching	OFF
Heavy	State D	ON	ON	ON D=0	Switching

Fig. 5.30 Operating state changing with the load current

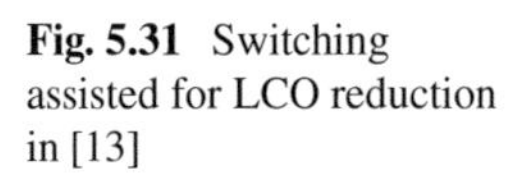
Fig. 5.31 Switching assisted for LCO reduction in [13]

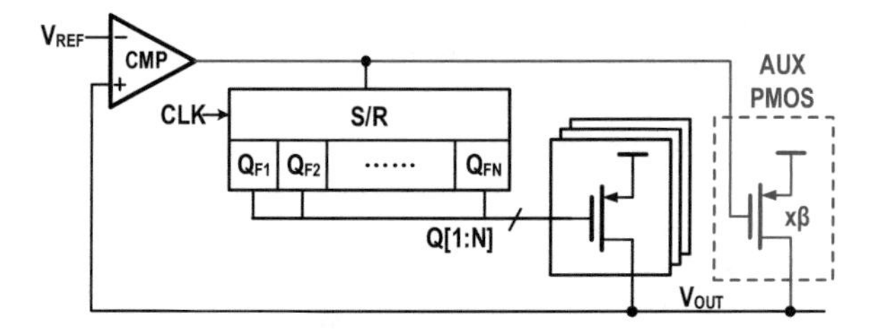

To ensure that the output current matches the load current, the first amplifier loop adjusts the value of V_{CP} such that the appropriate number of power switches is turned on, and the switching power switch operates at the appropriate duty cycle. Figure 5.30 depicts the power transistors' operating state changing with the load current.

Compared to the hysteresis mode switching structures in [4, 5], the approach in [12] sacrifices transient performance. When a transient occurs, the number of power switches turned on in [12] is proportional to the drop in the output voltage VOUT. All power switches will be turned on only when the drop reaches a certain value. In traditional hysteresis mode, once the comparator detects a drop (within a dozen mV), it can rapidly turn on all power transistors (Fig. 5.31).

Last but not least, several digital LDOs leverage the continuous adjustment characteristics of the switching control method to enhance LDO performance. For instance, Huang et al. [13] employ 1-bit power switch to reduce limited cycle oscillation (LCO), while Salem et al. [14] utilize 1-bit power switch to expand the output current range and improve output voltage accuracy. The digital switching hybrid control for LDO could be a possible way for the trade-off between high-current fast-transient response, good output accuracy, and low quiescent current consumption (Fig. 5.32).

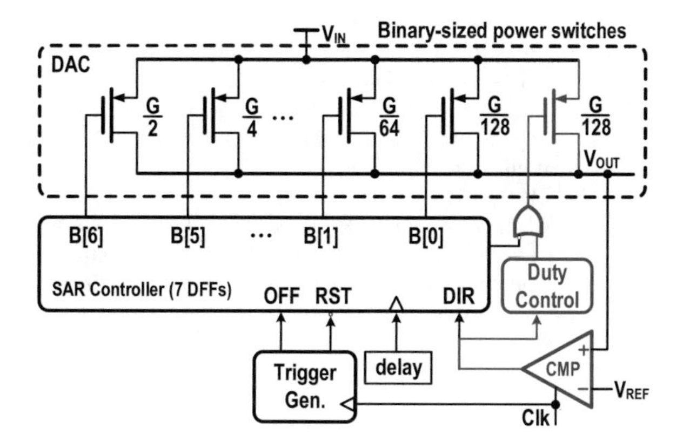

Fig. 5.32 Switching assisted for wide current range and high accuracy V_{OUT} in [14]

References

1. S. Kudva, S. Song, J. Poulton, J. Wilson, W. Zhao, C.T. Gray, A switching linear regulator based on a fast-self-clocked comparator with very low probability of meta-stability and a parallel analog ripple control module, in *Proceedings of IEEE Custom Integrated Circuits Conference (CICC)*, (2018), pp. 1–4
2. S. Saxena, S. Kudva, V. Srinivasan, M. Rodriguez, A distributed power supply scheme with dropout voltage in range 6mV–500mV and a low overhead retention mode, in *Proceeding of IEEE Custom Integrated Circuits Conference (CICC)*, (2024)
3. J. Bulzacchelli, Z. Deniz, T. Rasmus, J.A. Iadanza, W.L. Bucossi, S. Kim, Dual-loop system of distributed microregulators with high DC accuracy, load response time below 500ps and 85-mV dropout voltagE. IEEE J. Solid-State Circuits **47**(4), 863–874 (2012)
4. Z. Deniz, M. Sperling, J. Bulzacchelli, G. Still, R. Kruse, S. Kim, D. Boerstler, T. Gloekler, R. Robertazzi, K. Stawiasz, T. Diemoz, G. English, D. Hui, P. Muench, J. Friedrich, Distributed system of digitally controlled microregulators enabling per-core DVFS for the POWER8™ microprocessor, in *IEEE International Solid-State Circuits Conference (ISSCC) Digest of Technical Papers*, (2014), pp. 98–99
5. M. Perez, M. Sperling, J. Bulzacchelli, Z. Deniz, T. Diemoz, Distributed network of LDO microregulators providing sub-microsecond DVFS and IR drop compensation for a 24-Core microprocessor in 14-nm SOI CMOS. IEEE J. Solid-State Circuits **55**(3), 731–743 (2020)
6. X. Mao, Y. Lu, R.P. Martins, A scalable high-current high-accuracy dual-loop four-phase switching LDO for microprocessors. IEEE J. Solid-State Circuits **57**(6), 1841–1853 (2022)
7. Z. Wang, S.J. Kim, K. Bowman, M. Seok, Review, survey, and benchmark of recent digital LDO voltage regulators, in *IEEE Custom Integrated Circuits Conference (CICC)*, (2022), pp. 1–8
8. Y. Lee, W. Jang, H. Bae, J. Cho, H. Kim, 34.7A/mm^2 scalable distributed all-digital 6×6 dot-LDOs featuring freely linkable current-sharing network: a fine-grained on-chip power delivery solution in 28nm CMOS, in *IEEE International Solid-State Circuits Conference Digest of Technical Papers*, (2024), pp. 272–273
9. L. Zhao, Y. Lu, R.P. Martins, A digital LDO with Co-SA logic and TSPC dynamic latches for fast transient response. IEEE Solid-State Circuit Letters **1**(6), 154–157 (2018)
10. C.-H. Tsai, B.-M. Chen, H.-L. Li, Switching frequency stabilization techniques for adaptive on-time controlled buck converter with adaptive voltage positioning mechanism. IEEE Trans. Power Electron. **31**(1), 443–451 (2016)
11. X. Mao, Y. Lu, R.P. Martins, A 1-A switching LDO with 40-mV dropout voltage and fast DVS. IEEE Trans. Circuits Syst. II Express Briefs **67**(4), 725–729 (2020)

12. D.H. Jung, T.H. Kong, J.H. Yang, S. Kim, K. Kim, J. Park, M. Choi, J. Shin, A distributed digital ldo with time-multiplexing calibration loop achieving 40A/mm^2 current density and 1mA-to 6.4A ultra-wide load range in 5nm FinFET CMOS, in *IEEE International Solid-State Circuits Conference Digest of Technical Papers*, (2021), pp. 414–415
13. M. Huang, Y. Lu, S. U, R.P. Martins, A digital LDO with transient enhancement and limit cycle oscillation reduction, in *IEEE Asia Pacific Conference on Circuits and Systems (APCCAS)*, (2016)
14. L.G. Salem, J. Warchall, P.P. Mercier, A 100nA-to-2mA successive-approximation digital LDO with PD compensation and Sub-LSB duty control achieving a 15.1-ns response time at 0.5V, in *IEEE International Solid-State Circuits Conference Digest of Technical Papers*, (2017), pp. 340–342

Chapter 6
Distributed LDO

A fully integrated LDO is usually placed on one side or corner of the load chip to shorten the power supply distance and reduce IR drop (Fig. 6.1a). For high-current applications, especially when the physical domain is large (>1 mm2), the IR drop problem on the power supply network becomes very serious due to the parasitic impedance of the power delivery network. A larger IR drop will increase the guard band voltage of the digital system, thereby increasing the system power consumption. Furthermore, when a local transient occurs in an area far away from the LDO, the LDO may not be able to respond in time due to the long distance, resulting in a substantial local voltage droop and potential system malfunctions.

Distributed LDO involves splitting a large LDO into multiple smaller sub-LDOs and integrating them into the power delivery network, as shown in Fig. 6.1b. This distributed power supply network reduces on-chip current redistribution, thereby

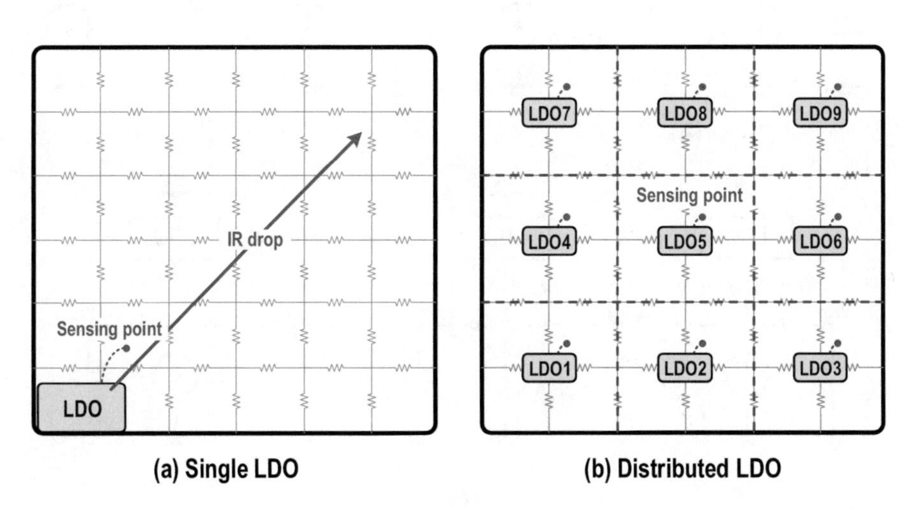

Fig. 6.1 (**a**) A single LDO and (**b**) a distributed LDO scheme

X. Mao et al., *Fully-Integrated Low-Dropout Regulators*, Analog Circuits and Signal Processing, https://doi.org/10.1007/978-3-031-84916-9_6

reducing the IR drop in the power grid on-chip. In addition, due to the multiple sensing points and the short distance to the load circuit, distributed sub-LDOs can respond fast to sudden local load transients.

A distributed LDO scheme offers many advantages over a single LDO design, but with more complexity. We are going to introduce it in detail in this chapter, including its performance benefits, design challenges, and parallel and dual-loop schemes.

6.1 Benefits of Distributed LDOs

Figure 6.2 depicts a simplified LDO power supply model, where C_L denotes the output capacitor, $I_{LOAD}[1]$ is the near-end load, $I_{LOAD}[2]$ represents the far-end load, and R_G is the parasitic resistance of the power delivery network. In Fig. 6.2a, a near-end feedback is employed; the voltage at $I_{LOAD}[1]$ is equal to V_{REF}, while the voltage at $I_{LOAD}[2]$ will decrease due to IR drop, which may affect the normal operation of $I_{LOAD}[2]$. Figure 6.2b utilizes a far-end feedback; the voltage at $I_{LOAD}[2]$ is equal to V_{REF}, and the near-end voltage will be higher, resulting in power dissipation.

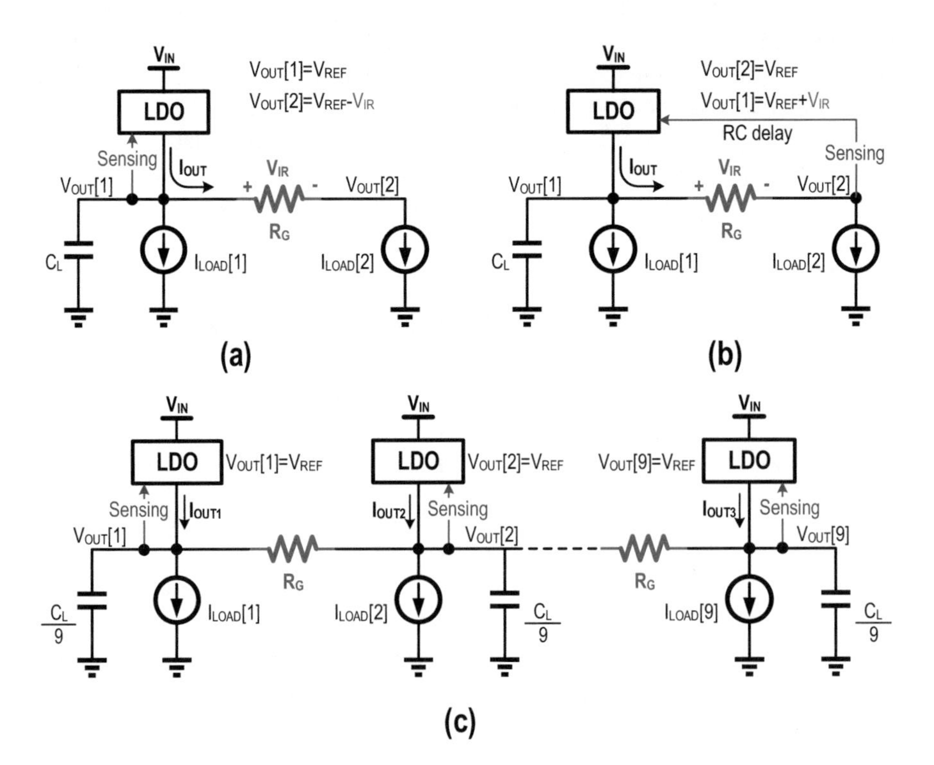

Fig. 6.2 (**a**) A single LDO with proximal feedback, (**b**) a single LDO with remote feedback, and (**c**) distributed LDOs with multiple local sensing points

Additionally, it will also adversely affect the transient response speed due to the larger propagation delay.

Figure 6.2c illustrates a distributed power supply scheme, which has the following advantages:

1. **Reducing IR drop**:

   Multiple LDOs can supply power to the nearby local loads, reducing current redistribution and consequently reducing IR voltage and I^2R losses on the power delivery network.

2. **Improved local transient performance**:

   When an LDO is close to the load, it can respond fast to local load changes and improve local transient performance.

3. **Thermal management optimization**:

   The power dissipation of the whole LDO is also distributed across multiple sub-LDOs, avoiding the hotspot issue associated with a single high-power LDO and facilitating the overall thermal management.

4. **Improved robustness**:

   Since multiple LDOs are employed to deliver power, and each LDO has a certain design margin, the failure of a single LDO may not compromise the overall system power supply.

5. **Improved system flexibility**:

   Scalability is one of the core features of a distributed LDO scheme. For different application scenarios, new requirements can be accommodated by increasing or decreasing the number of sub-LDOs without redesigning the LDO circuit.

6.2 Challenges of Distributed LDOs

Compared with a single LDO, distributed LDOs have the above advantages, but there are also many corresponding design challenges.

Integration Challenge

On-chip distributed LDOs would be integrated into the digital load (Fig. 6.3a). The digital load environment is inherently noisy, necessitating the LDO to possess robust noise immunity. In addition, routing analog signals such as the LDO's reference voltage and bias currents globally within the digital load has significant layout and analog-digital mixed-signal co-design challenges, especially for digital LDOs that require multiple analog reference voltages [1–3].

Due to embedded integration, the backend layout engineers need to pay much attention to the routing complexities. In particular, the digital circuit also needs to

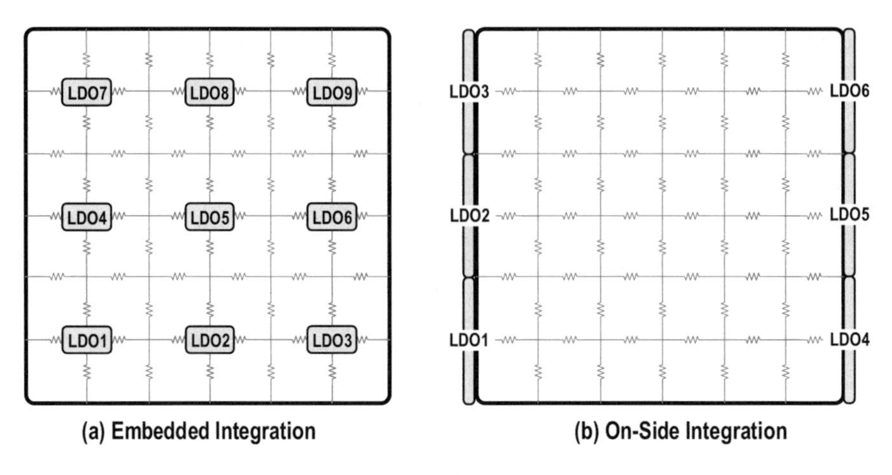

Fig. 6.3 (**a**) Embedded integration and (**b**) on-side integration

reserve free space inside to place the LDO, increasing the complexity of automated digital layout and routing. Therefore, a compact all-digital LDO devoid of analog global signals would be much more suitable for distributed applications.

A trade-off exists between power distribution granularity and integration complexity. Increasing the number of finely distributed LDOs aids in enhancing the power supply quality, but it also escalates the integration complexity significantly. As depicted in Fig. 6.3b, a compromise can be made by sacrificing certain performance and distributing the LDOs on both sides of the load instead of inside while leveraging package connections to reduce the impedance of the lateral power delivery network. This approach substantially mitigates the integration difficulties and presents a viable option for distributed integration.

Current Balancing Challenge

A distributed LDO system is composed of multiple sub-LDOs. Due to manufacturing deviations, each sub-LDO may have a different offset voltage. These sub-LDOs incorporate independent feedback and adjustment loops, so in theory, their output voltages will be different. Since the sub-LDOs share input and output grid, the output currents among the sub-LDOs are unbalanced.

The parasitic resistor R_G acts as a ballast resistor, and the unbalanced current is related to R_G and voltage difference, as illustrated in Fig. 6.4. A large R_G value results in small current unbalance. Conversely, when R_G is small, and the voltage difference is large, the high-output-voltage sub-LDO may operate at maximum capacity, while the sub-LDO with lower output voltage may be completely turned off. Severe current unbalances may cause local hotspots and electromigration issues. Furthermore, this unbalance compromises the ability of distributed regulators to respond to local disturbances, as a fully turned-on sub-LDO cannot provide additional current to the grid when the local voltage drops.

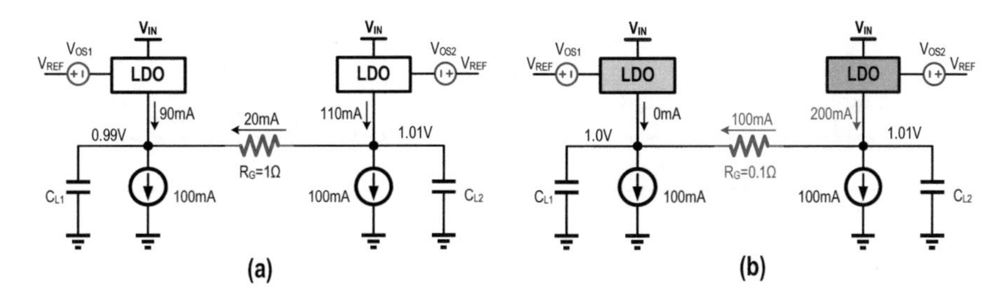

Fig. 6.4 Unbalanced current varies with (a) $R_G = 1\ \Omega$ and (b) $R_G = 0.1\ \Omega$

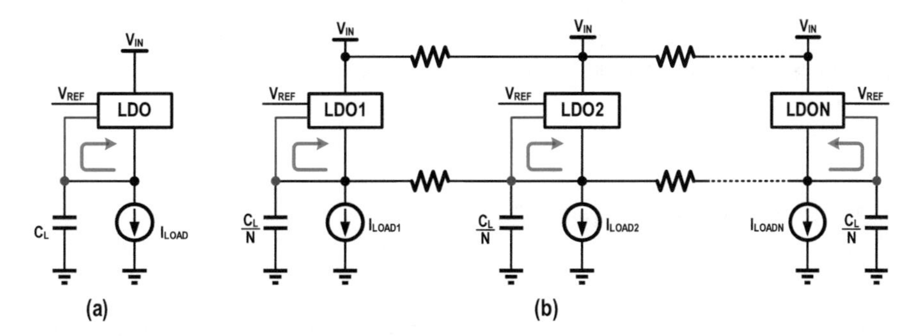

Fig. 6.5 (a) Single-loop LDO and (b) distributed LDO system

Stability and Scalability

As depicted in Fig. 6.5, compared to a single LDO, distributed LDOs can be considered as a multi-input, multi-output (MIMO) system [4]. A comprehensive theoretical stability analysis of such a system is highly complex. In addition to the typical factors influencing the LDO stability, we list three critical issues that may impact the stability of distributed LDOs.

1. Distributed LDOs share input and output networks. Adjustment of a local sub-LDO will also affect the other sub-LDOs accordingly. We need to consider whether the interaction of sub-LDOs would affect the system stability.
2. The output capacitance and parasitic impedance of the power network are distributed. Different load shapes and sizes, LDO integration methods, routing resources, and packaging forms result in significant variations in the parasitic impedance of the power delivery network. Thus, we also need to consider the impact of the parasitic impedance value on stability.
3. Assuming the total load capacity remains unchanged; theoretically, the more the sub-LDOs, the more evenly distributed output currents, the better the performances, and the more complex integration (Fig. 6.6). Therefore, we need to figure out how would the number of different sub-LDOs affect the system stability.

On the other hand, scalability is the core feature of distributed LDO. Adjusting the number of sub-LDOs to meet varying application requirements can significantly

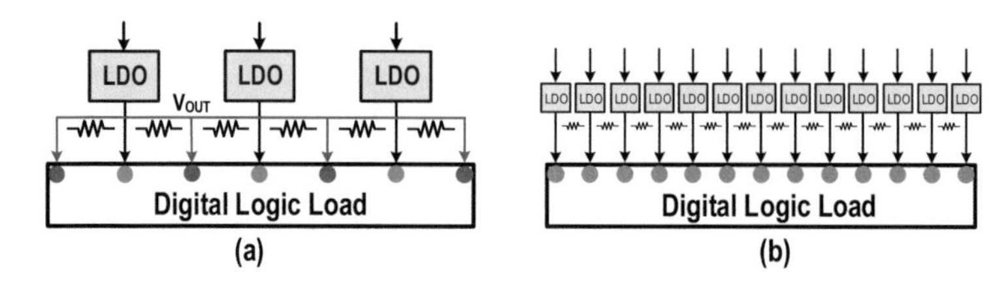

Fig. 6.6 (**a**) Coarse-grained distributed LDO and (**b**) fine-grained distributed LDO

reduce the development time and enhance design efficiency. Nonetheless, LDO stability is influenced by multiple factors including load capacity, output capacitance, and parasitic impedance. Practical application necessitates a grasp of scalability principles rooted in stability analysis before expanding upon this foundation can occur.

6.3 Parallel Distributed LDO

The most straightforward form of distributed LDO involves directly connecting multiple identical LDOs in parallel, also known as parallel distributed LDO. We will introduce several structures of parallel distributed LDOs in this section.

6.3.1 Event-Driven Digital Distributed LDO

Compared to the standalone single LDO, since each sub-LDO in a distributed LDO system features its own detection circuit and controller, a distributed LDO scheme consumes more quiescent current. Although conventional time-driven digital LDO structures can mitigate power consumption by lowering the operating frequency [5], this approach adversely impacts transient performance. Reference [6] presents an event-driven digital LDO that enhances current efficiency. In the steady state, its control circuit stops updating the state, thereby reducing the current consumption. More importantly, it can still improve the feedback loop latency at low power consumption and quickly respond to local transient changes.

Figure 6.7 illustrates the block diagram of the distributed LDO presented in [6]. Each sub-LDO incorporates an event-driven ADC, which outputs 5-bit thermometer code error information (LV[5:0]). Additionally, the architecture features independent proportional (P) controllers and integral (I) controllers. The P-controller manages three groups of PMOS arrays and two groups of 6-bit NMOS arrays, while the I-controller oversees a binary group of 9-bit power transistors.

For the work in [6], there is no intercommunication between the sub-LDOs. The reference voltage for each sub-LDO is generated by a DAC, and the clock is integrated within each sub-LDO. Figure 6.8 presents the schematic and detection

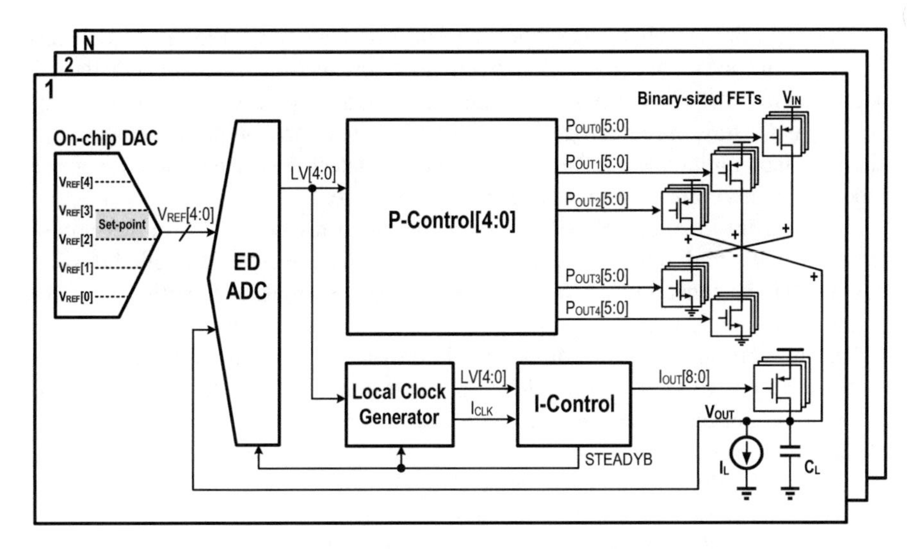

Fig. 6.7 The event-driven distributed LDO architecture in [6]

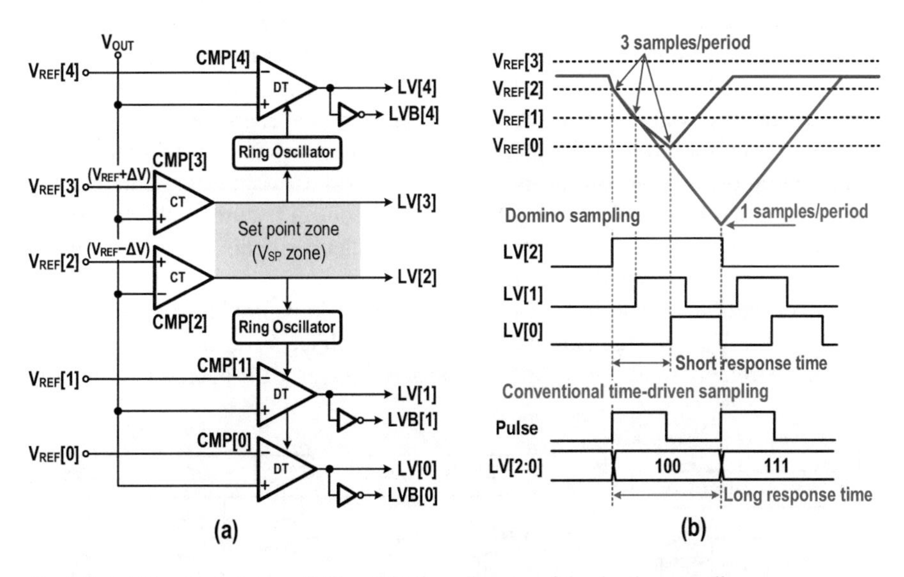

Fig. 6.8 (**a**) The event-driven ADC and (**b**) time diagram of the domino sampling

timing diagram of the event-driven ADC, which comprises two continuous-time (CT) comparators and three discrete-time (DT) comparators. The CT comparators CMP[2] and CMP[3], along with the reference voltages $V_{REF}[2]$ and $V_{REF}[3]$, implement a dead zone control. When the output voltage exceeds the range defined by $V_{REF}[2]$ and $V_{REF}[3]$, the DT comparators CMP[1] and CMP[4] are triggered. The operation of CMP[1] subsequently initiates CMP[0]. This sequential trigger detection mechanism is referred to as domino sampling in [6].

The event-driven scheme can improve detection latency with lower power consumption. In contrast to conventional fixed-clock time-driven detection, only the CT comparators CMP[2] and CMP[3] consume current during steady state in the event-driven ADC, thus minimizing the quiescent current consumption. When V_{OUT} changes, the continuous comparator can quickly detect the change of V_{OUT}, while time-driven detection requires at least one clock cycle.

The error information LV[4:0] output by the event-driven ADC is in thermometer code. In the traditional PID control scheme [7], the thermometer code information needs to be converted into binary code first and then multiplied with the control parameter to finally obtain the control code for the power transistors.

The proportional control component is crucial for determining the transient response speed. To minimize the delay in proportional control, the study [6] separates the proportional and integral control processes into two independent power supply paths. The I-control loop provides the steady-state output current, while the P-control loop addresses transient control currents. When the output voltage drops, the PMOS array in the P-control loop supplies a charging current; when the output voltage exceeds V_{REF}[3], the NMOS array provides a discharging current.

Figure 6.9 illustrates the conventional P-control and the domino P-control presented in [6]. In the domino P-control architecture, the power switches are organized into three PMOS arrays and two NMOS arrays. Notably, the output LV[4:0] from the ADC does not require binary encoding; instead, it directly controls the activation or deactivation of these arrays. The proportional coefficient is determined by multiplying K_P[1:0] with the preset value P_{ERROR}, which is a set of static values and is not on the critical path of the feedback control. Consequently, compared to the conventional P-control scheme, the domino P-control features a shorter critical path, thereby minimizing latency to the delay of a single trigger, which can improve the transient response speed of the LDO.

The conventional event-driven I-control has the sticking problem, as discussed in [8]. When the V_{OUT} voltage approaches the target voltage V_{REF}, the recovery speed becomes extremely slow. Synchronous control can increase the operating frequency

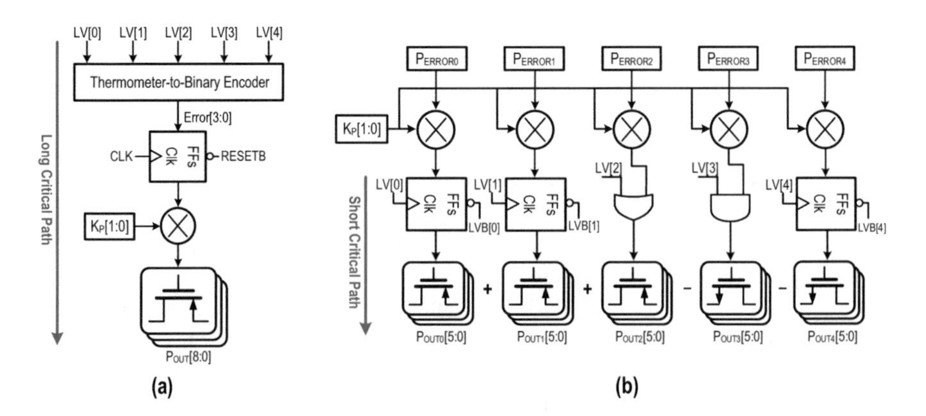

Fig. 6.9 (**a**) The conventional proportional controller and (**b**) domino P controller in [6]

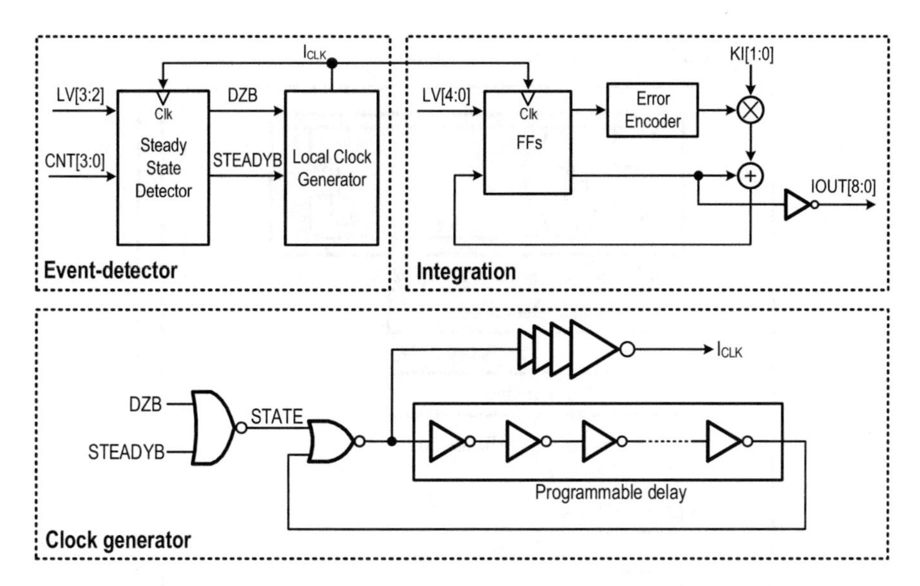

Fig. 6.10 The integral controller with a local clock generator in [6]

to reduce the settling time, but it will increase the power consumption and ripple in the steady state.

Figure 6.10 illustrates the integration controller and local clock generator in [6]. The steady-state detector determines whether the LDO is in idle mode or active mode based on the ADC output LV[4:0]. When V_{OUT} exceeds the set range, the LDO enters the active mode, prompting the local clock circuit to initiate operation. In this condition, the integration controller operates in time-driven mode. While meeting the stability requirement, a faster frequency helps to shorten the recovery time of V_{OUT}. Once V_{OUT} returns to the set voltage range and remains within it for N cycles, the LDO reverts to idle mode. Then, the local clock generator ceases operation, and the integration controller halts its output updates, which can eliminate the output ripple and stabilize V_{OUT} within the set voltage. This approach addresses the sticking problem while maintaining low static power consumption in the steady state.

6.3.2 All-Digital Parallel Distributed LDO

Since the distributed LDO is integrated within the digital load, the sub-LDO structure needs to be compact and anti-interference. A fully digital design is very helpful for distributed LDOs, and with the scaling of process, more area and performance benefits can be obtained. The study [9] presents a fully synthesizable parallel distributed LDO, with its block diagram illustrated in Fig. 6.11. This distributed structure comprises multiple sub-LDOs connected in parallel, each featuring its own power switches, digital supply voltage sensor (DSVS), and PID controller. Notably,

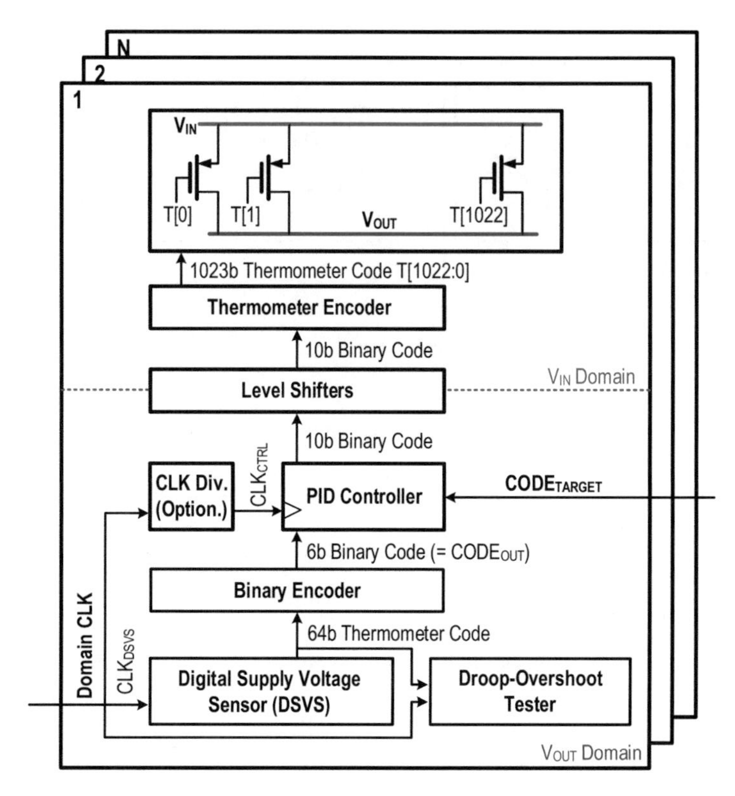

Fig. 6.11 The all-digital parallel distributed LDO in [9]

there is no communication between the sub-LDOs, nor is there a global controller. All circuits in distributed LDOs use standard library units and standard automatic layout wiring (APR) tools for comprehensive synthesis, layout, and wiring.

The DSVS presented in [9] employs an inverter chain time-to-digital converter (TDC) to quantize the output voltage. With a fixed clock cycle, variations in the supply voltage across the inverter chain result in differing delay times for the inverters, leading to corresponding differences in the output code values. The DSVS circuit is illustrated in Fig. 6.12. The input of DSVS is a 1.5 GHz clock, which is divided by two to generate a 750 MHz clock. This 750 MHz clock passes through the inverter chain every cycle. The flip-flop samples the inverter state at each rising edge of the clock. The multiplexer implements the polarity flipping of the code every cycle. DSVS can generate multiple output codes within a cycle (1.5 GHz). Additionally, the output of the DSVS utilizes AND gates to eliminate bubbles caused by metastability, simplifying the thermometer code to binary encoding logic.

However, the delay of the inverter chain in DSVS is sensitive to PVT variations. Additionally, frequency scaling can influence the output of the DSVS. To mitigate the effects of PVT variations, as well as frequency scaling, research [9] proposed a DSVS characterization and an active calibration flow. This approach employs the

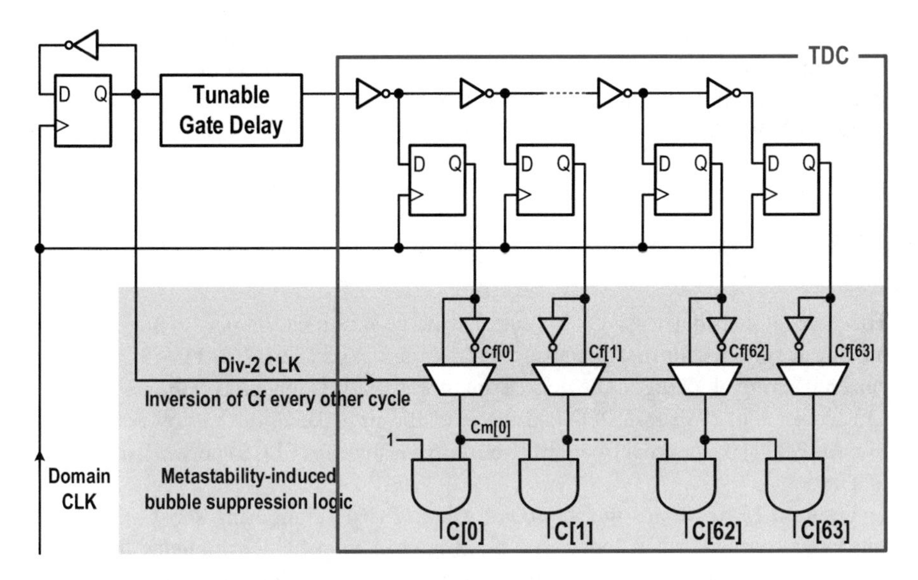

Fig. 6.12 The digital supply voltage sensor in [9]

trilinear interpolation [10] to update the target code value, $CODE_{TARGET}$, every <1 ms, based on the measured temperature, frequency, and target voltage. It should be noted that this flow requires sufficient distributed thermal sensors to detect the local temperature near the sub-LDO.

The DSVS outputs a 64-bit thermometer code value, which must be converted into a binary code value before being input into the subsequent PID controller. The power supply for the DSVS, binary encoder, and PID controller is provided by the output voltage V_{OUT}. The output of the PID controller is a binary code that is translated to the V_{IN} power domain through a level shifter. Subsequently, this binary code is converted into a thermometer code T[1022:0] to generate the control code for the power switches.

There is a current sharing problem in parallel distributed LDO. Take the study [9] as an example. The control code of the power switches can be expressed as

$$T = K_P \times e[n] + K_I \times \Sigma e[n] + K_D \times (e[n] - e[n-1]) \quad (6.1)$$

Here, $e[n]$ represents the difference between the V_{OUT} and V_{REF}. The coefficients K_P, K_I, and K_D denote the parameters of the PID controller. In the distributed LDO described in [9], each sub-LDO has its own independent DSVS and PID controller. Given that the sub-LDOs are located in various load regions, their DSVS output codes may exhibit relative quantization errors. According to Eq. (6.1), the difference in the control codes of the two sub-LDOs is given by

$$\Delta T = K_P \times \Delta e[n] + K_I \times \Sigma \Delta e[n] \quad (6.2)$$

It is evident that even a small quantization error can result in a large difference in the control code of the power switches due to the integration effect, leading to considerable current unbalance. In [9], the maximum unbalanced current may potentially reach 100%.

6.3.3 Parallel Distributed LDO with Current Sharing Network

The parallel distributed LDO structure has no communication between sub-LDOs, making it suitable for distributed integration and expansion. However, it faces the current sharing challenge. The degree of current sharing depends on the output voltage accuracy of the sub-LDOs and the parasitic impedance of the power network. The study [11] proposed a freely linkable current sharing (FLCS) network to address this issue.

Figure 6.13 illustrates the interconnection of two neighboring sub-LDOs. Each sub-LDO comprises two loops: the main control loop, which includes the ADC, regulation controller, and power switches (denoted by the blue line), and the FLCS loop (represented by the red line), which adjusts the input voltage D_{ADJ} for the main loop.

To achieve current sharing, it is essential to know the output current of the sub-LDO. For analog LDOs, a current detection circuit is usually necessary; however, in the case of digital LDOs, acquiring current information is straightforward, as the control code of the power transistor directly reflects the output current. To simplify the design, only the most significant bits (MSB) D<7:4> are utilized to represent the output current, which is sufficient to ensure a medium current sharing accuracy of (<12%).

LDO1 and LDO2 exhibit different input offset voltages, denoted as V_{OS1} and V_{OS2}. Without current sharing control, if V_{OUT1} exceeds $\mathrm{V}_{\mathrm{OUT2}}$, a current will flow from LDO1 to LDO2. In extreme cases, LDO2 may cease to supply power, resulting

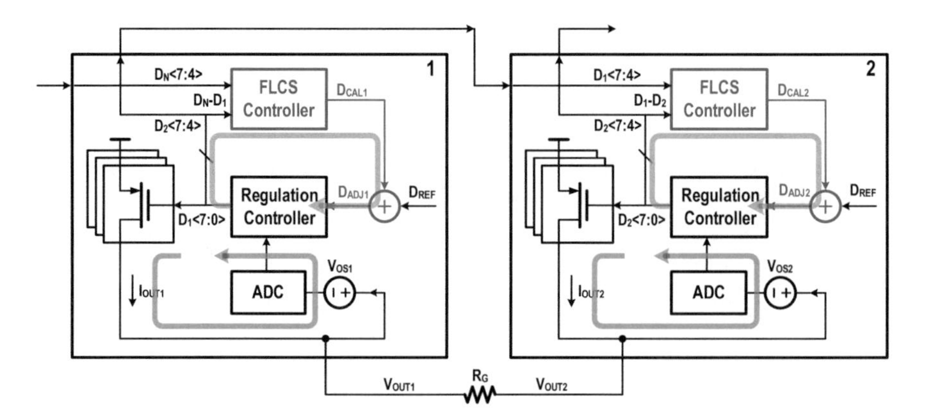

Fig. 6.13 Two neighboring sub-LDOs with FLCS in [11]

in all current being sourced solely from LDO1. To address the current sharing issue, LDO2 receives control information D<7:4>from LDO1 and compares it with its own D<7:4>. It then adjusts the LDO2 compensation voltage D_{ADJ}, which enhances the output current of LDO2 and reduces the output current of LDO1 accordingly. Through this collaborative approach between LDO1 and LDO2, current sharing will be finally achieved.

To ensure the stability of the current sharing loop, a gradual tracking method is employed, allowing I_{OUT1} and I_{OUT2} to progressively converge towards equality (Fig. 6.14). The adjustment speed for current sharing is not a critical parameter for distributed LDOs, and excessively immediate current sharing may cause oscillation issues.

Figure 6.15 presents the detailed block diagram of the distributed LDOs. The freely linkable current sharing controller comprises a digital comparator (CMP) and an integrator. Its operating frequency is set to 1/8 of that of the main loop, with the output changing by +1 or −1 each cycle. The damping factor α in the main loop

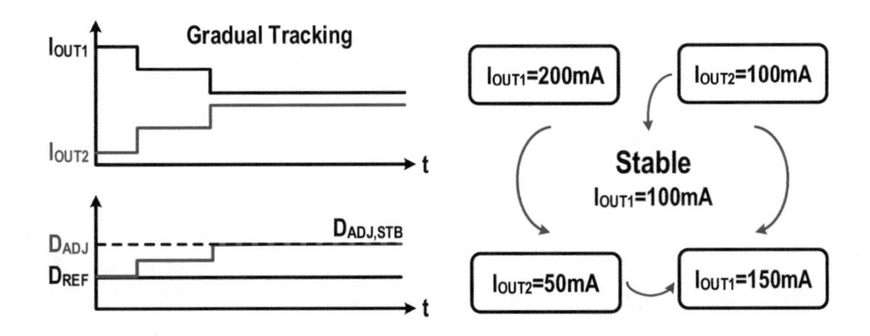

Fig. 6.14 Gradual tracking current balancing

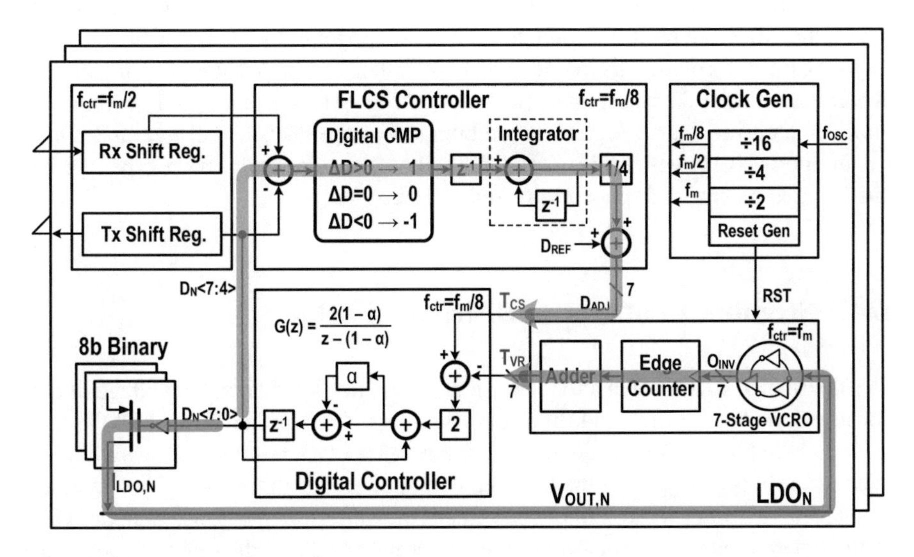

Fig. 6.15 The detailed block diagram of the distributed LDO in [11]

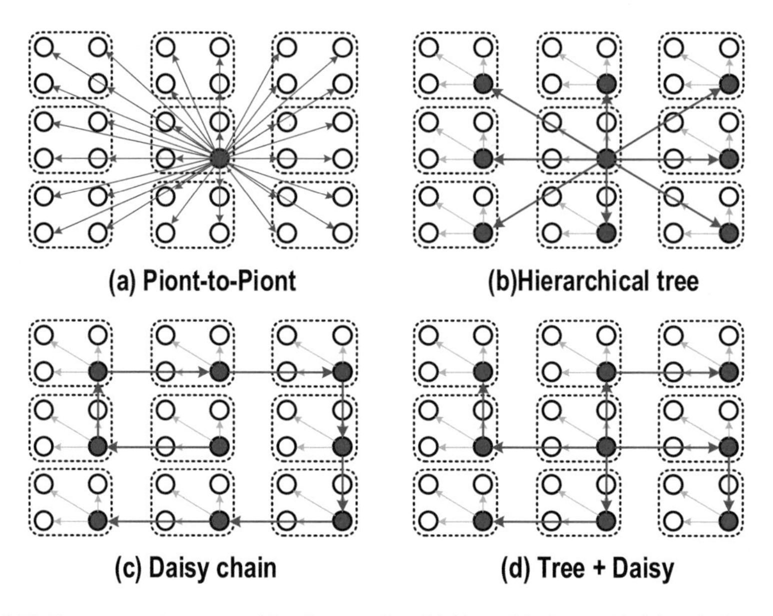

Fig. 6.16 Interconnection types: (**a**) point to point, (**b**) hierarchical tree, (**c**) daisy chain, and (**d**) hierarchical tree + daisy chain

controller influences both current sharing stability and voltage regulation performance. When α is set to 0, the freely linkable current sharing loop becomes unstable; conversely, if α approaches 1, the freely linkable current sharing loop stabilizes but at the expense of the main loop's voltage regulation performance. In [11], α is configured to 1/16 to optimize both stability and voltage regulation performance.

The current sharing network described in [11] supports various interconnection types, including the point-to-point configuration shown in Fig. 6.16a, which broadcasts control information from one LDO to the other LDOs. Additionally, it features a hierarchical tree distribution as illustrated in Fig. 6.16b, a daisy chain configuration depicted in Fig. 6.16c, and a combination of the hierarchical tree and daisy chain approach shown in Fig. 6.16d. The choice of interconnection type can be made based on specific application requirements.

6.4 Neighbor Cooperative Distributed LDO

In a distributed LDO, when a load transient occurs in a local area, the voltage of the local node will drop first, and the current will flow from the neighbor LDO to this node to help handle the transient. Then, the neighbor LDO output voltage will also drop accordingly, triggering the loop regulation of the neighbor LDO. This interaction creates a "water wave" effect that spreads outward from the transient node. This is the inherent passive regulation of the power network, and the diffusion speed of the "water wave" is influenced by the impedance of the network. To further improve

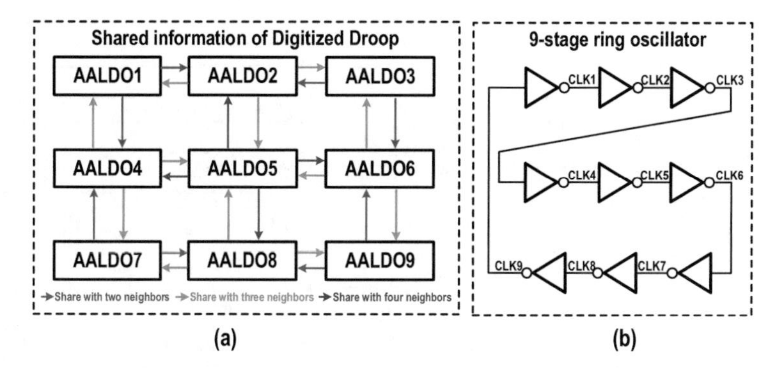

Fig. 6.17 (**a**) A 3 × 3 distributed LDO with droop information sharing and (**b**) ring oscillator

the transient response performance and mitigate voltage drops, the study [12] proposed a neighbor cooperative distributed LDO structure.

Figure 6.17 illustrates a 3 × 3 distributed LDO network of this structure. Sub-LDOs 1, 3, 7, and 9 have two neighbors, while sub-LDOs 2, 4, 6, and 8 each have three neighbors, and sub-LDO 5 has four neighbors. Each sub-LDO features an independent local feedback network and shares local voltage information with neighboring LDOs. When a load transient occurs, the neighboring LDOs actively adjust their control loops and increase their output current upon receiving voltage information from the affected area. This cooperative regulation can significantly improve transient performance.

The maximum time for a single phase to detect a voltage drop is one clock cycle. In [12], the nine sub-LDOs utilize phase-interleaved clock signals, allowing the detection time for a drop to be reduced to only 1/9 of a clock cycle. In the steady state, the clock frequency is set to 16 MHz. When the output voltage of any of the nine LDOs exceeds the specified range, the frequency increases immediately to 100 MHz, thereby reducing the LDO recovery time and enhancing transient performance.

Figure 6.18 illustrates the circuit architecture of the neighbor cooperative distributed LDO in [12], where the sub-LDO is implemented using an analog-assisted digital LDO scheme [13]. The 128 power MOSFETs and their buffers are organized into 16 groups, and the 8 buffers in each group share the same power supply voltage rail V_{BFH}/V_{BFL}. The voltage supply rail of each group of buffers can be switched between the digital V_{DD}/GND and the analog control voltage V_{ANA}.

The digital loop comprises two comparators and a digital controller. The comparators compare the local voltage V_{OUT} with the voltage boundary $V_{REF} \pm \Delta V$. The comparison results Q[1:0] correspond to three states: "11" for when the local voltage is above the boundary, "10" when it is at the boundary, and "00" when it is below the boundary. The finite-state machine (FSM) in the digital controller defines the state of the LDO based on the comparison results Q_N[1:0] of the local and neighboring LDOs. The LDO can operate in three modes: digital burst mode (DBM), digital-analog transition mode (DATM), and analog steady mode (ASM), as shown in Fig. 6.19.

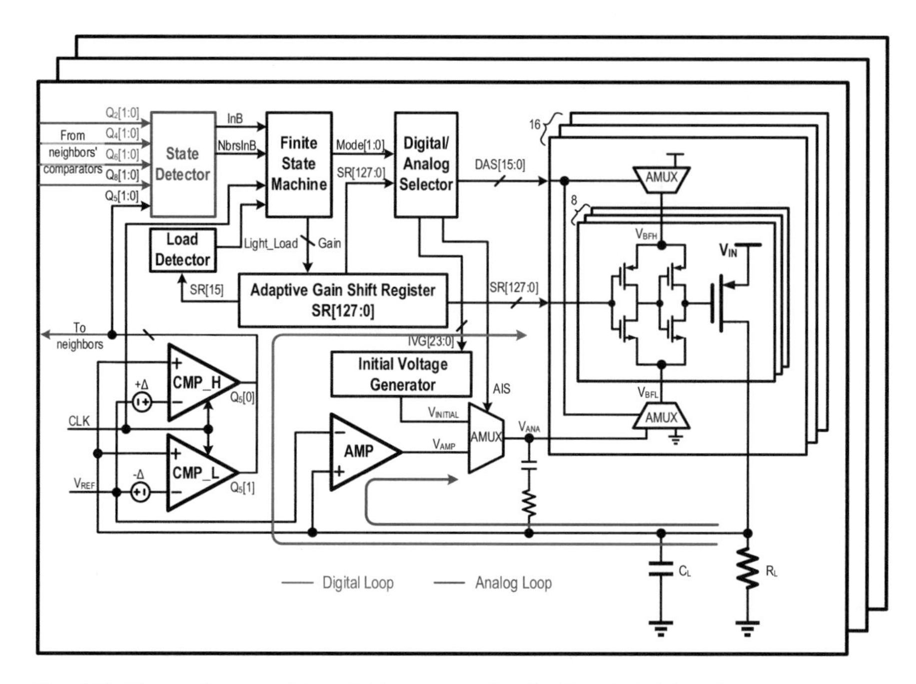

Fig. 6.18 The architecture of the neighbor cooperative distributed LDO in [12]

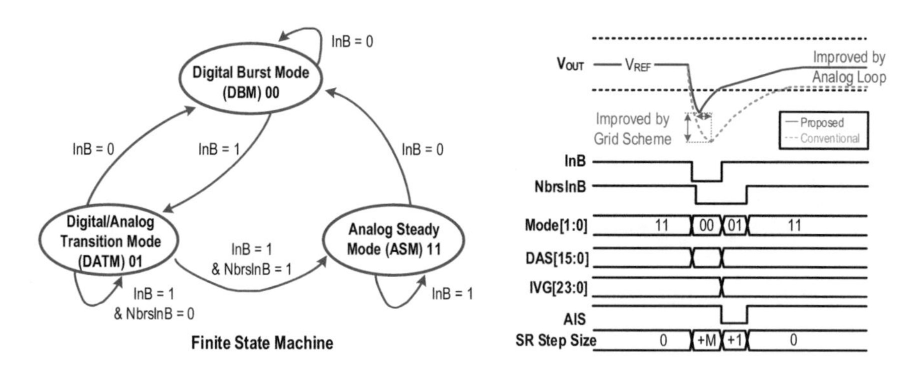

Fig. 6.19 The finite-state machine and operation timing diagram in [12]

When the V_{OUT} voltage of the local or neighbor LDOs falls outside the defined boundary (InB = 0 or NbrsInB = 0), DAS[15:0] selects V_{BFH} and V_{BFL} of all groups to connect to V_{DD} and ground, respectively. The controller operates DBM to achieve a fast response. The 128b adaptive gain shift register (AGSR) begins to shift left or right and can speed up the response by shifting multiple bits in one clock cycle during digital burst mode.

Taking LDO5 as an example, the gain of the shift register can be adjusted from −5 to +5 based on the local comparator result $Q_5[1:0]$ and the statuses of the four neighboring LDOs $Q_{2,4,6,8}[1:0]$. Figure 6.20 illustrates the step size in three modes.

Upon V_{OUT} entering the defined boundary (InB = 1), the controller will work in the digital-analog transition mode. Concurrently, the adaptive gain shift register

Q[1:0]	00						10	11					
Ligh Load	1	0						1	0				
	–	Num_Ngbrs$_{Q[1:0]=00}$					–	–	Num_Ngbrs$_{Q[1:0]=11}$				
SR Step Size		4	3	2	1	0			4	3	2	1	0
DBM	+1	+5	+4	+3	+2	+1	0	-1	-5	-4	-3	-2	-1
DATM	+1						0	-1					
ASM	0												

Fig. 6.20 The step size of AGSR in three modes

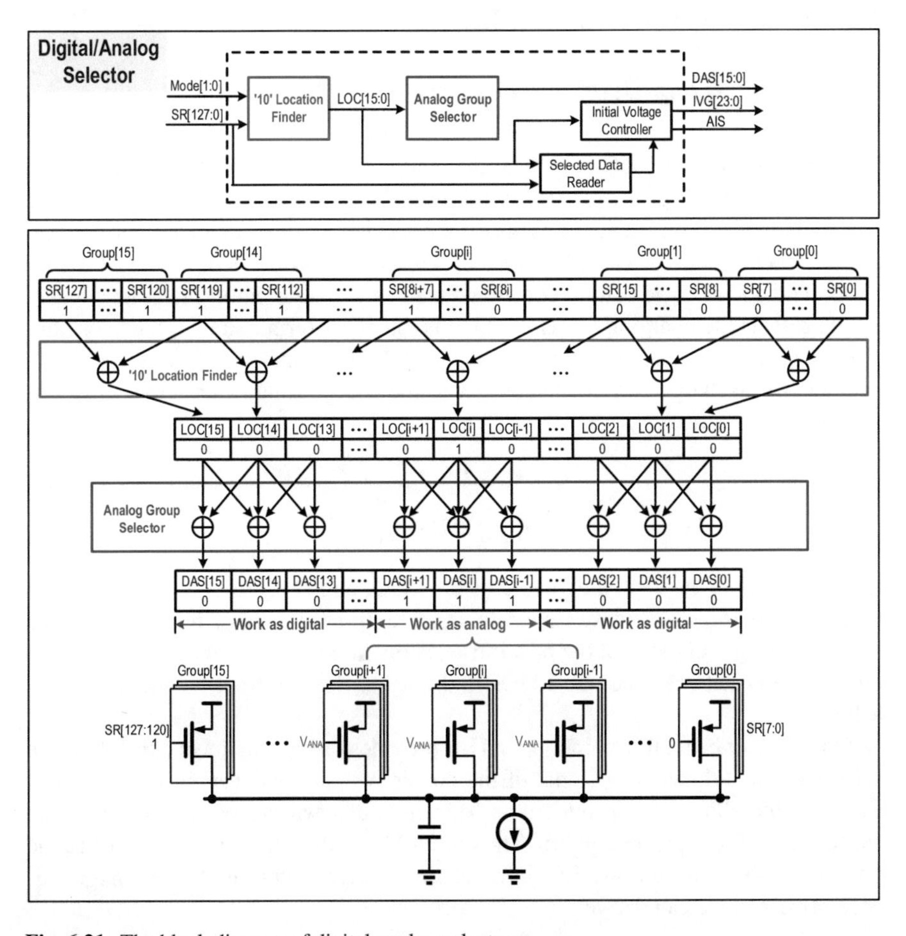

Fig. 6.21 The block diagram of digital-analog selector

(AGSR) halts its shifting operation. The initial voltage generator, controlled by IVG[23:0], generates the initial voltage, thus preparing the system for the smooth transition from digital to analog state.

The digital-analog selector is used to find the "10" boundary in the adaptive gain shift register and then selects this group along with the adjacent two groups to be controlled by the analog loop, as shown in Fig. 6.21. This selective control implies that only the gate voltages of these three selected groups are connected to the amplifier while the remaining power MOSFETs are still fully turned on/off by the digital loop.

When all four neighbors enter the boundary, the controller enters analog steady mode and the amplifier takes over control. With the help of the analog-assisted loop, the gate of the power MOSFET can be accurately controlled by the analog voltage, reducing the output ripple and improving the voltage regulation accuracy.

6.5 Dual-Loop Distributed LDO

6.5.1 Dual-Loop Distributed Switching LDO

Utilizing a switching LDO structure, the study [14] introduced the first dual-loop distributed LDO (Fig. 6.22a). The working principle of the switching LDO was discussed in Chap. 5. Here, we focus on the mechanism for achieving current balancing among its sub-LDOs. Figure 6.23 presents the equivalent model of a sub-LDO. The power stage can be conceptualized as a charge pump integrator, where I_{SW} represents the charging current, which is mainly determined by the power transistor strength and the input-output voltage difference, and the load current I_{LOAD} denotes the discharge current. According to the charge balance principle in steady state, we can express

$$I_{\mathrm{SW}} \times D \times T = I_{\mathrm{LOAD}} \times T \tag{6.3}$$

$$I_{\mathrm{LOAD}} = D \times I_{\mathrm{SW}} \tag{6.4}$$

where T denotes the switching cycle of the LDO, and D represents the duty cycle of the power transistor in steady state. According to Eq. (6.4), regardless of the manufacturing variations of the power transistors in different regions, as long as the duty cycle D between the sub-LDOs is the same, load current balancing can be achieved.

On the other hand, the main loop in [14] provides the up and down currents to the sub-LDOs, and the control signal of the power switches also serves as the control signal of the charge pump integrator. The up and down currents, along with the capacitor C_P, form a charge pump. The output V_{CP} of the charge pump serves as the reference voltage for the sub-LDO. In the steady state, according to the charge balance principle, we can derive

$$I_{\mathrm{DOWN}} \times D \times T = I_{\mathrm{UP}} \times (1 - D) \times T \tag{6.5}$$

$$\frac{I_{\mathrm{UP}}}{I_{\mathrm{DOWN}}} = \frac{D}{1 - D} \tag{6.6}$$

$$D = \frac{I_{\mathrm{UP}}}{I_{\mathrm{UP}} + I_{\mathrm{DOWN}}} \tag{6.7}$$

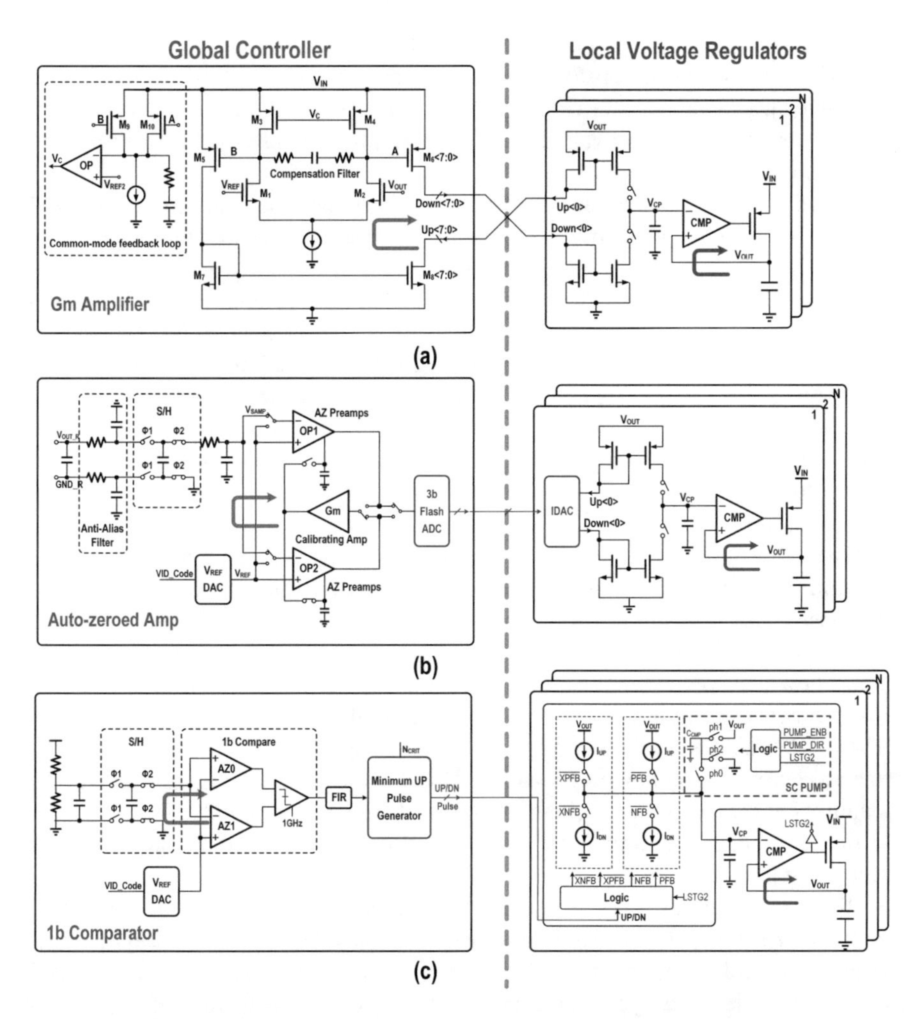

Fig. 6.22 Three dual-loop distributed switching LDO architectures: (**a**) with analog amplifier [14], or (**b**) with auto-zeroed amplifier [15], or (**c**) with auto-zeroed comparator [16]

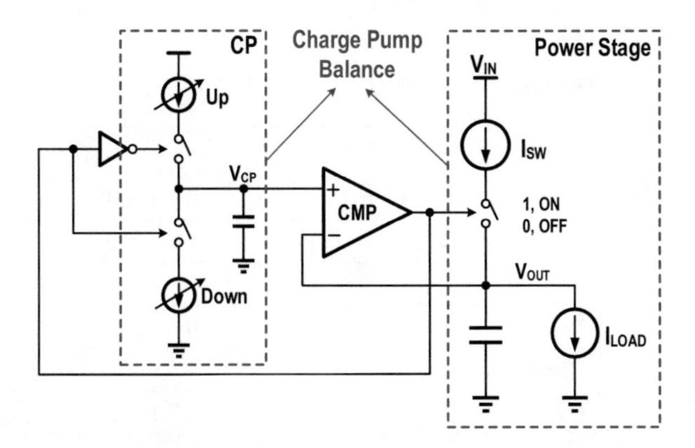

Fig. 6.23 The equivalent model of the local switching voltage regulator

As all sub-LDOs are governed by a shared global controller, they receive identical up and down current information. Consequently, all sub-LDOs operate with the same duty cycle D. Referring to Eq. (6.7), the same duty cycle ensures that all sub-LDOs generate equivalent output currents, thereby achieving current balancing.

In [14], the communication information between the global controller and local sub-LDO is up and down current signals. However, in processor environments, analog current information is not suitable for long-distance routing in a noisy environment. To solve this problem, the study [15] proposes a digital approach, as illustrated in Fig. 6.22b. The global controller employs an auto-zeroed amplifier. This design allows one amplifier to be operational in the loop while the other undergoes offset calibration, effectively mitigating offset voltage and enhancing output voltage accuracy. A 3-bit flash ADC quantizes the amplifier output into a 7-bit thermometer code, which is then transmitted to the sub-LDOs. Within each sub-LDO, a current digital-to-analog converter (IDAC) translates this code into the requisite up and down currents.

The work in [16] replaces the auto-zeroed amplifier with an auto-zeroed comparator. The comparator output undergoes processing through a finite impulse response (FIR) filter and the minimum up signal generator, resulting in a 2-bit up/down code. The code is subsequently transmitted to the sub-LDO. Table 6.1 illustrates how the up/down codes are gated with the signal LSTG2 (opposite to the driving signal of the power switches) to generate the charge pump control signal. The LDO structures proposed in [14, 15] require significant adjustment of the voltage V_{CP} during dynamic voltage scaling (DVS). However, the dual-loop structure's stability constraints limit the charging and discharging rates of the up/down current in the sub-LDOs. This limitation results in a slow adjustment of V_{CP}, consequently restricting the DVS rate.

To enhance the DVS rate, the study [16] introduces a switched-capacitor (SC) circuit to the sub-LDOs, which is activated during fast DVS operations. When V_{OUT} requires upward regulation, C_{PMP} alternates between V_{CP} and V_{OUT}, delivering a fixed charge to C_P in each cycle. Conversely, for downward regulation of V_{OUT}, C_{PMP} switches between GND and V_{CP} to expedite discharge.

In digital systems, load distribution is inherently nonuniform. Parallel distributed LDOs, as illustrated in Fig. 6.24a, attempt to equalize voltages across different

Table 6.1 The charge pump control signals

	LSTG2	UP	DN	$\overline{PFB}$	$\overline{NFB}$	$\overline{XPFB}$	$\overline{XNFB}$
Switches ON	0	1	0	0	0	0	0
	0	1	1	0	1	0	0
	0	0	1	0	1	0	1
Switches FF	1	1	0	1	0	1	0
	1	1	1	1	0	0	0
	1	0	1	0	0	0	0

regions, but encounter challenges related to current sharing. Various architectures for implementing dual-loop distributed LDOs are presented in Fig. 6.24b–e. Figure 6.24b depicts the most prevalent configuration, featuring a global controller governing distributed sub-LDOs with a single-sense point.

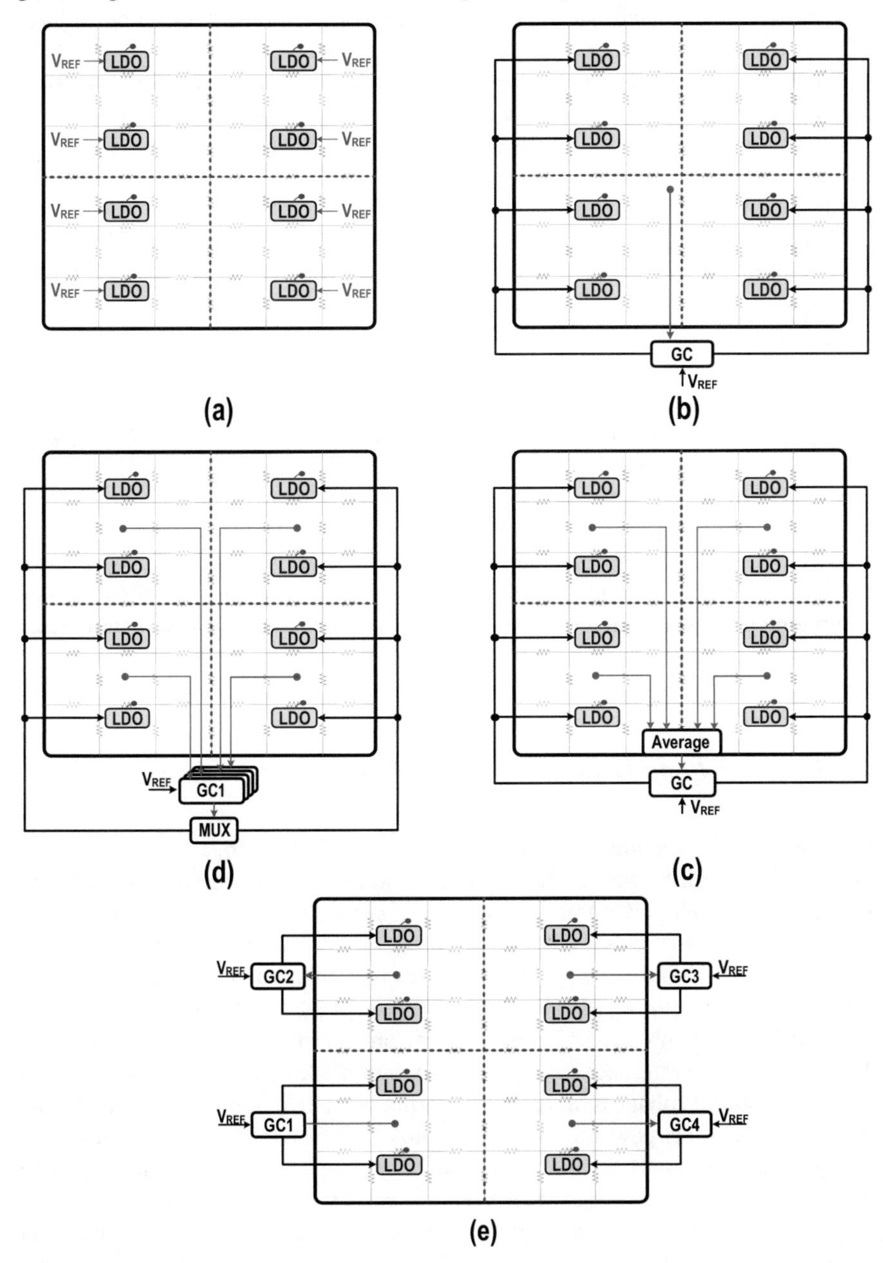

Fig. 6.24 Different sensing schemes in distributed LDOs: (**a**) parallel distributed LDO, (**b**) global controller with single-sense point, (**c**) global controller with average sensing, (**d**) global controller with multisense points, and (**e**) multi-sector scheme

The output voltage accuracy of dual-loop distributed LDOs is primarily determined by the global controller. However, in scenarios with unbalanced loads, the single-sense point global controller can only ensure voltage accuracy at the feedback point, leading to potential voltage disparities across different regions. Furthermore, load changes near the single-sense point will affect the output voltage DC value in other chip areas.

Figure 6.24c illustrates an alternative approach wherein voltages from multiple regions are detected and aggregated into an average voltage through a resistor network. This averaged voltage serves as feedback information for the global controller. This configuration mitigates the impact of localized load variations near the feedback point on the overall DC voltage.

Figure 6.24d illustrates multisense point detection, which dynamically selects the output of the global controller with the lowest feedback voltage as the actual control signal. This method ensures that the voltage at each detection point is equal to or higher than the target value, satisfying the V_{MIN} requirement. However, if the voltage of most chip areas is higher than the target value, it will also cause unnecessary power consumption.

Figure 6.24e presents a multi-sector configuration. The load is divided into multiple regions, each of which contains a complete dual-loop distributed LDO, that is, a global controller and several sub-LDOs. It is equivalent to the parallel connection of multiple dual-loop distributed LDOs. Within each sector, the sub-LDOs share current equally; however, current sharing between different sectors may present challenges. To alleviate this problem, the solution proposed in [16] implements a minimum duty cycle limitation. This approach prevents the LDO from completely shutting off, thereby maintaining continuous operation.

6.5.2 Dual-Loop Distributed Event-Driven Digital LDO

By employing a switching LDO structure, the works in [15, 16, 14] achieved a fast transient response. However, this approach resulted in relatively high power consumption due to all power transistors switching at high frequency. In contrast, the study [17] proposed dividing the power transistors into several units and implementing event-driven control. This method ensures that only a 1-bit power switch is in the flipping state in the steady state, effectively reducing the driving current.

Figure 6.25 shows the switching control and event-driven control. In event-driven control, the reference voltage is set to multiple levels. Under light-load conditions, the output voltage is high, and only power switch S_1 operates in the switching state, while the other power transistors remain off. As the load increases, the output voltage decreases; S_1 is fully turned on, activating path 2, and S_2 enters the switching state. With further increases in load, S_3 and S_4 are activated sequentially.

In the event-driven control depicted in Fig. 6.25b, the DC value of the output voltage varies significantly with the load over a wide range (>4ΔV). To further enhance output voltage accuracy, the study [17] incorporates a main loop.

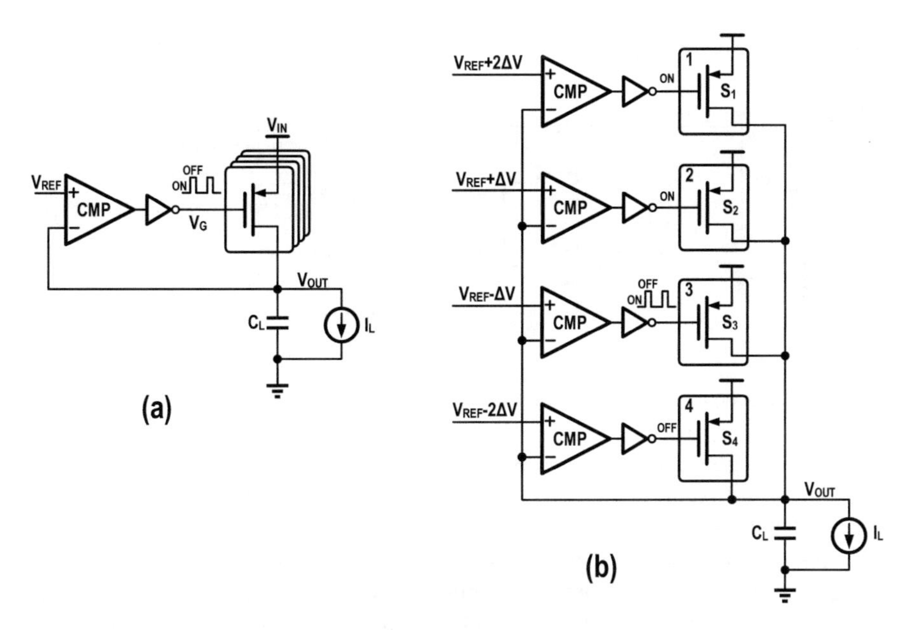

Fig. 6.25 (**a**) Switching control and (**b**) event-driven control

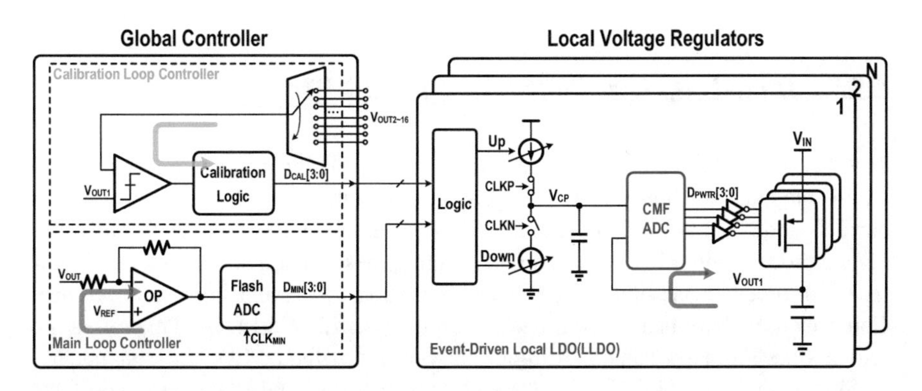

Fig. 6.26 The block diagram of the event-driven distributed LDO in [17]

Figure 6.26 illustrates the block diagram of the distributed LDO presented in [17]. It comprises three feedback loops: the main loop, the local loop, and the time-division multiplexing calibration loop.

The main loop employs an inverting amplifier to amplify the error between the voltage feedback signal V_{FB_MAIN} and the reference voltage. The amplified result is then processed through a flash ADC, converting it into a 4-bit digital signal $D_{MAIN}[3:0]$. This signal adjusts the up/down current of the charge pump in the local LDO (LLDO).

Figure 6.27 also shows the schematic of the LLDO. A current-mode flash ADC (CMFADC) serves as the event-driven control circuit in the LLDO. Its detailed

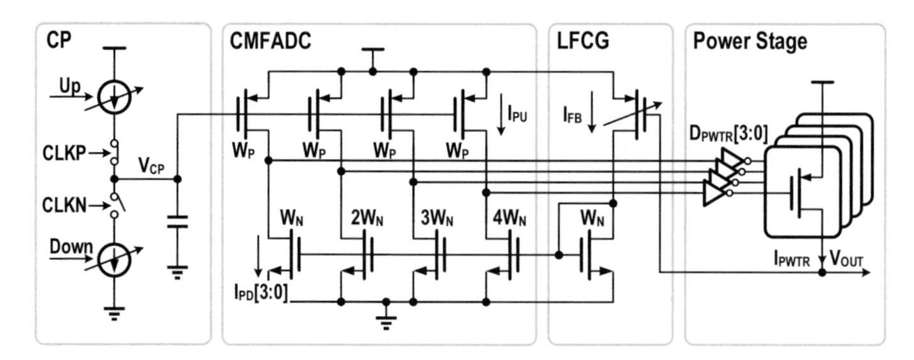

Fig. 6.27 The schematic of the LLDO in [17]

working principle is explained in Sect. 5.4. Due to mismatches and uneven load distribution, different LLDOs may produce varying output voltages. To mitigate these voltage differences, the study [17] implements a time-division multiplexing calibration loop. This loop sequentially selects the output voltage of each LLDO for comparison with LLDO1's output, generating a calibration signal $D_{CAL}[2:0]$. This signal is then fed into the respective LLDO to adjust its CP current, ensuring that its output matches that of LLDO1.

6.5.3 Dual-Loop Distributed All-Digital LDO

The dual-loop distributed LDOs described in [15–14] incorporate various analog circuit modules, such as charge pump-based integrators, analog comparators, and current-mode flash ADCs. These analog designs lengthen the design cycle and complicate process migration. Moreover, integrating these analog circuits within a digital environment necessitates careful consideration to prevent noise interference. Consequently, there is an urgent need to digitize distributed LDOs. This raises the critical question: How can an all-digital dual-loop distributed LDO be designed?

Shift register-based digital LDOs and ADC-based digital LDOs are two of the most classic DLDO structures, as illustrated in Fig. 6.28. The shift register-based DLDO boasts high comparator detection accuracy and features integral control, which enables high-output voltage accuracy. However, it can only turn on/off one power unit per cycle, resulting in a slow transient response. In contrast, the output voltage accuracy of an ADC-based DLDO is constrained by the resolution of the ADC.

To achieve both high accuracy and fast transient response, a high-resolution, high-speed ADC and a complex PID controller are necessary. The design requires careful trade-offs between power consumption, speed, accuracy, and complexity.

In the dual-loop LDO structure, the main loop typically employs integral control to ensure DC accuracy of the output voltage, while the fast loop generally utilizes

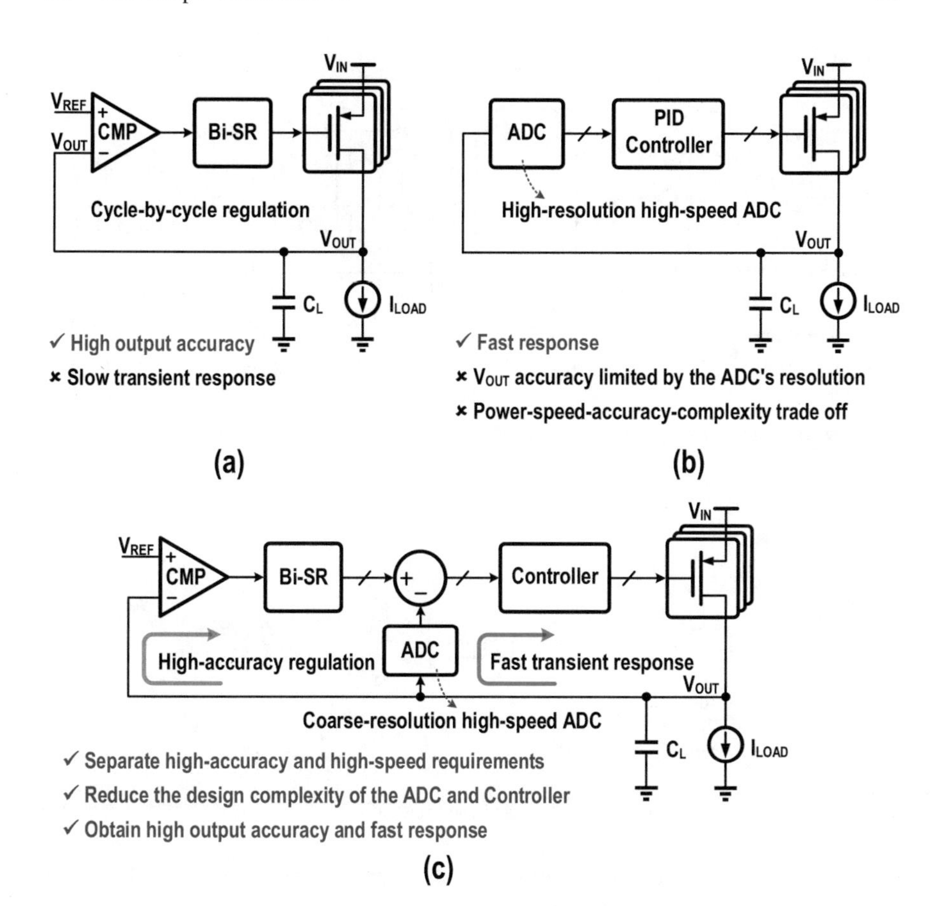

Fig. 6.28 (**a**) Shift register-based digital LDO, (**b**) ADC-based digital LDO, and (**c**) dual-loop digital LDO in [18]

proportional control to enhance transient response speed. By decoupling the high-resolution and high-speed requirements of the ADC and combining them with integral and proportional control mechanisms, we can achieve optimized performance.

As illustrated in Fig. 6.28c, the combination of a comparator and shift register forms an integral loop that ensures output voltage accuracy. The fast loop employs an ADC and proportional control to create a high-speed feedback loop, requiring only a coarse-resolution, high-speed ADC. Compared with traditional ADC-based LDO, this approach significantly reduces the design complexity of the ADC and controller.

The circuits for both the comparator and the ADC are depicted in Fig. 6.29. To achieve a fully digital design, the ADC utilizes the inverter chain TDC as a voltage detector, and the comparator is also implemented using standard cells. The TDC output is sensitive to PVT and frequency variations. Conventional time-domain quantization often requires calibration or use of paired quantizers to counterbalance

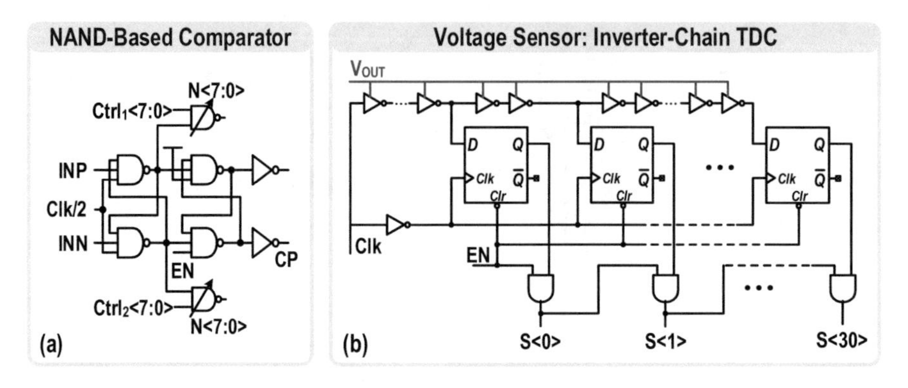

Fig. 6.29 (**a**) NAND-based comparator. (**b**) Inverter chain TDC

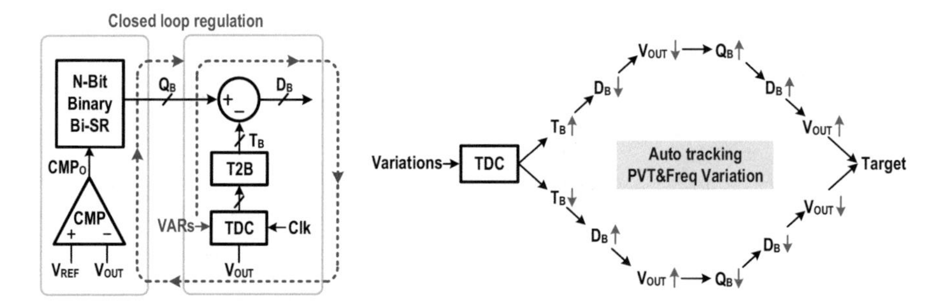

Fig. 6.30 The CMP-TDC structure with closed-loop regulation

the effects of PVT or clock variations. In the architecture described in [18], the integral loop can automatically track these variations and adjust the corresponding Q_B value to maintain output accuracy.

The blue dotted line in Fig. 6.30 illustrates the closed-loop tracking process. If the TDC output increases due to PVT or frequency changes, the differential D_B value decreases. After processing by the controller and power stage, V_{OUT} decreases.

Consequently, since $V_{OUT} < V_{REF}$, the comparator output remains high. This causes the shift register output Q_B to increase, the differential D_B to increase, and V_{OUT} to return to the target value. Since PVT and frequency changes are either slow (temperature, aging) or relatively fixed (process, frequency), and the integrating loop adjusts within nanoseconds, it can easily track these changes and ensure stability.

Figure 6.31 illustrates a single-point implementation of the all-digital dual-loop DLDO architecture. This design encompasses three key technical aspects: (1) all-digital dual-loop control, (2) coarse-fine tuning, and (3) asynchronous window control.

All-Digital Dual-Loop Control

The integral loop comprises a comparator and a 10-bit binary bidirectional shift register. The higher bits of the shift register output, QB < 9:4>, serve as the

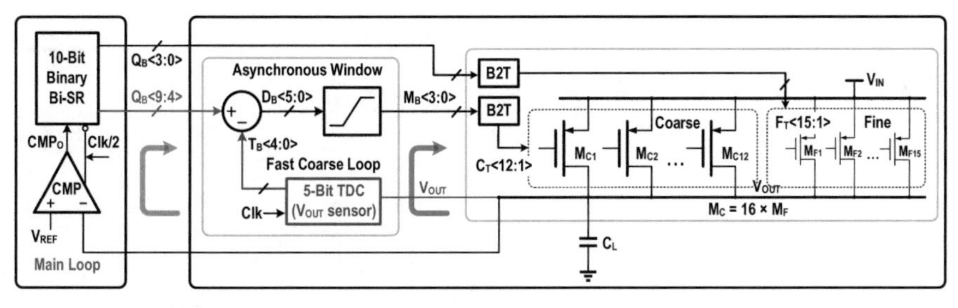

Q_B<9:4>: Binary code, higher bit output of Bi-SR.
Q_B<3:0>: Binary code, lower bit output of Bi-SR.
T_B<4:0>: Binary code, output of TDC.
D_B<5:0>: Binary code, D_B=Q_B-T_B
B2T: Binary code to thermometer code converter.
M_B<3:0>: Binary code, output of the asynchronous window.
C_T<12:1>: Thermometer code, control signals of the coarse switches.
T<15:1>: Thermometer code, control signals of the fine switches.

Fig. 6.31 The dual-loop all-digital LDO structure in [18]

reference code for the fast proportional loop. The fast proportional loop employs a time-to-digital converter (TDC) to quantize VOUT to TB < 4:0>, enabling the detection of output voltage changes within a single cycle. Asynchronous logic subtracts QB < 9:4 > from TB < 4:0 > to compute the difference, denoted as DB < 5:0>, where DB<5 > is the sign bit. The asynchronous window controls DB < 5:0>, and the resulting output directly manages the coarse power switches.

Coarse-Fine Tuning

The power switches are categorized into coarse and fine power switches, with a width ratio of 16:1. The integral loop utilizes $Q_B < 9{:}0 >$ to adjust all power switches. Figure 6.32 illustrates the dual-loop LDO's operational principle. As load current increases, $Q_B < 9{:}0 >$ increases. When all fine power switches are active ($Q_B < 3{:}0 > =$ 1111), further load current increases will raise the carry bit, increasing $Q_B < 9{:}4 >$ by 1. This action activates a coarse power switch and resets $Q_B < 3{:}0 >$ to 0000, ensuring smooth and precise regulation.

During sudden load transients, the asynchronous subtractor and control window, along with the TDC's capability to detect V_{OUT} drops within one cycle, enable the fast loop to proportionally activate coarse power switches and prevent further voltage drops. The integral loop regulating $Q_B < 9{:}0 >$ will subsequently adjust V_{OUT} to the target value.

Asynchronous Window Control

The asynchronous window control input is the difference between the integral loop output and the TDC output. Its output, $M_B < 3{:}0>$, directly regulates the coarse power switches. When $D_B < 5{:}0 >$ is within the window, $M_B < 3{:}0 >$ equals $D_B < 3{:}0>$. If $D_B < 5{:}0 >$ exceeds the upper limit (1100) or the lower limit (0000), all coarse adjustment switches are turned on or off accordingly:

$$M_B\langle 3:0\rangle = \begin{cases} 0000 & D_B < 0 \\ D_B\langle 3:0\rangle & 0 \le D_B \le 12 \\ 1100 & D_B > 12 \end{cases} \tag{6.8}$$

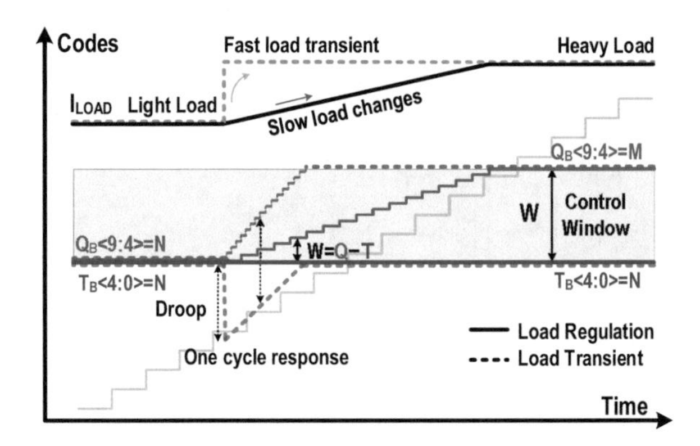

Fig. 6.32 Working principle of the dual-loop DLDO in [18]

The digital window size, W, affects loop stability and transient performance, determining the number of coarse power switches. For a fixed total strength of power switches, a smaller W value increases the strength of each coarse switch. During a load transient, this results in a larger output current for the same voltage drop, enhancing transient performance.

The integral loop regulates the power switches gradually, while the stability of the LDO is primarily determined by the proportional loop. The transfer function of the proportional loop is

$$H(s) = \frac{K_{\mathrm{TDC}} R_{\mathrm{L}} I_{\mathrm{MAX}}}{W\left(1 + sR_{\mathrm{L}} C_{\mathrm{L}}\right)} \times \frac{1 - e^{-Ts}}{sT} \tag{6.9}$$

where K_{TDC} denotes the TDC gain equivalent to 1/(TDC resolution), R_{L} represents the load resistance, I_{MAX} is the load capacity (representing the total strength of the power transistor), C_{L} is the output capacitance, and W denotes the control window size.

Figure 6.33 presents the Bode plot. With optimal parameter selection, loop stability remains independent of load changes. Increasing the window size W and output capacitance C_{L} enhances loop stability. A larger C_{L} improves transient performance, but an increased W leads to larger overshoot and undershoot during transients. In addition, a smaller I_{MAX} can also make the system more stable.

Building on the single LDO design, we duplicate the fast proportional loop as LVRs and spatially distribute them across the load. Figure 6.34 depicts the overall architecture of the dual-loop all-digital distributed LDO, which includes a global controller (GC) and nine LVRs. The GC comprises an integral loop, a ring oscillator, and all fine power switches (PSs).

Due to the spatial distribution of LVRs, TDC mismatches occur, resulting in varying $T_{\mathrm{BN}} < 4{:}0 >$ values. Despite having the same reference code $Q_{\mathrm{B}} < 9{:}4 >$, this mismatch causes current imbalances. To address this problem, a digital primary-secondary calibration is implemented to eliminate TDC mismatches. The output code T_{B1} from the TDC in LVR1 serves as the calibration code for the TDCs in the other LVRs.

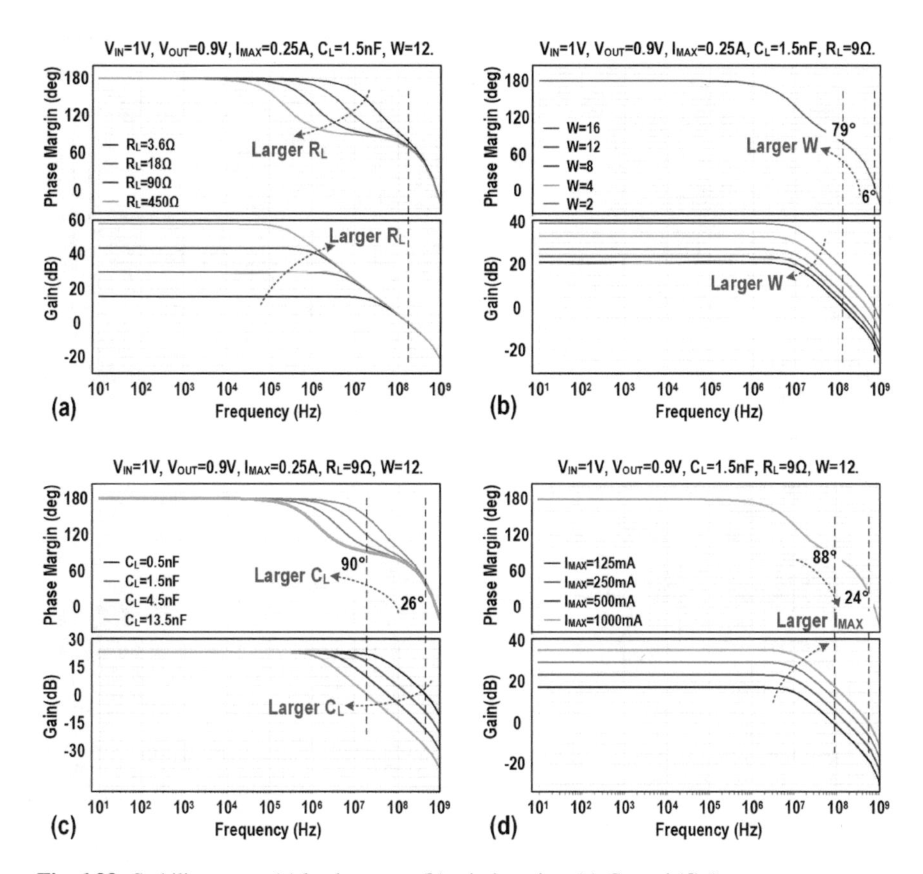

Fig. 6.33 Stability across (**a**) load current, (**b**) window size, (**c**) C_L, and (**d**) I_{MAX}

Figure 6.35a illustrates the simulation model for current sharing. Different DC offsets (V_{OS}) are added to each LVR to simulate quantization errors, resulting in varying TDC outputs. Additionally, all LVR outputs are directly shorted together to create an exaggerated current imbalance scenario. Figure 6.35b presents the simulation waveform of the current sharing calibration. Without calibration, the LVR output current depends on the offset voltage. Upon the arrival of the calibration signal TR's rising edge, the output currents of all LVRs become almost identical.

It is important to know how to analyze the stability of distributed LDO and guide its application. Using state-space equations to study the system stability of distributed LDO requires a lot of mathematical calculations. To simplify the analysis, a system model with two LVRs can be established to analyze the stability of a single LVR within the system, as shown in Fig. 6.36. Each LVR has a local load and local capacitor, and R_G is the parasitic resistance between the two LVRs.

When R_G is large, there is no interaction between the two LVRs, allowing each LVR to be treated as a separate regulator, as depicted in Fig. 6.36b. The stability analysis in this case is identical to that of a single LVR.

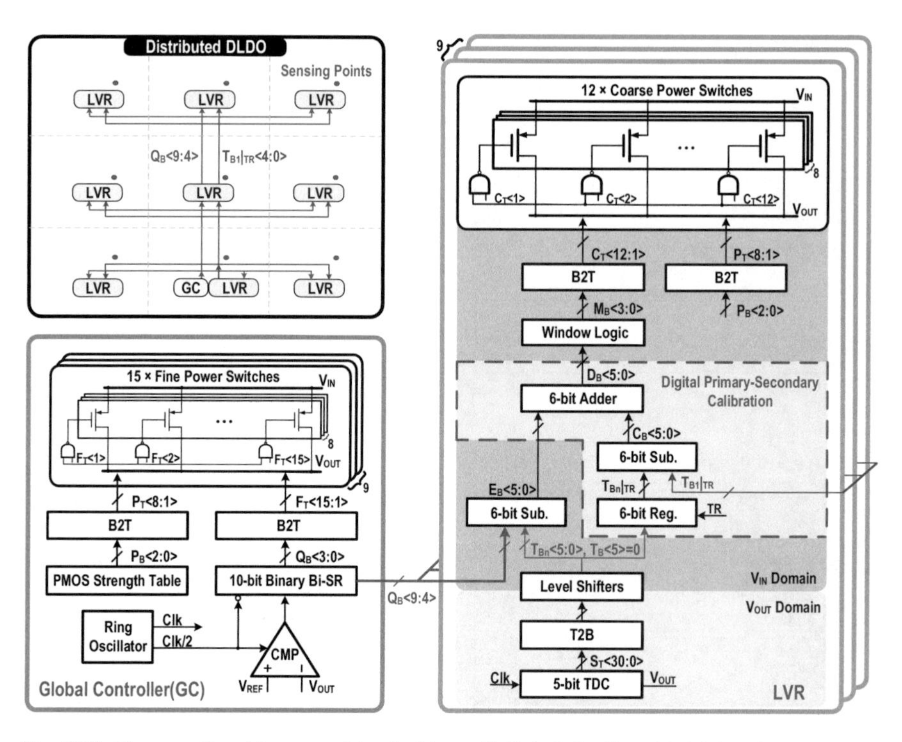

Fig. 6.34 The overall architecture of the dual-loop all-digital distributed LDO in [18]

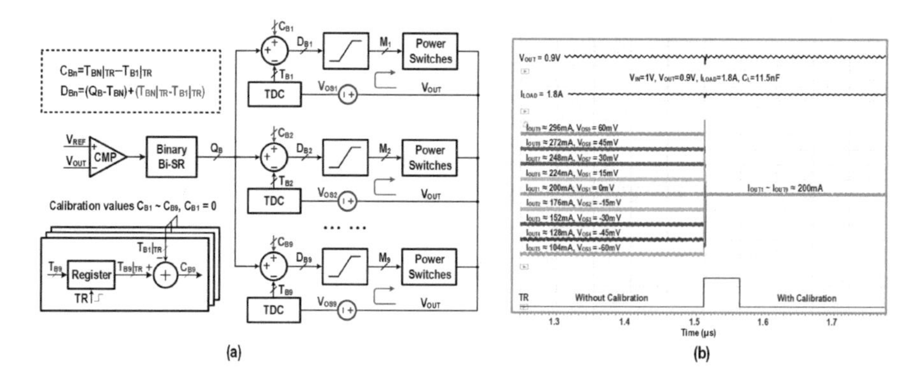

Fig. 6.35 (**a**) The simulation model for current sharing and (**b**) simulation waveforms

When R_G is small, the two LVRs, sharing the same controller and output response V_{OUT}, can be considered equivalent to a single LVR with doubled output capacity ($I_{MAX1} + I_{MAX2}$) and doubled output capacitance ($C_{L1} + C_{L2}$), as shown in Fig. 6.36a. The ratio of I_{MAX} to C_L remains constant, and stability is evaluated for various I_{MAX} and C_L values. This approach aids in analyzing the impact of different load capacities and output capacitances on stability when the application is extended.

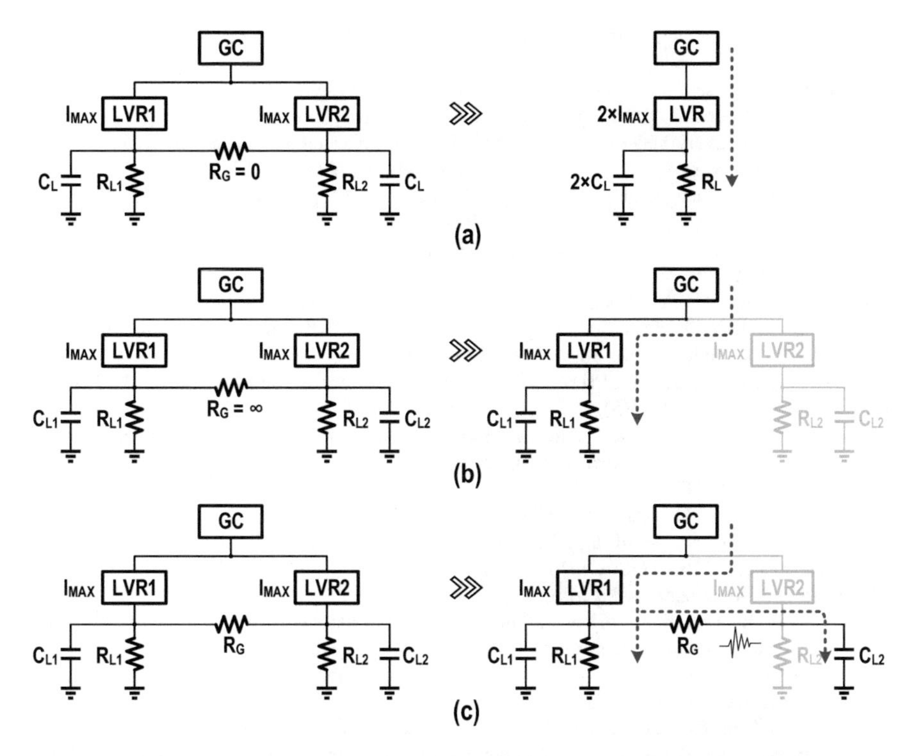

Fig. 6.36 Circuit model of two LVR connections with (**a**) $R_G = 0$, (**b**) $R_G = \infty$, and (**c**) R_G is a finite value

When R_G is moderate, each LVR assists the other in responding to its local load transient. The changes in R_{L2} and the regulation of LVR2 can be equivalent to the dynamic load current of LVR1. In this scenario, C_{L2} acts as the remote capacitance for LVR1, with R_G serving as the equivalent series resistance, as illustrated in Fig. 6.36c. A new transfer function can be derived, enabling the analysis of loop stability variations with different R_G values based on this transfer function.

References

1. R. Muthukaruppan, T. Mahajan, H.K. Krishnamurthy, S. Mangal, A. Dhanashekar, R. Ghayal, A digitally controlled linear regulator for per-core wide-range DVFS of atom cores in 14 nm tri-gate CMOS featuring non-linear control, adaptive gain and code roaming, in *Proceedings of 43rd IEEE European Conference on Solid-State Circuits (ESSCIRC)*, (2017), pp. 275–278
2. M. Zelikson, K. Luria, L. Gil, Y. Brown, V. Goldenbeg, D. Kasif, A Digital Low-Dropout (LDO) linear regulator with adaptive transfer function featuring 125A/mm^2 power density and autonomous bypass mode, in *IEEE International Solid-State Circuits Conference – (ISSCC) Digest of Technical Papers*, (2023), pp. 230–231

3. K. Ahmed, H. Krishnamurthy, C. Augustine, X. Liu, S. Weng, K. Ravichandran, J. Tschanz, V. De, A variation-adaptive integrated computational digital LDO in 22-nm CMOS with fast transient response. IEEE J. Solid-State Circuits **55**(4), 977–987 (2020)
4. S.B. Nasir, Y. Lee, A. Raychowdhury, Modeling and analysis of system stability in a distributed power delivery network with embedded digital linear regulators, in *Proceeding of 15th International Symposium on Quality Electronic Design*, (2014), pp. 68–75
5. S.B. Nasir, S. Gangopadhyay, A. Raychowdhury, A 0.13μm Fully Digital Low-Dropout Regulator With Adaptive Control and Reduced Dynamic Stability for Ultra-Wide Dynamic Range, in *IEEE International Solid-State Circuits Conference – (ISSCC) Digest of Technical Papers*, (2015), pp. 98–99
6. S. Kim, D. Kim, Y. Pu, C. Shi, S.B. Chang, M. Seok, 0.5–1-V, 90–400-mA, modular, distributed, 3 × 3 digital LDOs based on event-driven control and domino sampling and regulation. IEEE J. Solid-State Circuits **56**(9), 2781–2794 (2021)
7. K.H. Ang, G. Chong, Y. Li, PID control system analysis, design, and technology. IEEE Trans. Control Syst. Technol. **13**(5), 559–576 (2005)
8. D. Kim, S. Kim, H. Ham, J. Kim, M. Seok, 0.5V-V_{IN}, 165-mA/mm^2 fully-integrated digital LDO based on event-driven self-triggering control, in *IEEE Symposium on VLSI Circuits*, (2018), pp. 346–347
9. S. Bang, W. Lim, C. Augustine, A. Malavasi, M. Khellah, J. Tschanz, A fully synthesizable distributed and scalable all-digital LDO in 10 nm CMOS, in *IEEE International Solid-State Circuits Conference – (ISSCC) Digest of Technical Papers*, (2020), pp. 380–381
10. https://en.wikipedia.org/wiki/Trilinear_interpolation
11. Y. Lee, W. Jang, H. Bae, J. Cho, H. Kim, 34.7A/mm^2 scalable distributed all-digital 6×6 dot-LDOs featuring freely linkable current-sharing network: a fine-grained on-chip power delivery solution in 28nm CMOS, in *IEEE International Solid-State Circuits Conference – (ISSCC) Digest of Technical Papers*, (2024), pp. 272–273
12. Y. Lu, F. Yang, F. Chen, P.K.T. Mok, A 500mA analog-assisted digital-LDO-based on-chip distributed power delivery grid with cooperative regulation and IR-drop reduction in 65nm CMOS, in *IEEE International Solid-State Circuits Conference – (ISSCC) Digest of Technical Papers*, (2018), pp. 310–312
13. M. Huang, Y. Lu, U. Seng-Pan, R.P. Martins, An output-capacitor-free analog-assisted digital low-dropout regulator with tri-loop control, in *IEEE International Solid-State Circuits Conference – (ISSCC) Digest of Technical Papers*, (2017), pp. 342–343
14. J. Bulzacchelli, Z. Deniz, T. Rasmus, J.A. Iadanza, W.L. Bucossi, S. Kim, Dual-loop system of distributed microregulators with high DC accuracy, load response time below 500ps and 85-mV dropout voltage. IEEE J. Solid-State Circuits **47**(4), 863–874 (2012)
15. Z. Deniz, M. Sperling, J. Bulzacchelli, G. Still, R. Kruse, S. Kim, D. Boerstler, T. Gloekler, R. Robertazzi, K. Stawiasz, T. Diemoz, G. English, D. Hui, P. Muench, J. Friedrich, Distributed system of digitally controlled microregulators enabling per-Core DVFS for the POWER8™ microprocessor, in *IEEE International Solid-State Circuits Conference – (ISSCC) Digest of Technical Papers*, (2014), pp. 98–99
16. M. Perez, M. Sperling, J. Bulzacchelli, Z. Deniz, T. Diemoz, Distributed network of LDO microregulators providing sub-microsecond DVFS and IR drop compensation for a 24-Core microprocessor in 14-nm SOI CMOS. IEEE J. Solid-State Circuits **55**(3), 731–743 (2020)
17. D.H. Jung, T.H. Kong, J.H. Yang, S. Kim, K. Kim, J. Park, Distributed digital LDO with time-multiplexing calibration loop achieving 40A/mm2 current density and 1mA-to 6.4A ultra-wide load range in 5nm FinFET CMOS, in *IEEE International Solid-State Circuits Conference – (ISSCC) Digest of Technical Papers*, (2021), pp. 414–415
18. X. Mao, Y. Lu, R. Martins, A fully synthesizable all-digital dual-loop distributed low-dropout regulator. IEEE J. Solid-State Circuits **59**(6), 1871–1882 (2024)

Chapter 7
Conclusions on Fully Integrated LDOs

Low-dropout regulators play a vital role in all kinds of electronic devices. In this book, we have introduced and discussed basically all aspects of fully integrated low-dropout regulator designs, from fundamentals to advances, from classics to state of the art, from theory to applications, and from one to many.

7.1 Summary and Conclusions

For different applications, the specifications of LDO have different weights. In a noise-sensitive system, an LDO provides a clean low-noise supply and extraordinary power supply rejection (PSR) with certain dropout voltage. For microprocessors and digital systems, LDO is an easy way to enable the dynamic voltage and frequency scaling (DVFS) function to save the system efficiency; meanwhile, it provides ultrafast load transient response for high-current steps. For space-constrained and low-power applications, like biomedical implants [1], LDO is a very suitable choice for step-down of the high-input voltage for low-power circuit blocks, as it has no energy storage components (power inductors and capacitors), different from the switching-mode or switched-capacitor power converters.

Also, from the contents provided in this book, we can conclude that the control part and the power stage of an LDO can be hybrid solutions and be designed separately. We can have analog, digital, or time-domain control methods, while we may have analog, digital, and switching power stages. The control part may employ proportional-integral-derivative (PID) small-signal control methods and could also be equipped with large-signal control paths for a better trade-off of quiescent current consumption and transient response speed [2, 3].

Analog control is intrinsic, fast, and energy efficient. Therefore, analog control is suitable for the P and D control paths but may suffer from the gate pole of the large power transistor in high-current applications [4]. Also, analog control would be

X. Mao et al., *Fully-Integrated Low-Dropout Regulators*, Analog Circuits and Signal Processing, https://doi.org/10.1007/978-3-031-84916-9_7

suitable for low-power applications. Meanwhile, analog signals are not good for distribution over a large on-chip area, as they are much more vulnerable to environmental noise and capacitive coupling effects, when compared to their digital counterpart.

Digital control is very flexible, process scalable, and inherently good for low-voltage operation [5]. Thus, digital control is suitable for implementing the I path in the control loop and is suitable for the large signal control for high-current transient response. However, digital control is not very energy efficient, because it needs to convert the analog output voltage into digital domain with the power-speed-resolution trade-offs of the analog-to-digital converter (ADC). Also, the output current of a digital LDO suffers from the output accuracy of its power stage, which generates unavoidable limit-cycle oscillation (LCO) [6]. In addition, the digital control loop only deals with the feedback signals and has no response to the power supply ripples. Plus, the power stage of digital LDO is comprised of power switches which operate in triode/linear region and thus barely has power supply rejection capability.

Analog-digital hybrid LDO may combine the benefits of both analog and digital control loops and can be a good low-voltage solution with fast transient response and also certain power supply rejection. Analog control is fast and more energy efficient in small-signal domain, while digital control is instantaneously fast in large-signal domain. Analog-assisted digital and digitally assisted analog control schemes would always be interesting to be investigated [7–9].

Switching control, which uses the switch on-off timing duty cycle for regulating the output current and voltage, can be extremely fast [10], because the powerful switch driving stage may turn on or turn off the power stage instantly. But obviously, it would have large output ripples and also consume significantly large quiescent current [11]. Therefore, switching LDOs may only be suitable for high-current digital load with fast transient steps. Alternatively, high-current distributed LDOs welcome hybrid control methods [12].

7.2 Possible Future Directions

For digital systems, the total current consumption goes higher and higher at an exponential rate. So high-current, large-area, parallel/distributed power management solutions would be more important and demanding. For analog systems, ultralow quiescent current for low-power systems and ultrahigh PSR performance for low-noise systems would still be of great importance. For research, we need to divide ourselves into extremes.

In this fast-changing and -developing world, a fully synthesizable or an automatically designed LDO would definitely be highly favorable. There might be two types of auto-designed LDOs. One is the fully synthesizable digital LDOs with standard digital cells for better process migration and shorter time to market. This is relatively more "conventional" when compared to machine learning-based analog circuit design automation. With artificial intelligence (AI) and neural networks, all kinds of analog circuits, including LDO, can be generated with the well-trained model by feeding hundreds of thousands of existing analog circuits to the machine.

Analog circuit design is an art. But low-level arts can be easily replaced by silicon-based artificial intelligence. As carbon-based analog circuit designers/engineers, we need to investigate new applications, to incorporate with new technologies, and to be more focused on new design methodologies.

References

1. Y. Lu, W.-H. Ki, *CMOS Integrated Circuit Design for Wireless Power Transfer* (Springer, 2017) ISBN 978-981-10-2615-7
2. Y. Lu, M. Huang, R.P. Martins, PID control considerations for analog-digital hybrid low-dropout regulators (Invited Paper), in *IEEE International Conference on Electron Devices and Solid-State Circuits (EDSSC)*, (2019)
3. R.J. Milliken, J. Silva-Martinez, E. Sanchez-Sinencio, Full on-chip CMOS low-dropout voltage regulator. IEEE Trans. Circuits Syst. I: Regul. Pap. **54**(9), 1879–1890 (2007)
4. Y. Lu, Y. Wang, Q. Pan, W.-H. Ki, C.P. Yue, A fully-integrated low-dropout regulator with full-Spectrum power supply rejection. IEEE Trans. Circuits Syst. I: Regul. Pap. **62**(3), 707–716 (2015)
5. Y. Okuma, K. Ishida, Y. Ryu, X. Zhang, P.-H. Chen, K. Watanabe, M. Takamiya, T. Sakurai, 0.5-V input digital LDO with 98.7% current efficiency and 2.7-μA quiescent current in 65nm CMOS, in *2010 IEEE Custom Integrated Circuits Conference (CICC)*, (2010), pp. 1–4. https://doi.org/10.1109/CICC.2010.5617586
6. M. Huang, Y. Lu, S.-W. Sin, S.-P. U, R.P. Martins, W.-H. Ki, Limit cycle oscillation reduction for digital low dropout regulators. IEEE Trans. Circuits Syst. I: Regul. Pap. **63**(9), 903–907 (2016)
7. M. Huang, Y. Lu, S.-P. Uran, R.P. Martins, An analog-assisted tri-loop digital low-dropout regulator. IEEE J. Solid-State Circuits **53**(1), 20–34 (2018)
8. Y. Lu, Digitally assisted low dropout regulator design for low duty cycle IoT applications, in *IEEE Asia Pacific Conference on Circuits and Systems (APCCAS)*, (2016), pp. 33–36
9. S.B. Nasir, S. Sen, A. Raychowdhury, Switched-mode-control based hybrid LDO for fine-grain power management of digital load circuits. IEEE J. Solid-State Circuits **53**(2), 569–581 (2018)
10. J.F. Bulzacchelli et al., Dual-loop system of distributed microregulators with high DC accuracy, load response time below 500 ps, and 85-mV dropout voltage. IEEE J. Solid-State Circuits **47**(4), 863–874 (2012)
11. X. Mao, Y. Lu, R.P. Martins, A scalable high-current high-accuracy dual-loop four-phase switching LDO for microprocessors. IEEE J. Solid-State Circuits **57**(6), 1841–1853 (2022)
12. X. Mao, Y. Lu, R.P. Martins, A fully synthesizable all-digital dual-loop distributed low-dropout regulator. IEEE J. Solid-State Circuits **59**(6), 1871–1882 (2024)

Index

X. Mao et al., *Fully-Integrated Low-Dropout Regulators*, Analog Circuits and Signal Processing, https://doi.org/10.1007/978-3-031-84916-9